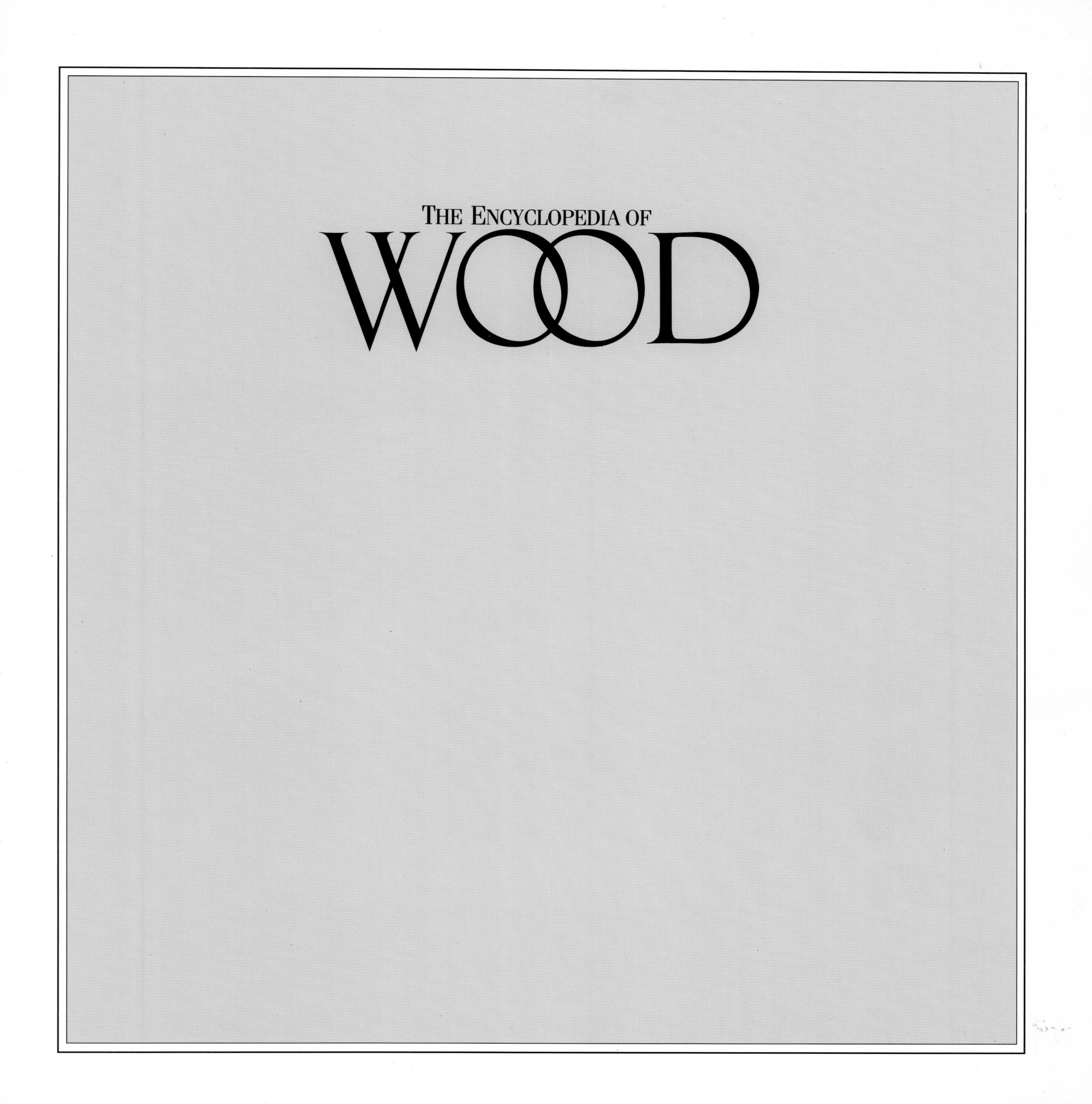
THE ENCYCLOPEDIA OF
WOOD

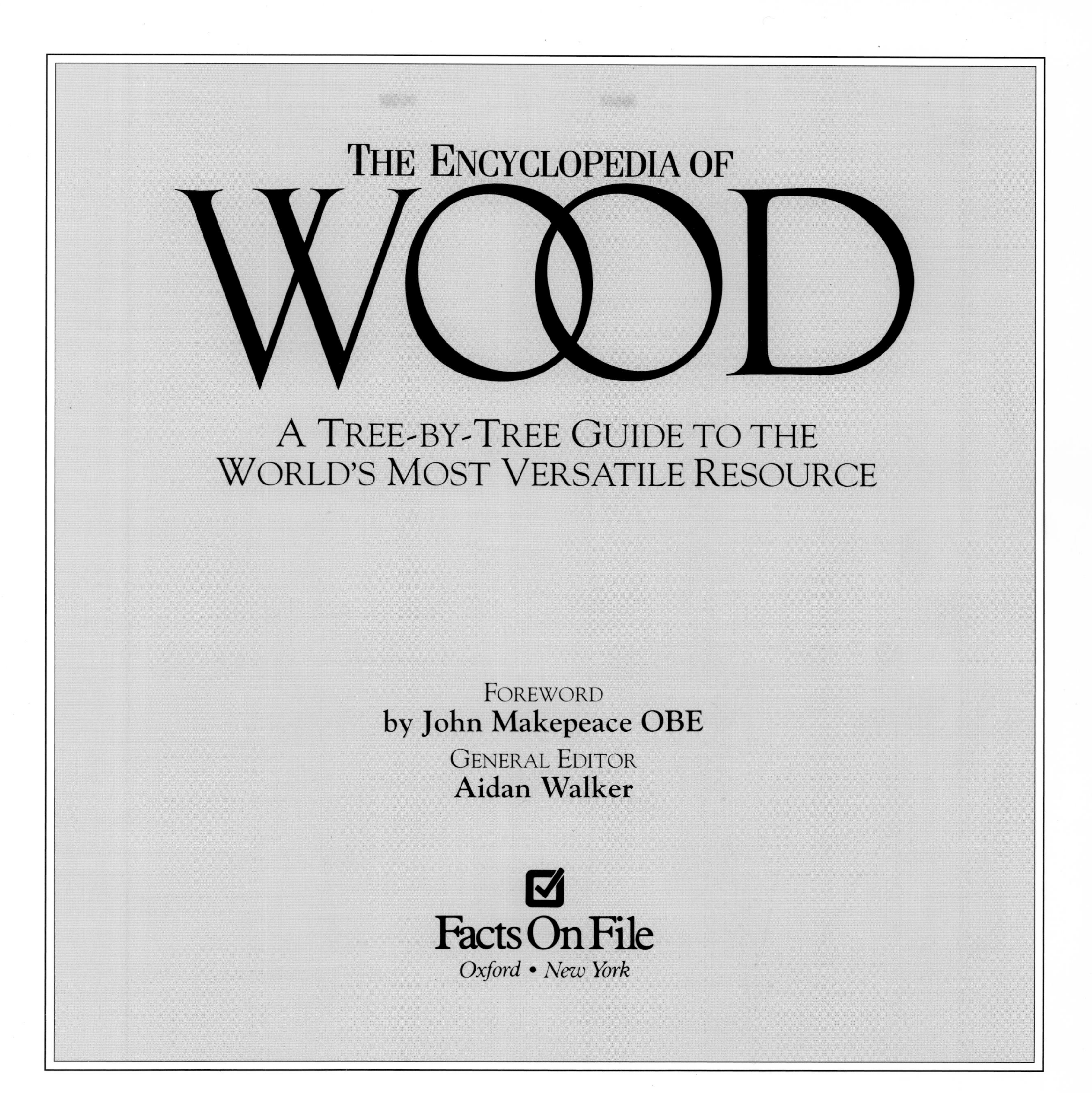

THE ENCYCLOPEDIA OF WOOD

A TREE-BY-TREE GUIDE TO THE WORLD'S MOST VERSATILE RESOURCE

FOREWORD
by John Makepeace OBE
GENERAL EDITOR
Aidan Walker

Facts On File
Oxford • New York

THE ENCYCLOPEDIA OF WOOD

For information contact:
Facts On File, Ltd.
Collins Street
Oxford OX4 1XJ
United Kingdom

or Facts On File, Inc
460 Park Avenue South
New York NY 10016
USA

British Library Cataloguing in Publication Data

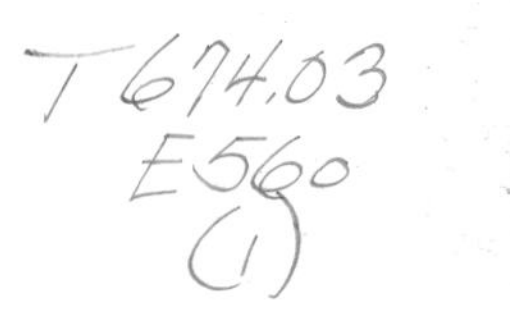

Lincoln, William et al
Encyclopedia of wood,
1. Wood, - Encyclopaedias
I. Title
620. 1'2'0321

ISBN 0-8160-2159-7

Library of Congress Cataloguing-in-Publication Data

Encyclopedia of wood: a directory of timbers and their special uses/
by William Lincoln... [et al.].
p. cm.
Includes index.
ISBN 0-8160-2159-7
1. Wood--Dictionaries. I. Lincoln, William.
TA419.E53 1989 89-33439
674'.03--dc20 CIP

This book was designed and produced by
Quarto Publishing plc.
The Old Brewery,
6 Blundell Street,
London N7 9BH

General Editor Aidan Walker
Senior Editor David Game
Designers Frances Austen, Penny Dawes and Andy Turnbull
Illustrators David Kemp, Kevin Maddison and Janos Marffy
Picture Manager Joanna Wiese
Picture Researchers Linda Proud and Rose Taylor

Art Director Moira Clinch
Editorial Director Carolyn King

With thanks to Ingrid Clifford and Messrs. Phoenix (Hahn), Edmonton, London.

Manufactured in Hong Kong by Regent Publishing Services Ltd
Printed by South Sea Int'l Press Ltd., Hong Kong
10 9 8 7 6 5 4 3 2 1

The ethereal beauty of this chair
in English holly, designed by John Makepeace,
suggests the living form of the tree

CONTENTS

142 THE WORLD OF WOOD
by Luke Hughes

John Makepeace, OBE; a furniture designer and educator whose work is based on a passionate relationship with wood

FOREWORD

by John Makepeace, OBE

It is encouraging to see a book like this which puts timber in its broad human context. In itself, the book is both an expression of the world's hopes associated with timber, and a statement of its fears and problems. The intense pressure to specialize in all sorts of activity, which is a unique factor in modern life, has fostered areas of learning, technology and practice that are discreet; communication suffers to the point of breakdown. But the solution to problems frequently arises from cross-fertilization between disciplines, and one of the welcome characteristics of this book is that it recognizes and deals with not only conservation, but the even more subtle relationship between trees and man as well as the technical, botanical and commercial aspects.

Perhaps the enormous amount of information available in any given knowledge area is a pressing factor in often divisive specialization. Overwhelmed with an ever-growing mountain of data, the challenge for us all is to keep sight of the totality and the relatedness of its parts. Awareness of the cause and effect of our actions can only help to clarify our sense of purpose in the complex world in which we live.

When we come to timber and the way people deal with it – and in it – the phenomenon of separation is as noticeable, and as lamentable, as ever. Specialists, from the forester and wood scientist, through the manufacturer and merchant, up to the conservationist and consumer, see others' interests as quite different from their own. Policies aimed at short-term political and commercial profit with little concern for medium-to-long term results; poor standards of forest management; mediocre quality in the wood-related product, whatever it be; a high proportion of waste; all contribute to marginal economics, with little research and investment into the future of our trees and forests, and thus of commercial timber.

The Great Hall at Parnham House (LEFT), the Elizabethan manor which is home to Makepeace's School for Craftsmen in Wood, his own furniture design and making workshop, and the charity, the Parnham Trust. Much of his work is characterised by a strong 'organic' feel; he will often design an entire piece to be made from a single tree. Indeed the enormous table in the foreground, one of Makepeace's designs, echoes the structure of a tree, with its top joining in curved-edge segments and its heavily carved trunk-like base.

Robert Ingham, Principal of the School for Craftsmen in Wood (LEFT)*, stresses the importance of accuracy in machined components. The prototype building of Hooke Park School for Advanced Manufacturing in Wood,* (BELOW LEFT) *a few miles from Parnham, where a conservationist design philosophy goes hand in hand with forest management training.*

The theme of forest clearance, followed by agricultural and industrial development sustained by heavy reliance on imports of timber, is widely evident. Of the 10 million hectares of forest cut each year in the world, one million are replanted, largely with fast-growing mono-cultures. Their role is to supply pulp industries dependent on huge volumes of timber to feed capital-and energy-intensive processes in the manufacture of low-value products.

If timber growing is seen, as it must be, as one of the solutions, it is still vital to undertake it according to principles of conservation. They require a broad approach to encompass the pressing environmental, social, industrial and economic needs associated with plantations. Conservation groups and popular opinion are heightening interest in the planting of native species, especially broad-leaved hardwoods, which goes some way towards relieving pressure on tropical timbers and creating a high-value resource. Broad-leaved forests are also more botanically diverse and kinder to flora and fauna than conifers. Yet, to reap the full potential benefit, it is crucial that young trees should be carefully managed.

Natural forest is inherently diverse. The primary concern in managing planted forest is the promotion of growth and quality in a crop, by removing 9 out of every 10 trees during the forest cycle, so that the prime trees have space to develop their crowns and their girth. The trees removed, known as thinnings, represent 50 per cent of the annual harvest.

It is obvious that the best use must be made of resources which have the combined advantage of being both local and renewable. The need for imports is reduced – eventually eliminated – and a forest regeneration process, akin to the natural cycle, gets under way. There appear to be at least three ways in which currently wasteful practices can be improved:

- Techniques must be recovered and re-developed for making the surface of commonly available but visually plain woods more attractive, by using textures and colour. This long-standing tradition is well overdue for revival.
- Many substantial trunks and limbs, which do not conform to the conventional standards of straightness normally required, are rejected. These pieces often contain a useful proportion of timber with interesting grain patterns or curvature; imaginative makers can exploit these characteristics to structural and aesthetic advantage. The introduction of portable sawmills and the lower overheads associated with them make this a realistic possibility.
- Thinnings offer a very substantial and renewable resource worldwide. Their effective utilization generates income to improve woodlands, and their removal promotes the growth of remaining trees, which can then become a long-term source of larger-section material.

As the world finally begins to recognize the vital role of trees in its health and survival – a process of recognition to which books like this one contribute – the supply of thinnings from the planted forests will become a huge new resource. Despite the fact that they have been used locally for building and small 'human-scale' industry for centuries, they are a 'new material' in large-scale commercial terms, and as such their properties are but poorly understood. Research in this field is leading to new low-capital technologies suitable for vigorous 'green' industries to operate in balance with local resources and local needs.

It is a happy coincidence that having established the School for Craftsmen in Wood in 1977, The Parnham Trust launches Hooke Park College for Advanced Manufacturing in Wood in the same season as the publication of this book.

The College aims to generate a network of manufacturing businesses which utilize sustainable indigenous timber, especially thinnings, in the production of quality products and buildings. The broadly based course, which covers the spectrum from forest through design, manufacturing systems and enterprise development, is intended to attract environmentally sensitive entrepreneurs into industry. The innovative technologies developed for the construction of the College demonstrate the scope for collaboration between foresters, scientists, engineers, designers and the construction industry.

The choice of material is a crucial decision. No longer can designers, architects and makers specify the use of timber without regard for its renewal. This adds a global dimension to the painstaking endeavours of all those who use timber, all those who are in business with it or who simply respond to its endless fascination. Wood lovers and wood users – thankfully they are almost always one and the same – are gaining a new world consciousness, considering the sources of their supply, becoming (sometimes painfully) aware of the effects of their demands on peoples living on the other side of the world. Apart from serving people who have a relationship with wood with the information and detail they need, it is a book like this which feeds and educates that worldwide 'wood awareness'.

John Makepeace.

The vaulted framework of the training workshops at Hooke Park (LEFT) is made from forest thinnings — young trees that would otherwise be considered as waste. All-round training: a Visual Studies session at Parnham (BELOW) from which some of England's most distinctive furniture designers emerge.

French polishing a traditional barometer case at London specialists Garner and Marney (ABOVE). The most highly prized of all finishes for its deep, warm lustre, french polish requires great skill to apply.

THE CRAFTSMAN

As a furniture maker I am in a dilemma. I love growing trees and hate to see them felled, but I also love wood. I love to stroke it, to admire its infinite variety of grain, colour and texture. I am endlessly fascinated that no two boards, even from the same species, are ever exactly the same. I consider myself fortunate that I work daily with a material that is both beautiful and constantly alive and moving, and one that invariably improves with age and usage. So different from lifeless steel or unchanging plastic, it is the movement in timber, its sheer unpredictability, that provides the challenge, the fascination and at times the frustration that, allied to its beauty, must make wood the most exciting and wonderful material with which to design and work.

A lifetime is never long enough to exhaust the potential of wood, nor to master its idiosyncrasies, for there is such a tremendous variety of species, each with different working properties – which may vary even within one tree. Indeed, in a single board one area may yield timber perfect for chair legs; the rest is useless for legs, yet perfect for door panels and thus we can use each piece of wood without waste.

Working with this living material is a lifelong challenge, but sadly, for many modern people, wood is something that comes cut, bundled, even sanded and polythene-wrapped for convenience. Or it has been neatly chopped up, crushed and glued into chipboard, medium-density fibreboard or hardboard, where dreary uniformity is the price to be paid for a material that is stable, convenient – and dead. This book will provide a great service if it does no more than encourage more people to investigate the real world of living wood, the ultimate three-dimensional material that can be carved, shaped and fashioned by human hands.

Trees are vital to humanity's existence on earth. They provide the landscape and the environment that make 20th-century life tolerable. Their by-products provide essential building materials, fuel, the basic elements of countless everyday objects used all over the globe. More subtly, they have also given us, since civilization began, a sense of communication with our natural surroundings; they house spirits as well as animals and birds, they are sources of medication, of meditation and of solace. It is trees and the trade in them which sustain the economies of many of the poorer nations of the world; they are those countries' national as well as natural resource, critically undervalued. Only when the world puts a higher value on the timber crop can more selective and restricted felling be promoted and financed.

Trees are vital to the ecology of the world, as we can see from the increasingly acute problems caused by deforestation; it is now having a global effect, changing the weather and environment and causing disasters such as the flooding in Bangladesh, soil erosion in the Himalayas, famine in Sub-Saharan Africa, drought and the extinction of wildlife. The crime is that it is all so unnecessary. Fossil fuels, themselves the subject of conservationist concern, are

A rowing 'eight' on the stocks in the workshops of London riverside boatbuilders George Sims (RIGHT). The ribs are ash, renowned for its flexibility. Boat building is a time honoured craft whose skills are still treasured even when faced with competition from synthetics.

extracted through holes in the ground. Their removal does not change the face of the earth or the air we breathe (though of course their exploitation does); unlike timber, they are not renewable.

My own concern for the future, shared by artists, craftsmen and timber users worldwide, was brought very much closer to home with the storm of October 1987 in England, which uprooted literally millions of trees. Compared to hurricanes of the tropics, of course, it was merely a strong breeze; still, it left southern England devastated. Even though – or perhaps because – that in no way compared with the utter devastation of vast areas of virgin tropical rain forest, it made people far more aware of the importance of trees as an amenity to be cherished. Despite the good intentions of those who wished the trees they had known and loved – some of which had stood for

The main timbers of the porch of this Tudor church at Shoreham (LEFT), in the ancient English county of Kent, are from a single split oak tree; it bears a telling likeness of spirit to sculptor Jim Partridge's single-tree 'gateway' to Hockley Wood, Essex (BELOW). The ravages of Dutch elm disease felled this 500-year old tree (BELOW LEFT), a sad monument for generations to come.

A cabinetmaker at London's Garner and Marney (LEFT) concentrates on minute brass hinges and screws. Though some parts of the furniture industry are highly machine-intensive, craftsmen and clients alike know there is no substitute for the skill and care of experienced hands; when it comes to wood, true quality was never automated.

A table in elm (ABOVE) by the author of this section of the book, Alan Peters. The top is from a single plank, a phenomenon unheard of today; the through-jointing of the legs serves decorative and practical purposes, in that the wood needs a stabilising influence as it shrinks and swells according to atmosphere.

more than 200 years – to be used, the wood could not command an economic price. A great deal of beautiful hardwood was hauled away for firewood.

This experience finally convinced me that seasoned hardwood timber prices, held down as they are by cheap imports from the rain forests, far from being too high, are in fact far too low. Timber, as a percentage of the cost of building a house or item of furniture, costs far less than forty years ago or indeed in past centuries, when every effort was made to economize on a valuable resource. Little was wasted. Why else were houses and barns often built of reclaimed ships' timbers? For what other reason has Chinese furniture for centuries often been made from reclaimed rosewood? In contrast, today it is more economic to take a prime forest, a prime tree, or a new plank, and burn what would once have been treasured and used.

The waste of the tropical rain forest is a continuing man-made tragedy, first highlighted early this century when the kauri forests of New Zealand and the Spanish mahogany of Central America were logged to commercial extinction. It seems there has been an endless determination to use up the world's natural resources without thought for tomorrow; now the tide is beginning to turn, but it still cannot be stressed too heavily that trees are a renewable crop and as such, except in certain circumstances of particular amenity value, they are best felled selectively at maturity when the timber is at its prime for commercial use. Indeed, this has been the time-honoured practice for centuries in much of Britain and Europe. The real, long-term tragedy of the 20th century is that we have not planted the right species in the right places and in the right quantities to provide for this constant renewal.

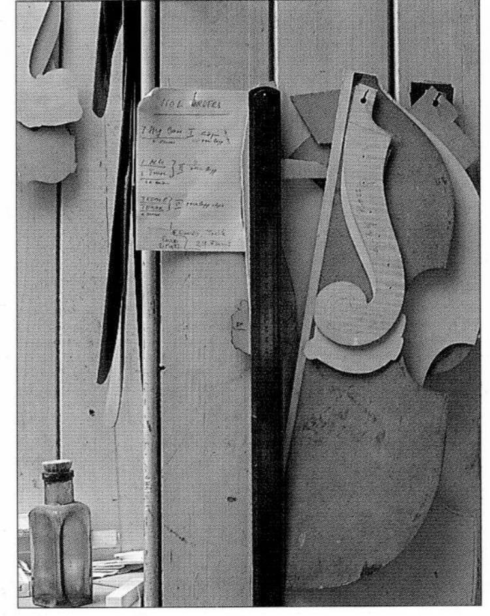

*In Edward Withers' lutemaker's workshop in London's Soho (*ABOVE*), instrument templates wait for re-use; Devon maker David Savage chose fragrant, lightweight cedar for this rock-collector's chest (*BELOW*).*

*Contrast this woodcarver, at work in his Dorset studio (*RIGHT*), with the Far-Eastern one-man production line on the opposite page. The range of carving equipment grows continuously, but the quality of output depends more on the craftsman's inspiration than on his tools and technique.*

The tiers of a Japanese temple's wooden roofs (ABOVE) rise to the sky. Japan has developed a highly sophisticated building joinery system based entirely on wood. This woodcarver (LEFT) in Chiang Mai, Thailand, with a couple of chisels and his feet to hold the work, turns out a great volume of devotional statues for tourists, religious and otherwise; a Korean temple's curvilinear decoration and pillars (BELOW LEFT) from single tree trunks contrast with the intricately squared work of the Japanese.

HOPES FOR THE FUTURE

Amongst the comparatively small community of wood lovers and users like myself, at least, a positive note can be struck; there are encouraging trends. Nowadays in the developed countries, more and more furniture and woodware is being made in small workshops than at any other time since the turn of the century. It is these workshops that exploit local timbers that have until recently not been considered economically viable. Species not normally associated with furniture, for example sycamore, brown ash, acacia, cherry, holly, burr oak, macrocarpa; small trees, interesting parts of trees, trees that are semi-rotten, timber commonly regarded as the waste products of our woodlands and sawmills – all can reveal wonders of beauty when transformed and revealed by sensitive hands. Now we have the exciting phenomenon of an influx of timber suppliers, in direct response to woodworkers' demands, handling and promoting timber that 10 years ago would have been burned, left to rot or, at best, used for pit props. The portable bandmill has given small-scale timber users the opportunity to convert hitherto 'uneconomic' timber on the spot where the tree falls and haul it already planked, making efficient use of a limited resource. English furniture designer John Makepeace is creating a school specifically to teach the integrated skills of forest management, product design and flexible manufacturing, using forest waste as raw material.

Against the dual background of deforestation on a massive scale and an increasingly vocal public calling for sanity, such localized phenomena and initiatives must not be interpreted as having an immediate and widespread physical effect. Their importance is more one of attitude, of an awareness that is the first essential for conservation at international level. I sincerely trust that this book, which tells us so much about the beauty and wonder of wood as a material as well as the beauty of the trees that provide it, will stimulate more and more people to reverse the trend of the past 100 years. It must encourage everyone to make more responsible use of this valuable natural resource, and to press for and participate in an unselfish programme of hardwood planting for future generations to use and enjoy.

Lungs of the earth:
trees and forests are crucial to
humanity's well being

THE LIVING TREE

No single form of plant life means so much to humanity as the tree. Highly complex bio-systems, providers of oxygen for us to breathe, shade for us to rest in, nuts and fruit for us to eat, habitats for birds, insects (and spirits), wood to burn and wood to build; the inter-relationship between humankind and tree life is as old as humanity itself.

Trees have the longest lifespan of all higher forms of life. The oldest survivor is the Bristlecone pine *Pinus longeava*, which grows in the White Mountains of California; it is 5,000 years old. It was probably a seedling when the Egyptians built their pyramids.

ANATOMY OF THE TREE

Trees have the longest lifespan of all higher forms of life. The oldest survivor is the Bristlecone pine *Pinus longeava*, which grows in the White Mountains of California; it is 5,000 years old. It was probably a seedling when the Egyptians built their pyramids.

Conifer trees, which we call softwoods (their Latin name is *Gymnospermae*) emerged 275 million years ago, eventually covering two-thirds of the earth's surface. These magnificent evergreens towered above other land plants for a further 90 million years before the first birds, snakes and dinosaurs appeared.

The *Angiospermae*, which we call hardwoods, are the broad-leaved, flowering and fruit-bearing trees which first emerged about 140 million years after the conifers. Upheaval of the earth's crust, and subsequent severe climatic changes, drastically reduced them and extensive treeless savannah plains and prairies took their place.

Today, the evergreen conifers form large forests in the cool northern regions, including fir (*Abies*), spruce (*Picea*), larch (*Larix*), cedar (*Cedrus*) yew (*Taxus*), and the true pines (*Pinus*).

SOFTWOODS AND HARDWOODS

These names refer to structural differences, not ones of softness and hardness. Some hardwoods, such as balsa and poplar, are much softer than botanical softwoods like yew or pitch pine. *Gymnospermae* – the softwoods – are naked seeded trees, which means the seeds are not enclosed in a pod. They are cone-bearing, with evergreen needles or scale-like leaves, and grow in the cool temperate northern regions. Conifers supply the bulk of the world's commercial timber.

The hardwoods — *Angiospermae* — hide their seeds in fruit seed cases. These broad-leaved trees produce fruits and flowers and are deciduous, which means in temperate zones they shed their leaves every autumn. There are two kinds of hardwood; those that carry their seeds in single-lobe cases, such as the palm tree, are called *monocotyledoneae*, those whose seed cases have two lobes are *dicotyledoneae*. The majority of our valuable commercial hardwoods come from this latter group.

The *dicotyledoneae* also sub-divide into two types. Polypetalous species possess both a calyx and a multi-petalled corolla, and produce beautiful flowers and fruits. Apple, almond, pear, peach, and cherry are polypetalous, as are horse chestnut, maple and mahogany. Those in the apetalous group have inconspicuous flowers, lack the corolla and often the calyx, but they are the strongest and most durable woods. Timbers such as hickory, birch, alder, beech, oak, elm and walnut come from apetalous trees.

HOW A TREE GROWS

Some tree's flowers are scented to attract pollinating bees and birds, others have catkins. Male catkins, commonly long, hanging, tassel-shaped clusters of flowers, release clouds of pollen, which is collected on the exposed stigmas of bud-shaped female catkins. The development of fruit and seed follows. Many catkin-bearing trees yield large nuts, while conifers produce big woody cones that contain numerous winged seeds. Pines flower when they are 10 years old and beeches, for instance, after more than 25 years.

Some seeds have broad-bladed wings or tufts of hair to aid wind dispersal, or to enable them to float in water; some have hooks to grab insects, are sticky to repel birds, or scented to attract bats. The seed ripens within the fruit and the birds pluck the fruit, then drop

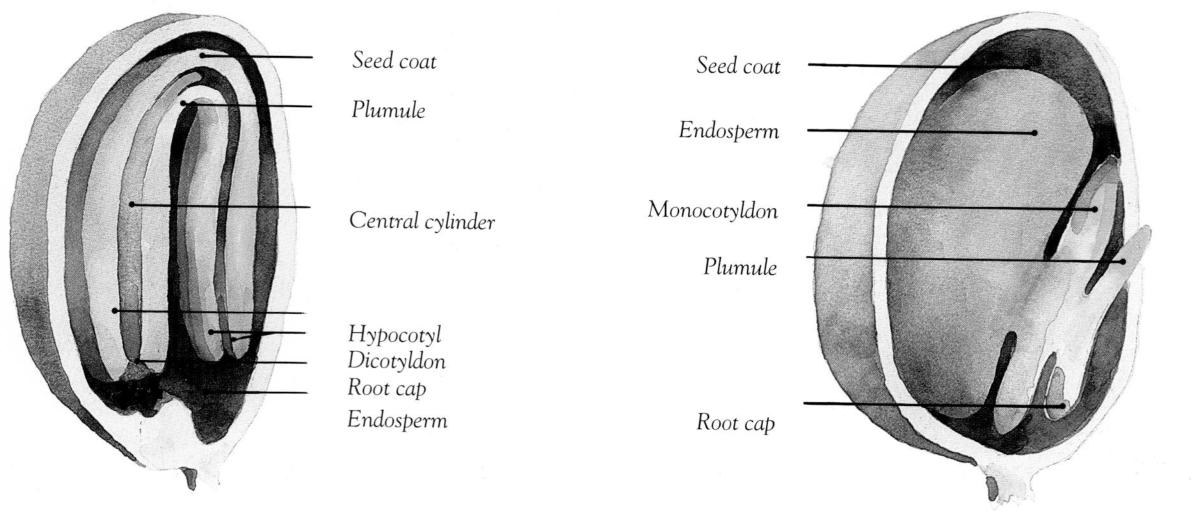

Power packs: the tiny seeds from which great trees can grow, providing shade, fruit and even timber. Angiospermae produce hardwoods; their seeds (FAR LEFT and LEFT) are enclosed in casings of either one lobe (monocotyledon) or two or more (dicotyledon). Dicotyledonae themselves divide into two groups; trees which produce flowers and fruit and those with few or no petals. The timber of the latter type is generally stronger and longer-lasting.

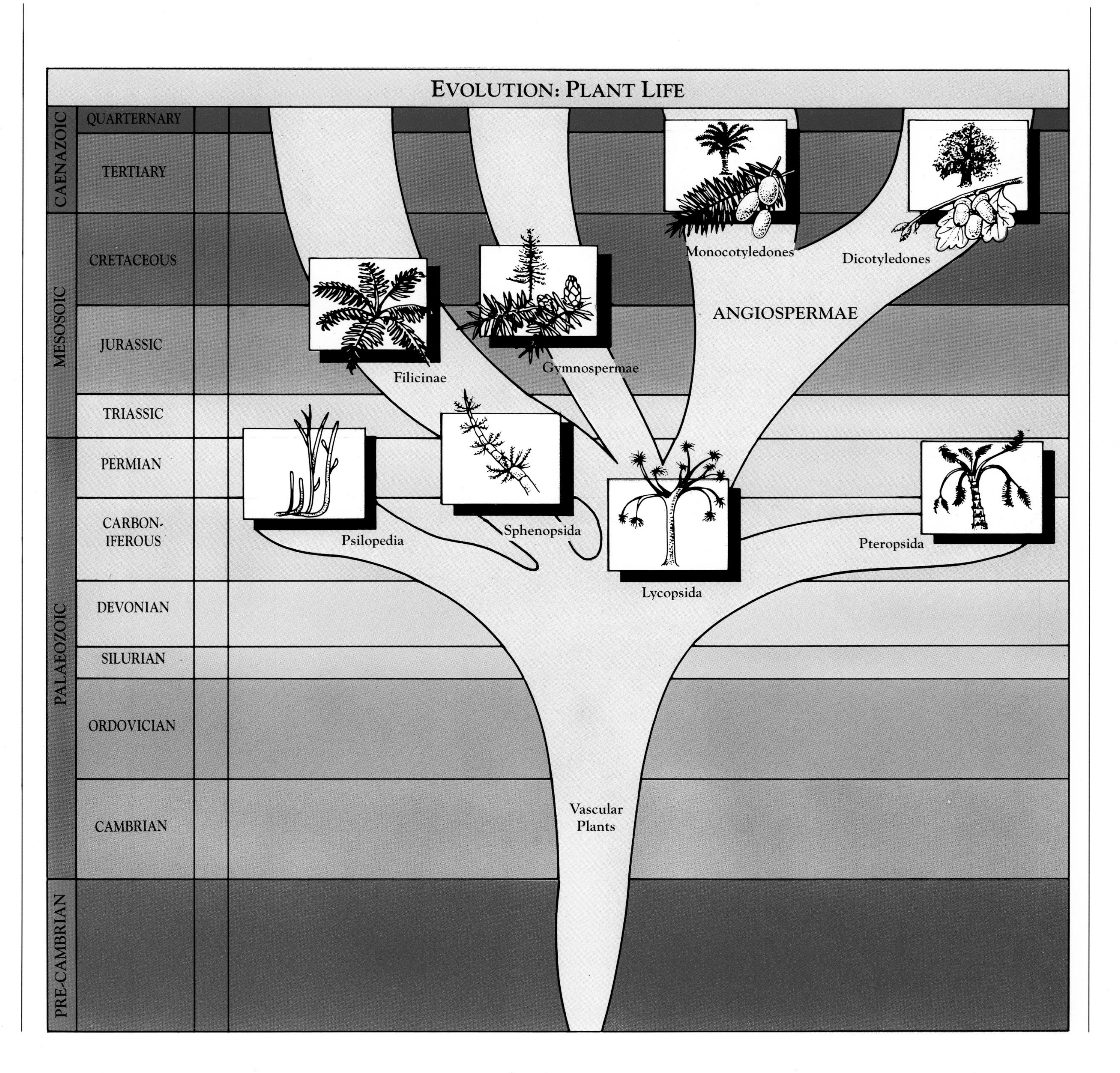
EVOLUTION: PLANT LIFE
CAENAZOIC
MESOSOIC
PALAEOZOIC
PRE-CAMBRIAN
QUARTERNARY
TERTIARY
CRETACEOUS
JURASSIC
TRIASSIC
PERMIAN
CARBON-IFEROUS
DEVONIAN
SILURIAN
ORDOVICIAN
CAMBRIAN
Monocotyledones
Dicotyledones
ANGIOSPERMAE
Filicinae
Gymnospermae
Psilopedia
Sphenopsida
Lycopsida
Pteropsida
Vascular Plants

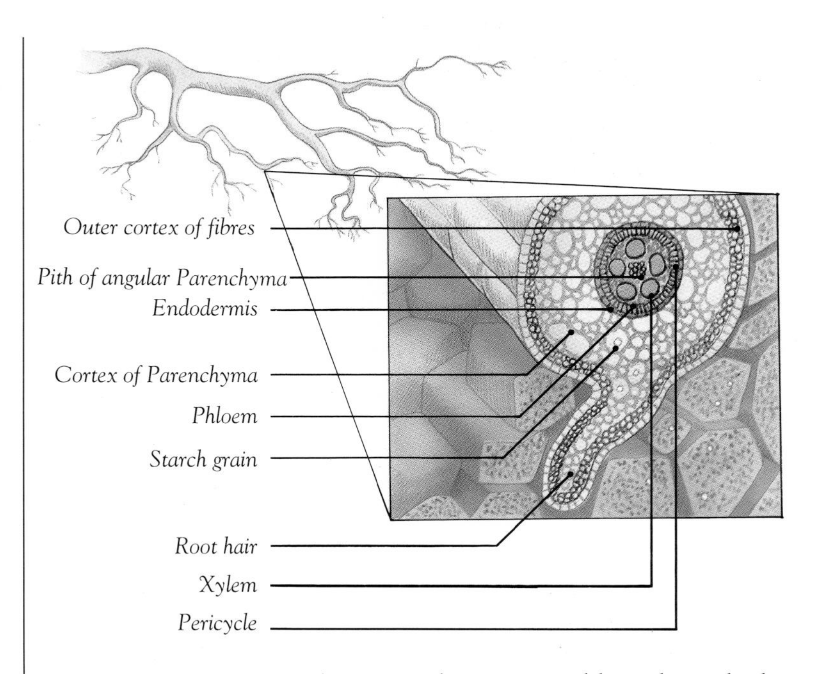

A root (LEFT). The hairs absorb water and mineral salts; a cortex of large thin-walled parenchyma cells surrounds the single layer of box-like endodermis cells, inside which is the pericycle, another single layer. This contains the two food systems, xylem carrying water and phloem carrying organic food. The pith is right in the centre. The stem (BELOW) shows the crucial single-cell layer of growth, the cambium; xylem (sapwood) is inside this, and phloem (inner bast and outer bark) outside it. Sapwood becomes hard, inert heartwood as growth develops. Vascular rays carry food.

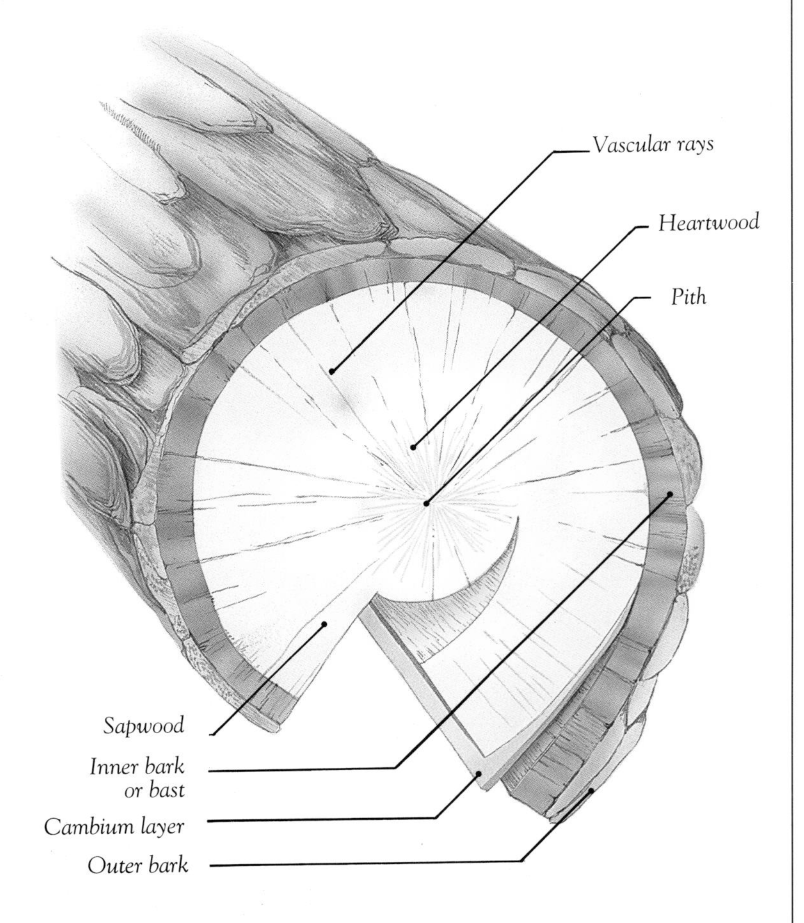

the seeds to the ground. Nut-producing trees, like oaks with their acorns, rely on squirrels, birds or other animals to spread the seeds.

For the majority of seeds, successful growth is impossible. They may land on hard, rocky or infertile ground, thick turf, grazing or farming land tilled by the plough. A few strike good soil, not covered in grass or weeds. Very few seeds find a niche on the forest floor.

The living embryo seedling lies dormant until conditions for sprouting are right and the soil is warm and moist. The bulk of the seed which surrounds the embryo consists of stored food – fats, starch and protein which it needs to start the growth process. But this food supply will soon run out; the seed needs light to make its own food. Water is absorbed, and after a few days the latent plantlet splits the seed case and rapidly grows a root tip and a shoot tip.

By the action of the control chemical auxin, the phototropic shoot tip grows upwards to the light, while the geotropic root tip grows into the soil to perform its function of anchorage, support, food storage and absorption of water and nutrients.

The seed leaves are storage organs. Some, like oak, remain within the husk beneath the ground, and the first shoot grows out from a bud between them. With other trees such as maple and all pines, the seed leaves are raised into the air on an expanding stalk while the husk that enclosed them falls away. At the tip of the shoot is the growing point (apical meristem) which governs the growth in length of the shoot.

Root tips also possess a growing point and develop short-lived root hairs. Via these, inorganic ions permeate freely through the cell walls and air spaces of outside root cells. Water enters by osmosis, the flow from a solution of low salt concentration (the ground water) to high salt concentration (the root hair cell). The water carries in solution salts and elements essential for life, including nitrogen, phosphorus and potassium, and lesser amounts of iron, magnesium, calcium, sodium, sulphur and other trace elements.

Within a few weeks the root system develops, either in the form of fibrous spreading roots growing near the surface, or taproots which can grow six metres (19ft) deep in mature trees. Firs, maple and beech have fibrous roots, but pine, hickory and oak have tap roots. These systems amount to 10% of the tree.

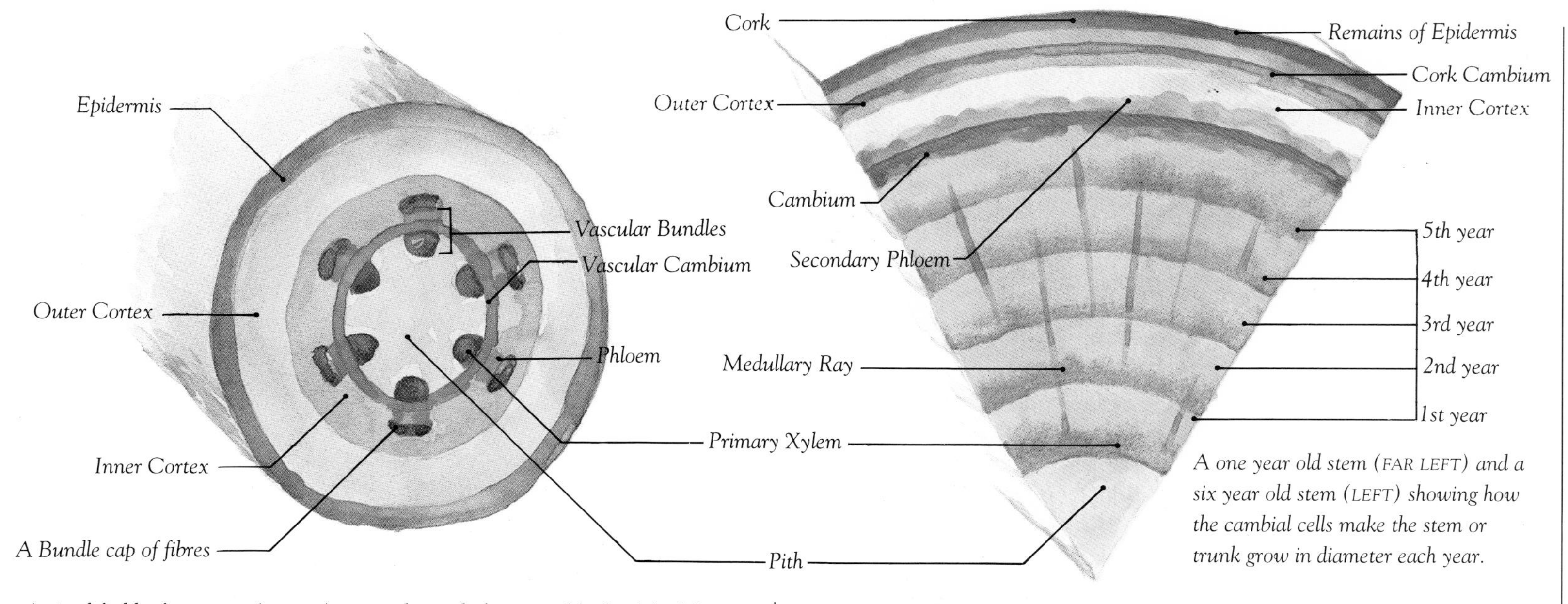

A one year old stem (FAR LEFT) and a six year old stem (LEFT) showing how the cambial cells make the stem or trunk grow in diameter each year.

A simplified leaf structure (BELOW). Both surfaces contain stomata – pores; epidermal guard cells control water flow. The spongy mesophyll lies beneath the vertical 'palisade' of thin-walled parenchyma cells, which contain green chloroplasts for photosynthesis.

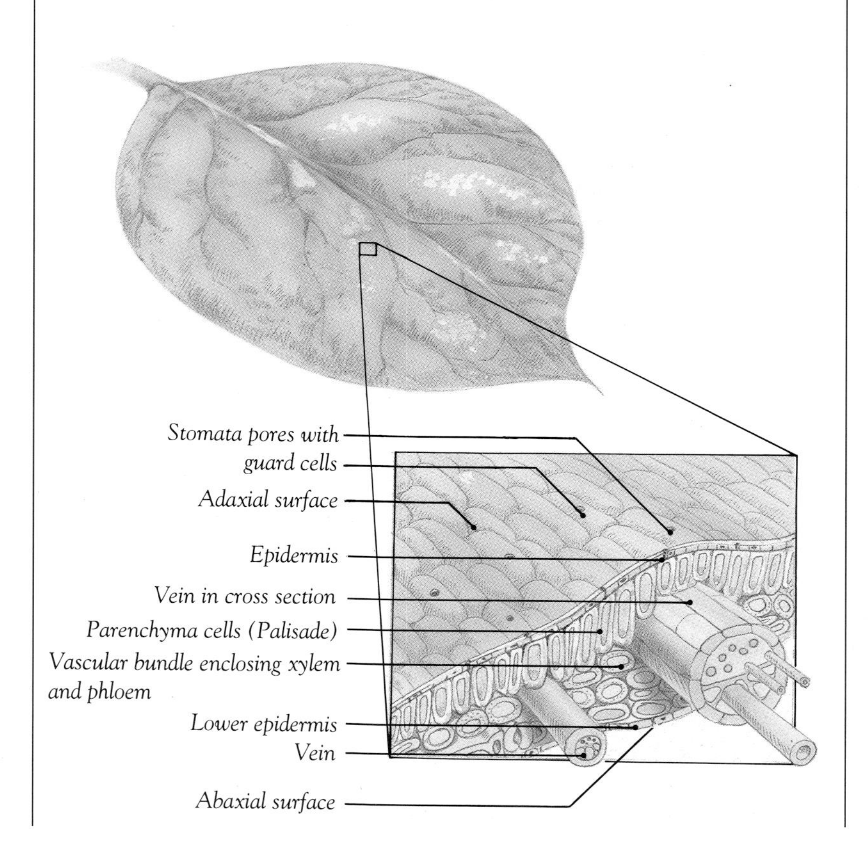

During the first year of its life, the sapling develops a layer of internal generative cells called the vascular cambium, which forms a continuous sheath completely enclosing all living parts of the tree. This single, simple layer, one cell thick, is the crucial element in the tree's growth. During periods of activity the cambial cells cause the trees to increase in diameter; they multiply and divide into new cells, forming xylem on the inside of the sheath, or phloem on the outside, making a layer of 'bast'.

Large volumes of air, consisting of nitrogen, oxygen, and a tiny fraction of carbon dioxide, are filtered into the organism through the leaves. Departing air carries with it large amounts of water. This constant evaporation, known as transpiration, is passed through perforations in the leaf surface called stomata. In a constant stream, lost water is replaced by more, carried from the roots via the xylem. On a summer's day a large tree may take up 100 gallons of water from the roots, and lose 90% of that through the leaf stomata by evaporation.

In the leaf cells are particles called chloroplasts containing chloropyhll. The solar energy absorbed by the leaves activates this photosynthetic green pigment, which gives the leaves their colour, to create carbohydrates from carbon dioxide and water. Oxygen is formed as a by-product. These combine with the water and mineral nutrients drawn up from the roots in the complex chemical process known as photosynthesis, which produces sucrose sugar, a soluble carbohydrate than can flow freely through the leaf veins. Thus the leaves feed vital energy back down to all parts of the growing tree to create new tissue.

The tree now has a two-way system. Dissolved mineral salts ascend from the roots through the xylem tissues of the sapwood, and the energy-giving sugar leaf-sap travels down to the roots and throughout the tree in the outer phloem tissues, or bast. The resulting wood is a chemical compound of cellulose and lignin and various minerals which contribute to its density and hardness. The tree also develops a protective coating of cork cells that we call bark, which is waterproof so that the underlying tissues do not dry out. It has pores called *lenticels* to admit air and shield the nutritious bast beneath it from gnawing birds and beasts; it also acts as a barrier to fungal spores and a buffer to temperature changes.

Growth Rings As the sap rises in the spring and ends in the autumn in temperate climates, this interrupted growth cycle causes a distinctive wood layer to form which is called a growth ring. These are not 'annual rings' by which to count the age of a tree, because cold winters or periodic droughts can interrupt the growth cycle. In the tropics the growth may be continuous, and the wood may appear not to have any growth rings at all.

Each growth ring has two distinct zones. Earlywood cells formed in the spring have thin walls and large cavities, while latewood cells formed in the summer have small cavities with thick walls. It is the contrast between these two layers which enables us to identify the tree with the naked eye, when viewing the end section of a piece of timber. A colour difference between earlywood and latewood can also be seen on the longitudinal or tangential surface in some woods.

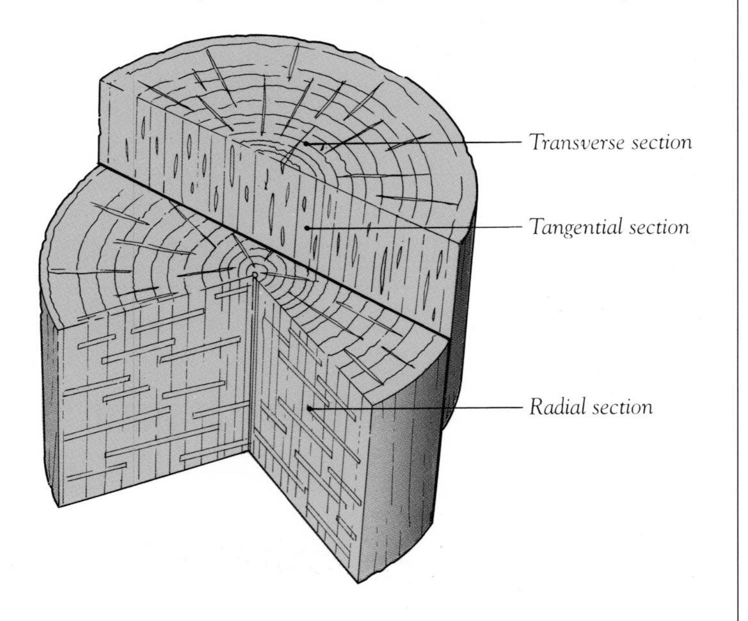

The trunk of a tree showing the various ways in which it can be cut (ABOVE)*. A transverse section is a straight cross-section across the trunk: a tangential will give an idea of vertical structures; and a radial section shows vascular rays to best advantage.*

The wood scientist studies the number of growth rings to the inch and the proportion of earlywood to latewood cells to determine the wood's strength, toughness and durability. The width of the growth rings varies according to the moisture absorbed by the tree, and will show the annual variations in climate, and all the hardships such as forest fires, insect or fungal attack endured during the tree's lifetime.

Growth rings will also show the presence of 'reaction wood', which forms with a special structure and chemical composition to help the tree overcome mechanical stresses – near the base of the trunk where the weight is carried, or if it is growing on a hillside, for example, the rings will be wider spaced on the 'uphill' side, indicating an extra pull to make the tree grow upright; such 'tension wood' occurs more usually in hardwoods. Softwoods exhibit an equivalent reaction, by producing 'compression wood' with much tighter growth rings, on the downhill side of the tree for a pushing effect. Strong branches, high winds, stones or foreign objects that become included in the body of the tree, all produce reaction wood; it is characteristically difficult to work, becoming unpredictable under the saw or blade because of the release of stresses, and showing disproportionate longitudinal shrinkage.

As the tree increases in diameter, the cambium layer just below the bast grows further away from the inner sapwood cells which become inactive, cease to conduct sap and die. Only the parenchyma, the food storage cells, are left alive as the wood transforms into a darker colour and becomes what we know as heartwood, which provides strength and support to the tree. It is far more valuable commercially than the weaker sapwood.

CELL STRUCTURE

All wood is formed of tube-like cells which differ in appearance according to their three main functions. Vessel cells carry the sap throughout the tree; supportive cells provide the tree's strength, and food storage cells appear in the form of soft tissues that comprise up to a fifth of the tree's total volume. The arrangement of these three types of wood cells into tissues, and their visual appearance form distinctive patterns in every tree and are used as an aid to positive identification by the wood anatomist.

Under a microscope, softwood cells appear to have a honeycomb structure. There is a special type of cell, the *tracheid*, which functions as both a vessel and a supportive cell. Tracheids are thin and blunt-tipped, with pitted side walls and large open cavities with rounded ends. They overlap each other so the pits of adjoining tracheids connect, enabling the sap to flow between them. Earlywood tracheids are thinner walled, with larger internal cavities and more pitting for better sap conduction; latewood tracheids develop much

GROWTH RINGS

Endgrain of a trunk of plantation softwood showing the growth rings (RIGHT); the tree below has much denser latewood, a result of slower growth.

Endgrain of oak, Querqus Robor (BELOW). Note the difference between the dark heartwood and light sapwood, and the distrubance in the growth rings (tension wood) round the branch, seen as a knot.

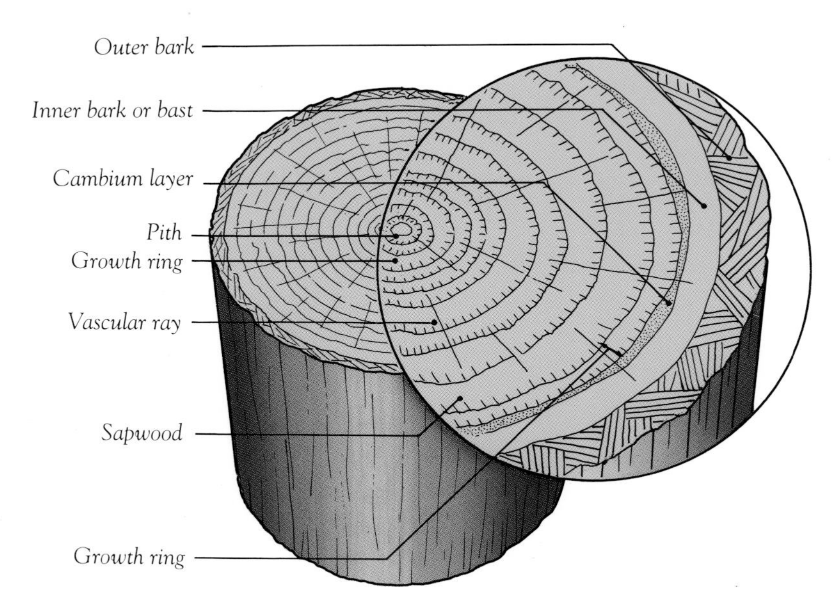

A section of trunk showing cambium, phloem, xylem, heartwood and pith (LEFT). Poor conditions cause slow growth and close rings; faster growth can be detected because the rings are farther apart and the proportion of dense 'summer wood' or latewood is smaller. Quick-grown timber is generally less structurally strong than slow-grown.

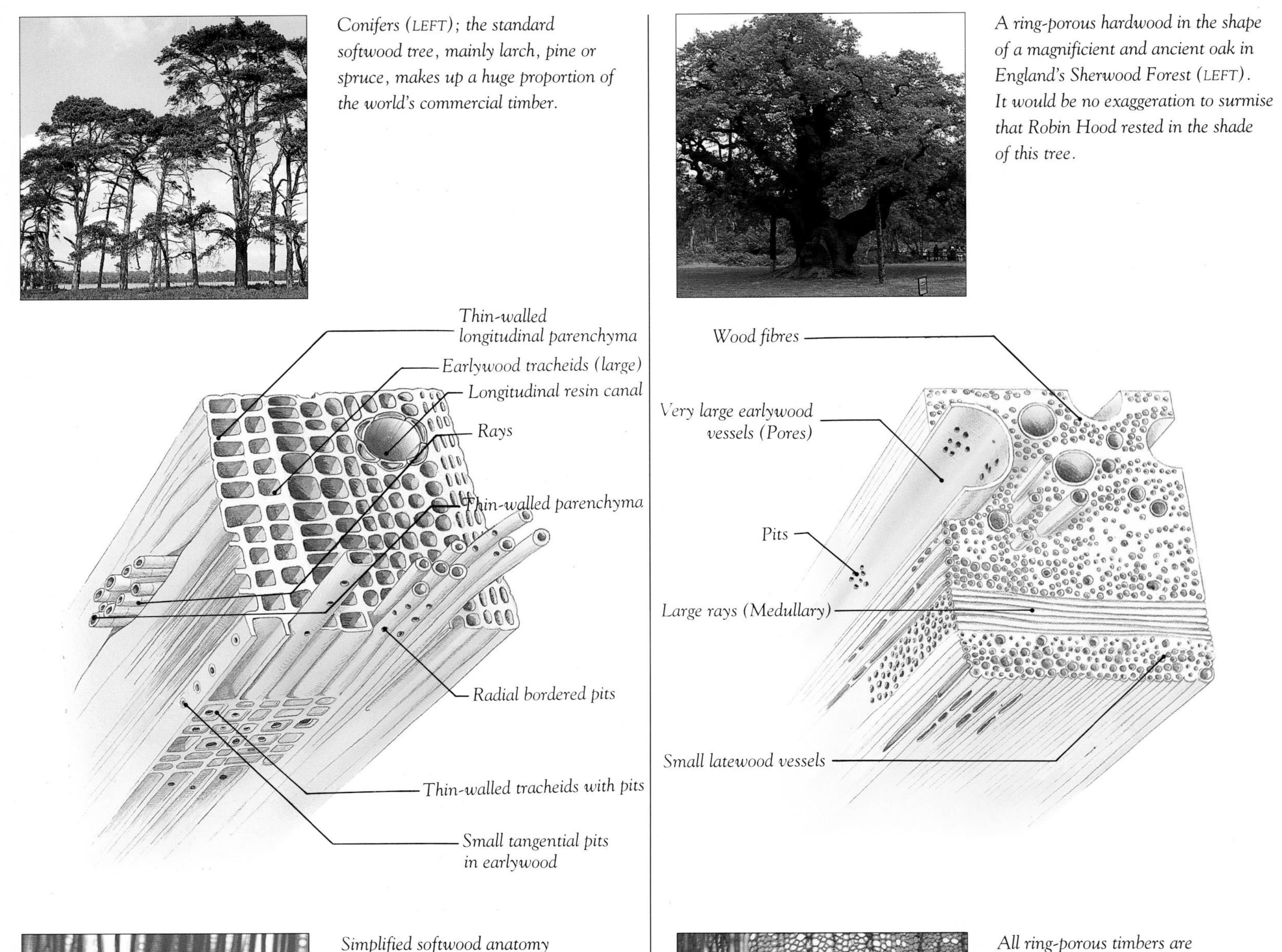

Conifers (LEFT); the standard softwood tree, mainly larch, pine or spruce, makes up a huge proportion of the world's commercial timber.

A ring-porous hardwood in the shape of a magnificient and ancient oak in England's Sherwood Forest (LEFT). It would be no exaggeration to surmise that Robin Hood rested in the shade of this tree.

Simplified softwood anatomy (ABOVE): the thick-walled tracheids which give strength can be seen at the right, the thinner-walled earlywood tracheids at the left. The fine medullary rays, for food storage, travel across the tracheids; the large opening is a resin duct. A micrograph of a pine tree cut tangentially (LEFT). The bordered pits on the tracheids can be clearly seen.

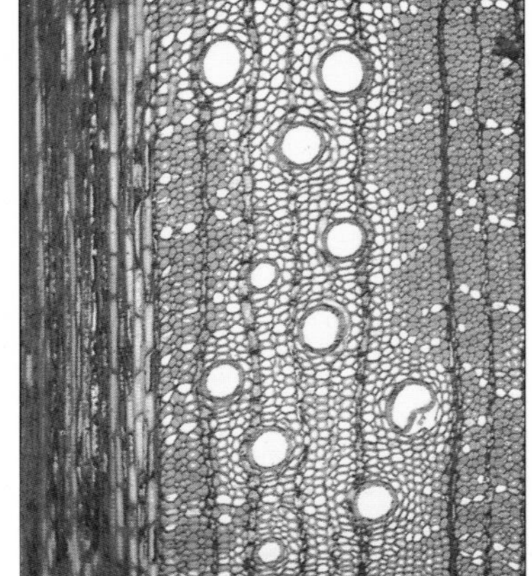

All ring-porous timbers are characterised by the marked difference between earlywood, with its large pores, and the small pores of latewood (ABOVE). The contrast itself forms the growth ring. Strength is gained from the large number of fibre cells in latewood; parenchyma tissue, laid in rays at right angles to the rings, holds food. Pores or vessels are shown in the micrograph (LEFT) of a cross-section of oak summer wood.

The diffuse porous maple (LEFT), familiar to Americans and Canadians as a source of sweetness and to dancers and sports lovers on both sides of the Atlantic as the best flooring timber there is. The consistent cellular structure is the secret of this hardwearing reputation.

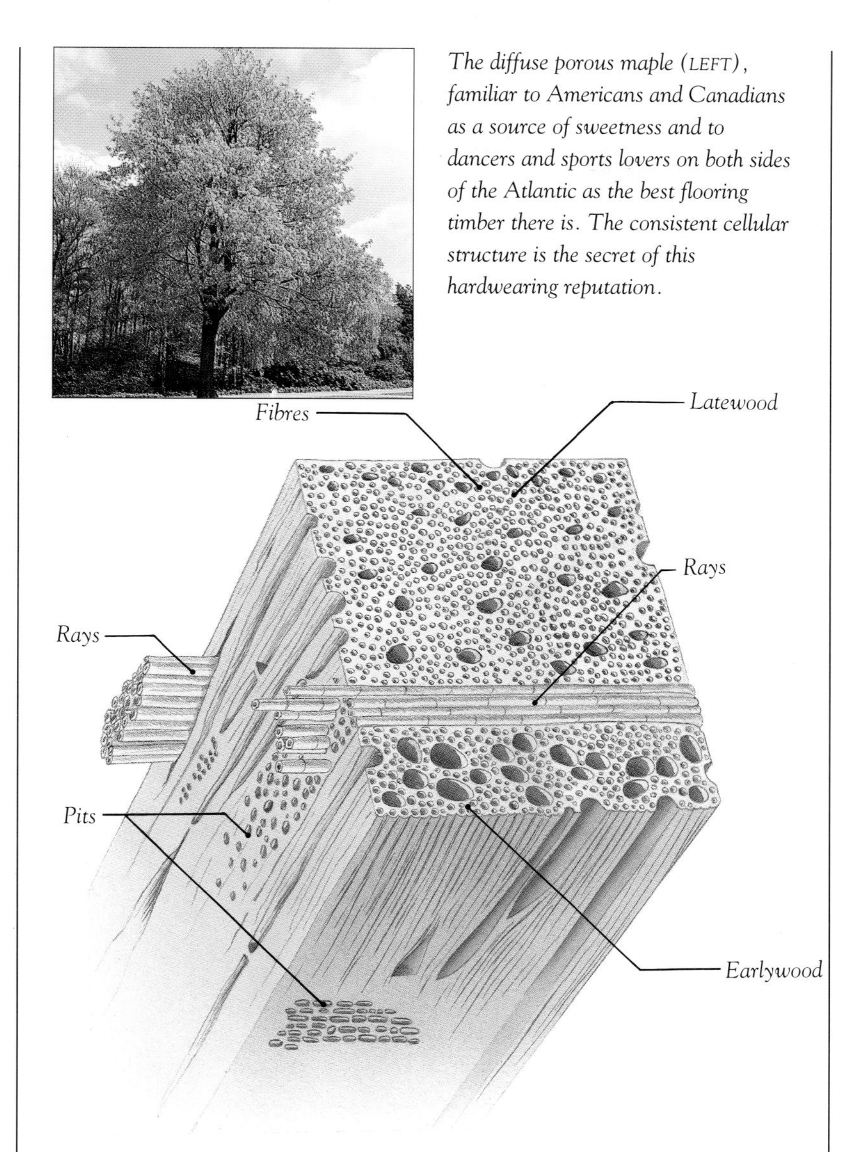

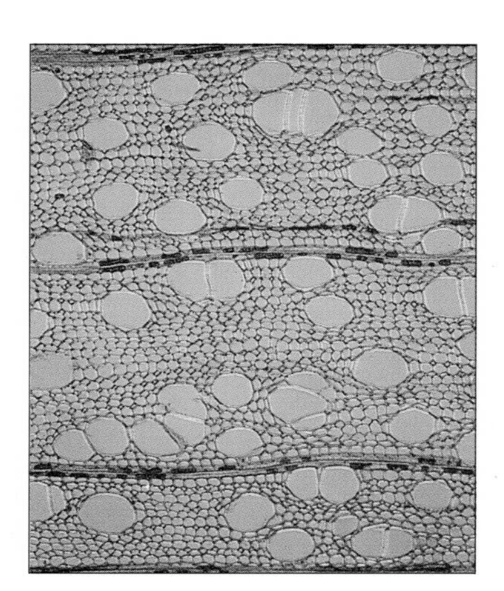

Diffuse-porous woods (ABOVE) show little or no gradation in size between early- and latewood pores or vessels, which are surrounded by fine strength-giving fibres. Tropical timbers, which grow continuously all year, show almost no growth ring pattern at all. Storage tissue, not so pronounced as in ring-porous hardwoods, lies in the form of rays across the pores and fibres. A cross-section micrograph of maple displays the regular size of the pores (LEFT).

thicker walls with smaller pits in their sidewalls and very small cavities. Their function is to provide strength rather than sap transport. In all softwoods, sap conduction and mechanical support are performed by both earlywood and latewood tracheids, which measure about 3mm (1/8in) long.

Hardwoods, in contrast, have two entirely different types of cell for sap carrying and mechanical support. The 'vessel' cells have thin walls, are open-ended and piled one on top of the other like short lengths of drainpipe. In cross section, vessel cells are visible as pores which on endgrain surfaces can be seen as very tiny holes. In fine-textured wood, pores can only be seen with a magnifying glass.

Where the earlywood pores are much larger than the latewood ones and form well defined growth rings, the tree is known as 'ring-porous', for instance ash or elm. When the pores are scattered across the growth ring with little difference between early and latewood pores, the wood is described as 'diffuse-porous'; beech, birch, cherry and willow are examples. Pore arrangements in hardwoods are described as 'solitary' if they are isolated from each other; 'clustered' if they are of a given number, or in 'chains', in which case the direction of the chain becomes a diagnostic feature.

The mechanical support of hardwoods is performed by fibres, which are long, narrow, minutely pitted, thick-walled cells, sharply tapered at each end. They make up the bulk of the woody tissue and are similar in structure to the latewood tracheids of softwoods, except that they are only about 1mm long.

The vessel cells of hardwoods do similar work to the earlywood tracheids of softwoods, and the hardwood fibres are the equivalent of the softwood latewood tracheids. The major difference between the woods is the complete absence of pores in softwood growth rings, and the presence of pores in hardwoods.

Food Storage This function in both softwoods and hardwoods is carried out by thin, brick-shaped cells of soft tissue called 'parenchyma', which form into rays running radially to the tree's vertical axis, horizontally and at right angles to the growth rings. They originate at the cambium and run towards the pith at the centre of the tree – the wood formed in its first year.

These rays vary from extremely fine to wide and broad. They might be of the same width, as in chestnut, or of two different widths, as in oak; they may combine to form an aggregate ray formation, as in alder. There are many different groups of parenchyma, and its absence or presence, its type, form and distribution are valuable diagnostic features for wood identification.

Many softwoods, such as spruce, larch, Douglas fir, and the pines, have resin ducts, surrounded by secretory parenchyma cells which exude resin into channels. Sometimes resin ducts are found in hardwoods such as meranti, seraya and the mahoganies.

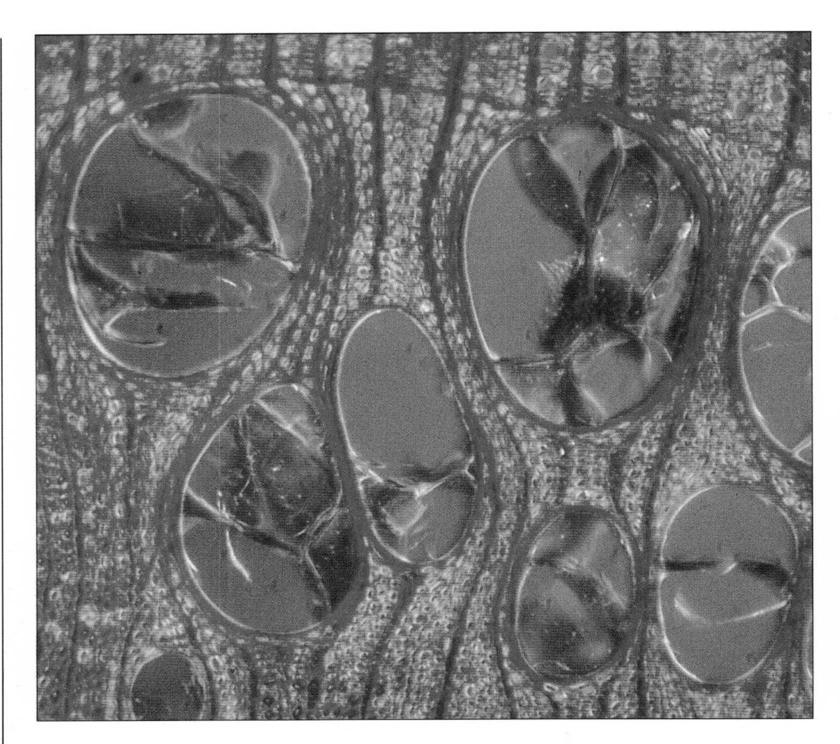

The photomicrographs on this page are taken with polarising filters, which differentiate in colour between the various forms of tissue. A cross-section through white oak (LEFT), showing the large vessels in spring wood; the vessels in this sample of sugar maple appear blue (BELOW). Cross-section through the cambium layer of basswood (RIGHT).

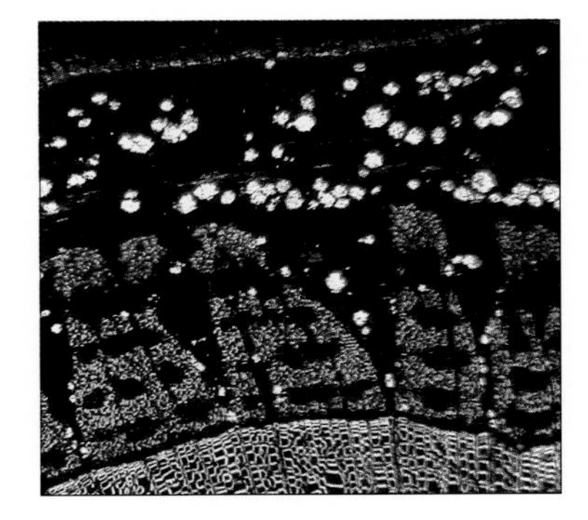

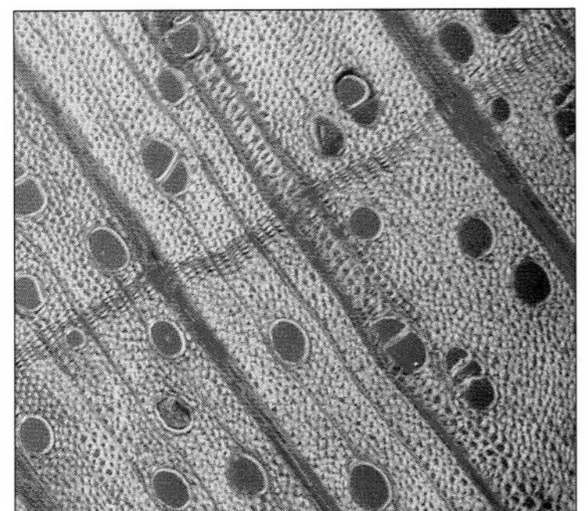

A light micrograph of the centre of an oak tree (BELOW LEFT), showing the inert pith fibres at the centre. One season's growth is shown in the two sets of larger earlywood pores. Longitudinal section through the sap-conducting pipe-like cells of elm (BELOW); cross-section of Platanus occidentalis *showing the consistently sized vessels (BOTTOM).*

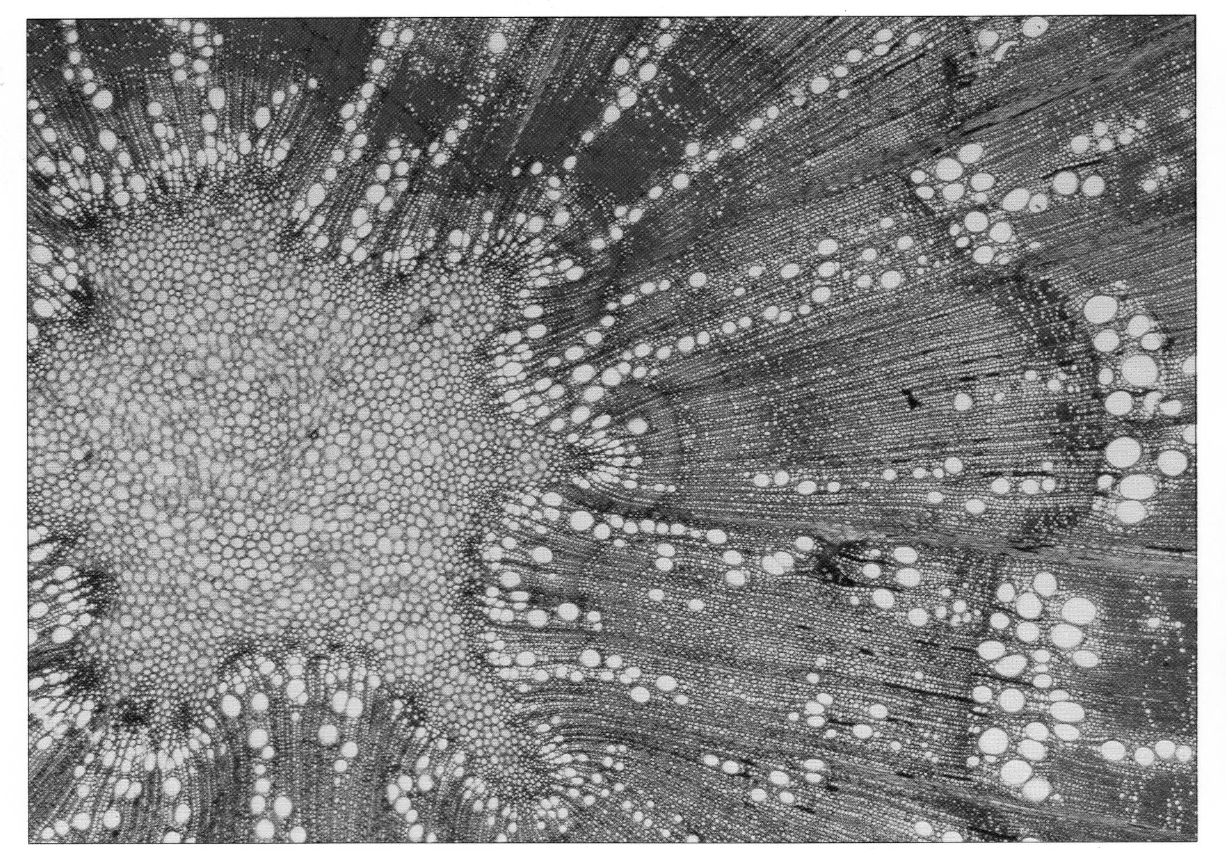

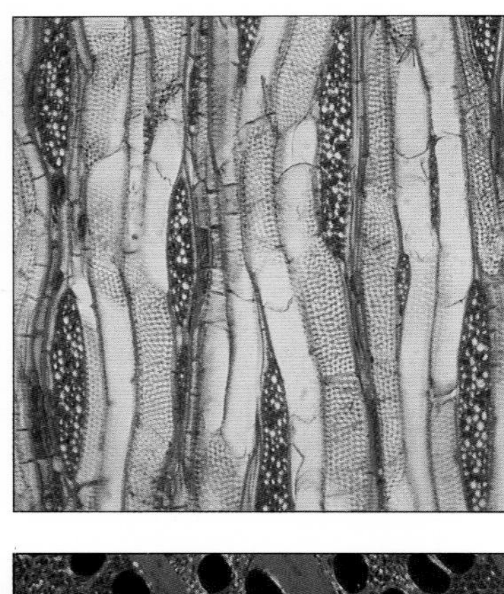

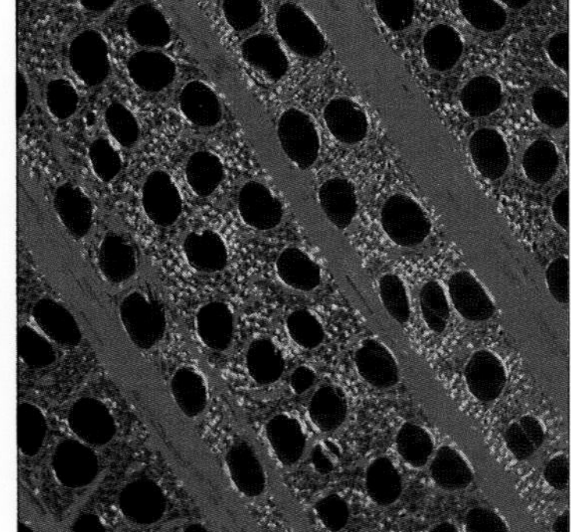

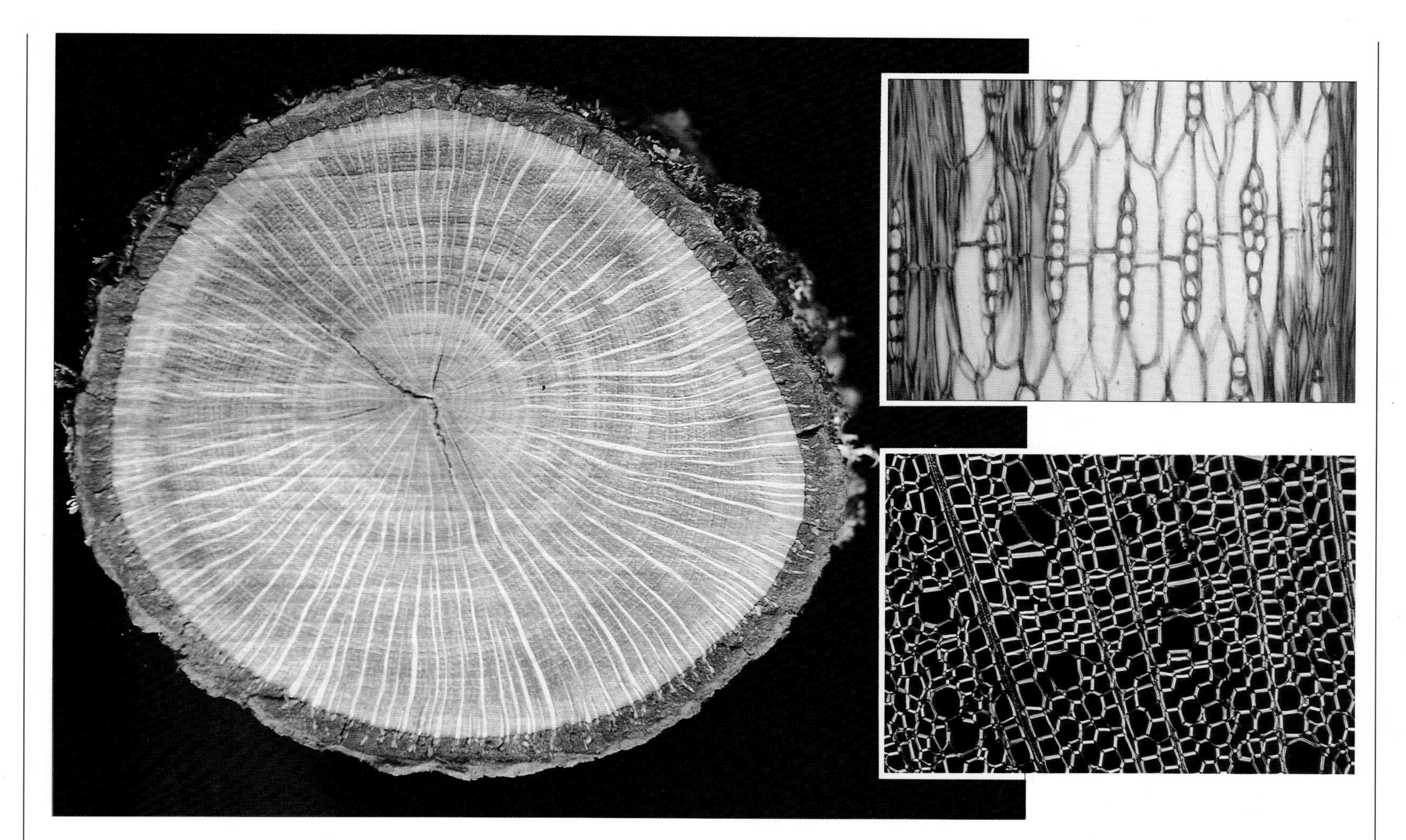

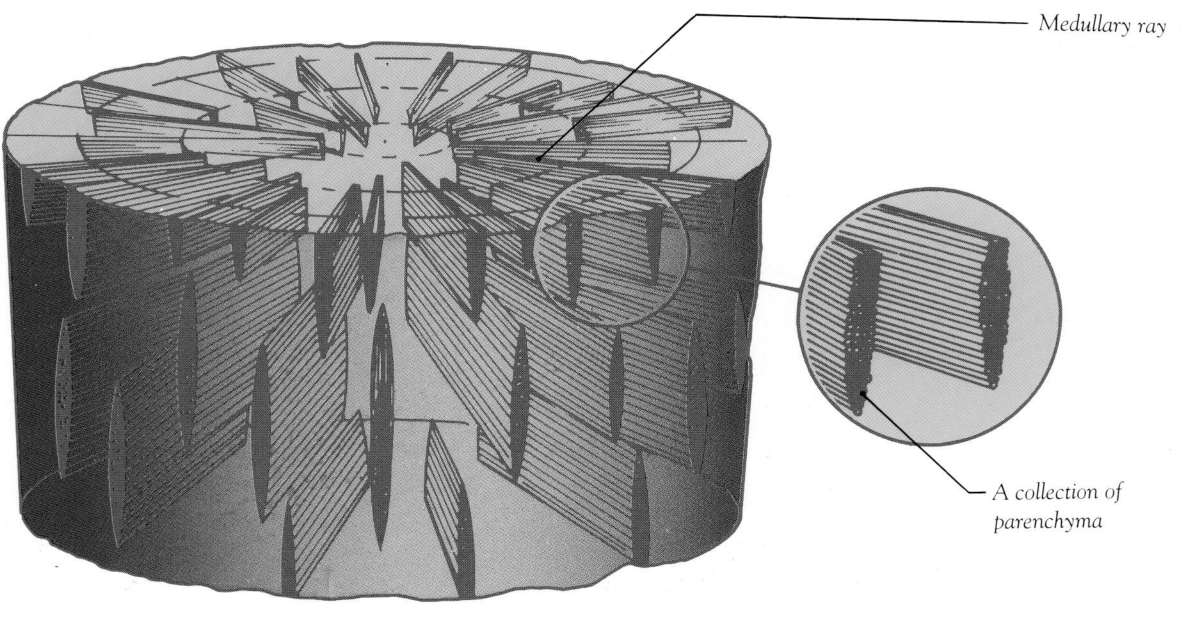

Medullary or vascular rays are collections of parenchyma, cells which store food, characteristically arranged radially (RIGHT) across the vertical lie of fibres and vessels (pores). Rays often produce a very characteristic appearance, as in this excellent sample of oak (ABOVE), whose 'silvering' is highly sought after in furniture and panelling. A tangential section through mahogany (TOP RIGHT), looking at the open ends of the tiny rays bunched between the vessel cells; a lime twig in cross section showing the rays as thin solid lines running between the red-coloured xylem cells (CENTRE RIGHT), laid in surrounding non-radial parenchyma.

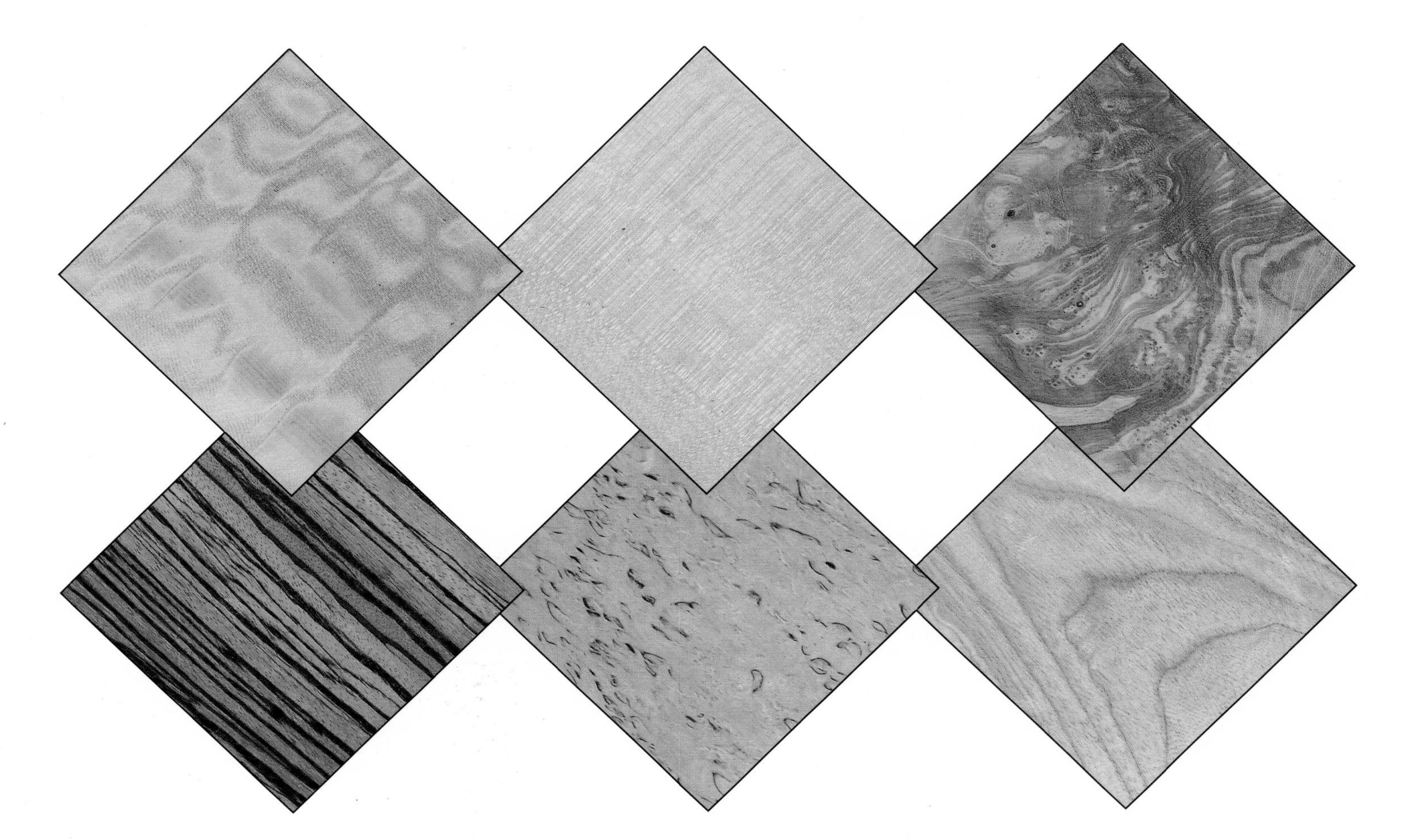

Quilted Japanese horse chestnut (TOP LEFT), *lace figured sycamore* (TOP CENTRE), *olive ash stumpwood* (TOP RIGHT). *Dark-striped zebrano* (BOTTOM LEFT), *disease-flecked masur birch* (BOTTOM CENTRE), *flat-cut Japanese ash* (BOTTOM RIGHT)

GRAIN AND FIGURE

If you say a piece of wood has beautiful grain, you are more likely to be talking about the surface pattern. Grain, strictly speaking, refers to the lines visible on a cut board that show the intersection of the growth rings and the plane of the board itself.

The natural arrangement of the wood fibres in relation to the main axis of the tree produces eight types of grain. Straight grain is where the fibres run parallel to the vertical axis of the tree; irregular grain is caused where the fibres contort around knots, swollen buttresses, or crotches.

Cross grain occurs where the fibres are not parallel to the main axis of the tree, and wavy grain where the fibres form short waves in a regular pattern. Curly grain forms where these waves are in irregular sequence. Spiral grain develops when the fibres form a spiral around the circumference of the tree. Diagonal grain results from a flat cut board of spiral grain. Interlocked grain occurs when the fibres change direction at intervals from a right-handed spiral to a left-handed one and back again.

The pattern on the surface of wood is correctly known as the 'figure', and results from the interaction of combined natural features and the way the log is cut to achieve an effect. Natural features include the scarcity or frequency of growth rings; colour and tone variations between early and latewood cells; pigments and markings in the structure; contortions around knots and butts, and so on.

A fine striped figure can result on a radially-cut board when there is a marked variation in density between early and latewood cells. Wavy-grained timber provides 'fiddleback' figure or, when combined with spiral growth, a block-mottled figure as in avodiré.

'Roe' figure is obtained when both interlocked and wavy grain combine as in afrormosia. Sometimes reverse spiral growth occurs which causes interlocked grain, and when this is radially cut it produces a beautiful close ribbon stripe as in pencil-striped sapele. A variation in pigmentation can yield a dark stripe as in zebrano, and other combinations of grain result in such attractive figures as blistered, quilted, plumpudding, pommelle, moiré, and snail quilt.

Wood that is radially cut, the faces of the boards parallel with the rays, produces a strong or deep ray figure as in 'raindrop' figured oak, lacewood from the plane tree, 'lace' figured sycamore and Australian silky oak.

Burrs are wart-like other phenomena growths which affect some trees. They give the appearance of tight clusters of dormant buds, each with a dark pith caused by stunted growth, that failed to develop into branches. For turners especially, burrs are the most highly treasured of all woods.

Curls Logs are cut from a point just above the root buttress, to a point below the first limb or fork of the tree. When this is cut through, it forms an attractive 'curl' or 'crotch' figure, sometimes called 'feather', elliptical in outline with a strong central plume.

Texture This is governed by the variation in size of the early-and latewood cells. Diffuse-porous woods like boxwood, with narrow vessels and fine rays, are fine-textured; ring-porous woods like oak, with wide vessels and broad rays, are coarse-textured. Mahogany is medium-textured.

Lustre This is the ability of the wood cells to reflect light and is related to texture. Smooth, fine-textured woods are more lustrous than coarse-textured ones.

Odour Resinous pines and many other woods have a strong natural odour. Camphor wood is used to line the interior of clothes closets, and cigar-box cedar to make humidifiers.

Lucinda Leech's strong-grained ash sideboard (LEFT) and John Makepeace's burr oak and burr elm serving table (BELOW) both make use of visually interesting grain or figure. The twisted oak (BELOW LEFT) will show spiral grain.

WORLD FOREST TYPES

The main types of forest are determined by climate, which means location in relation to the equator, and altitude. They are generally delineated in 'band zones', but of course there is great overlap: conifers, for instance, grow from just below the polar regions right into areas of Mediterranean climate.

Coniferous Forests Above a latitude of 60° north, the great coniferous evergreen forests or taiga encircles the polar regions. It stretches from Siberia across Northern Europe and Scandinavia, into Canada and Alaska, to within 3,200km (2,000 miles) of the North Pole, where the forest gives way to treeless tundra.

Cone-bearing trees also grow in temperate zones. The southern pines of the USA survive the winter by restricting water loss through their exceptionally long and tough, narrow needle-shaped leaves with blue waxy surfaces. Conifers can withstand drought because they adjust to a climate where precipitation falls as snow and the ground water freezes for a great part of the year. The lack of water is just as severe as it would be in the drought conditions of a hot desert. Xerophytic or water-conserving conifers can grow in the frozen north, on limestone mountains, or in the Mediterranean. Douglas firs grow from Mexico to Alaska, especially on the eastern slopes of the Sierra Nevada where the giant redwoods flourish.

Temperate Hardwood Forests Broad-leaved forests stretch across the temperate zones of Europe, Asia and America, broken only by the high mountain ranges or arid prairies. Oak, beech, ash, birch, and maple occur in pure or mixed stands, with other isolated trees, in the most northern regions.

There is ample summer warmth and sufficient rainfall, even in cold winters. These deciduous trees survive by shedding their leaves in autumn, causing the circulation of sap to stop moving. It doesn't fall as is commonly believed. These trees cannot draw water from the soil when the temperature falls bellow 5°C (40°F). If they retained their thirsty leaves they would die of drought through transpiration of water they could not replace.

Mediterranean Forests To the south of the temperate region, mixed forests of conifers and broad-leaved evergreen trees occur.

Sub-tropical Forests Below the temperate zones there are the sub-tropical forests of Europe, Central and West Africa, India, Asia, South Africa, Central and South America, and Southern and Western Australia. Each has its own characteristic forest types,

Mixed temperate hardwoo in Tuscany.

Map showing the distribution of Coniferous, Mixed Hardwood, Light tropical, Temperate Hardwood and Mediterranean forests.

KEY;

- *Coniferous forest*
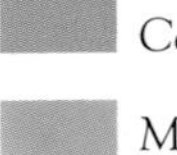
- *Mixed Hardwood and Coniferous forest*
- *Light Tropical forest*
- *Temperate Hardwood forest*
- *Mediterranean forest*

Mixed hardwood, predominantly beech, in the New Forest, England.

The coniferous forests of the far north.

The sub-tropical forest in the Abel Tasman National Park, New Zealand.

Mixed coniferous and deciduous Mediterranean forest in Portugal.

designed to resist summer drought rather than winter cold. Long annual dry seasons oblige the broad-leaved trees to conserve their water supply in tough, thick, leathery, waxy-surfaced leaves, which are sharp and spiny to resist animals.

Savannah Forests Almost a third of the land surface of the world is arid, with high daytime temperatures and very cold night-time ones. Such deserts occur in Australia, India, Southern Arabia, Africa and in both North and South America. Not all deserts are flat; some are mountainous like the Painted Desert in the USA and provide 'waddies' for the desert trees.

Savannah-type forest occurs in North America where there is heavy rainfall: on the open prairies of Canada east of the Rockies, and on the plains of the United States as far south as the Gulf of Mexico. Large areas of savannah forest occur right across Africa and down to South Africa, where scattered species can be found similar to the familiar flat-topped acacia of the African savannah.

Tropical Rain Forests In South America, West, Central and East Africa, and from India, Malaysia and Indonesia throughout South-East Asia to Papua New Guinea, very high temperatures and high annual rainfall encourage a dense growth of lush forest in a wide belt round the equator. Beautiful and essential to the earth's eco-system, these forests are now under severe and immediate threat. There is an extraordinary variety of trees, hundreds of different species per square kilometre, often up to 30m (98ft) tall, with long clear boles up to 20m (65ft) before the first branches. Even taller trees, known as emergents, project above the main canopy up to 50m (164ft). Each major region has its own specialities.

Palm trees also occur in these forests, namely coconut palms, sugar palms, betal, oil and date palms. In tropical forests leaf fall and flowering and fruiting are not seasonal but continuous. These mature 'high forests' are typical of tropical rain forests, where rainfall exceeds 1,500mm (60in) per year with no prolonged dry spells.

Montane Forests The 'band zoning' of forest from the North to South poles is paralleled by changes in altitude. One can travel from the hot desert scrub, too dry for tropical forest, up through broad-leaved trees to coniferous forest, up above the treeline, through alpine tundra into the snow. The treeline varies according to the annual rainfall, shelter, temperature, and soil conditions. Montane forests occur in Mexico, Peru and Chile, East Africa, China and Tibet, and in Western Australia.

Mixed hardwood and Coniferous Forests occur in the Northern Temperate zone. Further north coniferous trees predominate and conversely there are more hardwood trees in the southern areas.

The mangrove forests are located in river deltas in tropical areas.

Map showing the distribution of Savannah, Tropical Rain and Montane forests.

High montane forest in Rwanda, Africa

Dense and luxuriant tropical rainforest

Scattered Savannah forest in Port Moresby, Papua New Guinea.

THE RAINFORESTS

At the present rate of destruction, most of the world's vast areas of tropical forest will have disappeared by the end of the first quarter of next century – in about 35 years' time. With them will go half our plant and animal species, many of them never even identified or named. Billions of people will be affected by changing climatic conditions, soil erosion and lost water resources.

THE CURRENT SITUATION

Rainforest covers a world total of about 3½ million square miles, 7% of the earth's land surface. Since the beginning of this century, say researchers, 50% of the existing rainforest has been lost, and another 25-30% will have gone by the year 2000. It is currently being destroyed at the rate of 80-200 million hectares a year, or between 1 and 2% of the original total area; put another way, and taking the low figure, that amounts to 100 acres a *minute*.

Some of these forests have existed in much the same form for millions of years. This long history has contributed to their extraordinary diversity, in which there may be over 200 species of tree in a single acre of ground. Madagascar alone supports 2,000 tree species; Canada and the continental USA together boast a total of 700 species, yet just 10 one-hectare plots of Borneo forest have yielded the same number.

One of the most frightening factors is the speed at which the remaining forest is being removed. For instance in Peninsular Malaysia over the last 20 years, one-third of the forest area has been cleared, not by extracting the same quantities annually but as a result of increasing the rate exponentially from 2,248,000m^3 in 1960 to 10,428,000m^3 in 1981. An increase in rate is not the overall picture, however, so it is fair to say that there is a 1980s world average of some 40,000 square miles 'conversion' a year – which does't mean total

The declining rainforests, (BELOW). *The accelerating rate of disappearance is as fearful as the disappearance itself – it is a geometric progression, like the population growth which causes the main bulk of the destruction. It is easy to blame logging and the international trade because greed and environmental vandalism are identifiable, but the profit seekers are not the prime culprits.*

Clear felling north of Manaus (LEFT) in the Amazon basin. Such heartless destruction looks dramatic, but Brazil is no longer the biggest exporter of tropical timbers; most enter the trade from South-East Asia. The Amazon forest disappears largely for farming and firewood. A logging track in Papua New Guinea (BELOW). Timber companies open up the forests to other destructive influences, but at the same time the forest is continually regenerating.

destruction; it signifies degradation of forest from its natural state. Even against that 40,000 square miles 'average' must be placed the 1987 figure for forest burning in Brazil alone – 30,000 square miles.

As an example of the effect of the destruction on the life of a country, Nigeria presents a drastic picture; figures from the Food and Agriculture Organization (FAO) of the United Nations, which plays a central, if controversial, role in information gathering, state that in the years between 1976 and 1980, 285,000 hectares a year were being deforested. The export value of forest products declined from £22.1m in 1973 to £0.7m in 1984, while imports of timber increased from £49.7m to £203m over the same period. Thus timber imports rose in 11 years from just over double the value of wood-related exports to more than 200 times as much!

It is difficult to believe that a natural disaster on this scale is actually taking place now. Yet the first signs are already being felt, such as the 'greenhouse' effect – the gradual warming of the atmosphere caused by a build-up of carbon monoxide, which traps the sun's heat. When the vast forests become contributors to that build-up by burning, rather than playing their natural role of consuming the gas, the ecological implications are dire.

Financially hardwoods are an important export for many Third World countries. Worldwide demand for tropical wood has increased dramatically and is currently worth 8,000m US$ per year, although Britain imports significantly less than 50 years ago. Japan, on its own one of the world's biggest markets, operates many of the logging companies in Asia. It imports raw material direct, while the USA's imports are more manufactured and semi-manufactured goods. Despite reserves of native wood, the USA. imported 2,200m US$ worth of tropical timber products in 1984, and expects requirements to have doubled by the end of this century.

Rainforest has a fragile structure. The soils are poor, and all the contents of the forest depend on constant recycling of nutrients and a complex interdependence between species. Whilst it can recover from selective culling of its 'crops', too great a disturbance upsets the balance to a point where regeneration becomes impossible.

Sources of many useful products have been found in the rainforest. In addition to timber, local communities harvest rattan, rubber, nuts and fruits, whilst agriculturalists and scientists have developed chocolate, coffee and medicinal drugs: just a few of the everyday products originating in the 'jungle'. Destruction removes the chances of discovering more of these important substances and destroys the habitat of many animal species as well.

STATISTICAL DIFFICULTIES

Vast and remote areas, primitive conditions, difficult communications and political problems all make it hard to compile accurate information on what is actually going on. Much research is necessary, as without information it is impossible to formulate coherent plans to utilize and conserve forest resources. Researchers and conservationists now, however, are beginning to question the expenditure of time and money on information gathering while the forest disappears so fast; quick action is becoming an increasingly urgent priority.

Logging in New South Wales, Australia (LEFT); this stringy bark is growing in a temperate forest. A track will be needed to pull it out.

Bulldozing selectively felled logs in Bougainville, Papua New Guinea (BELOW). Damage is spread to the surrounding eco-structure.

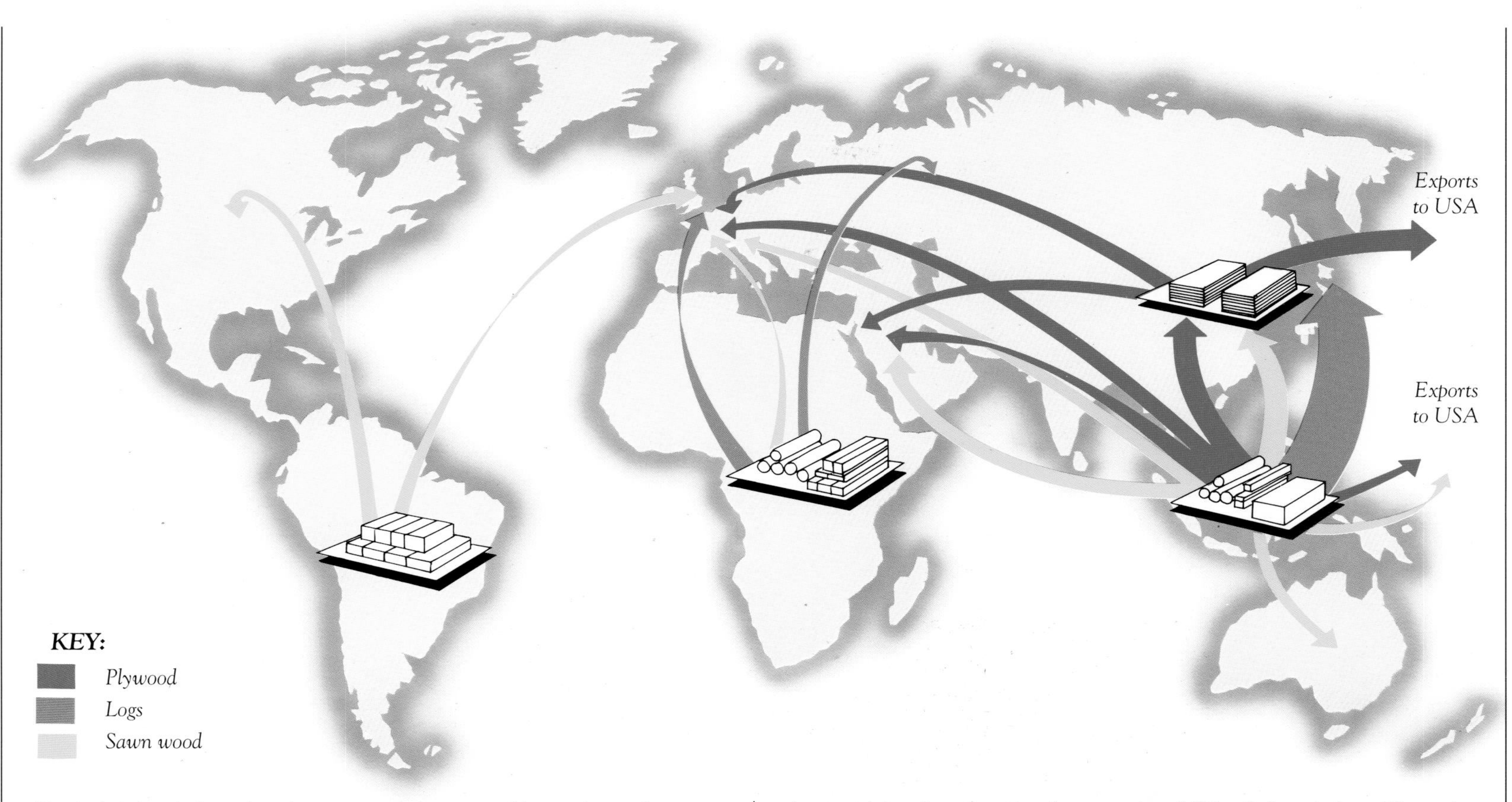

Tropical timber, in its various forms, moves principally from the Third world to the developed countries where it is consumed in ever-increasing quantities (ABOVE).

Recently satellite surveys have helped to assess the actual quantity of forest remaining. The FAO's 1982 survey is a major source of information, but there are arguments about what constitutes 'deforestation'. In India the government claimed that the forest cover was about 22%, until a satellite survey showed the real amount was 10%. This sort of discrepancy has become evident in many other once forested areas.

However, at least this information is being compiled on the spot. Many of us only see statistics gathered at a distance, filtered, interpreted and presented with inbuilt bias. For example, the government recommendation on logging in the Amazon is to extract not more than 10% of the timber so as to allow for natural regeneration. Local companies claim that in any case it is not economic to take more than this because 90% of the trees are too small, too big or the wrong species, so they conform to the requirement anyway. However heavy bulldozers and tractors need large areas for access, thus increasing the amount *removed* for a given amount *extracted*. In addition, up to 50% of the forest can be damaged in the process of extracting 10% of the timber. Thus the total loss can be seen to be much higher than 10%. It is practices such as this which must be controlled.

But there is a considerable difference between all the world's rainforest being logged out and all the economically accessible timber being utilized. Geographically, and thus economically, much of the world's tropical timber is impossible to get at using present-day methods of extraction.

In most places where logging is taking place, only a tiny proportion of species is extracted for commercial use. The same type of tree may vary depending on the precise conditions in which it grew. In addition, different places are threatened by different social, political or environmental problems, so generalized solutions are inappropriate. Rainforests throughout the world vary enormously and so do the people depending directly upon them.

THE COMPLEX ISSUES

Timber use is often identified as a main cause of deforestation, but logging is only one of the many reasons for land clearance. Rapidly increasing populations, coupled with increasing awareness of the 'advantages' of a cash economy, put constant pressure on land. Agricultural needs for subsistence and the possibilities of cash crops

Gathering firewood in a plantation in Rwanda, Africa (ABOVE); *here, the taking of small trees is controlled. Soil erosion on the Himalayan foothills in the Trisuli valley, Nepal* (RIGHT). *Intensive deforestation, including teak cultivation, has left the bare slopes to shed their soil on to the plains and silt the Ganges.*

such as coffee, cocoa and palm oil, mean that large areas of forest land are cleared, frequently by burning rather than using the timber. This is dramatically demonstrated in the Amazon, where huge areas are burnt to create impoverished grasslands for short-term beef production for export to the USA. This is primarily because the forest itself is undervalued, seldom seen as a viable source of revenue in the long term. Since 1962 more than 25% of Central America's rainforest has been turned to grass for beef to go to the US; one-third of Costa Rica's forests have been cut for cattle since 1960, and beef exports have risen by 700%. But Costa Ricans eat half as much beef as they did – less per person than a North American cat.

Using timber obtained by responsible selective methods can provide an income for landowners indefinitely and is a viable alternative to complete forest clearance for agriculture. In any case, the poor forest soils can only sustain tilling or grazing for a short period, so after two or three years more land must be cleared.

Quite apart from its use as a source of timber, conserving the forest has many other useful functions ecologically and environmentally. Several recent disasters caused by large-scale flooding and drought in different parts of the world have been blamed on deforestation. The trees act as an environmental regulator, gradually absorbing and releasing moisture like a giant sponge. Remove the tree cover and violent storms flood an area, washing away the precious top soil in one rapid and uncontrolled burst followed by periods of no water at all.

Consistent Bangladeshi flooding, bringing great loss of life and destruction of property, is blamed on the rising Ganges – six to 12in a year – which is a product of soil erosion from the slopes of the Himalayas. Ten tons of soil per acre per year are lost from the slopes; crop failure and hunger are natural consequences, even before the swollen Ganges hits Bangladesh. And it all starts with the loss of Nepal's forests; approximately half gone since the early 1950s.

Then there is the downward spiral; wood is scarce, so animal dung and crop residue are burnt, thus further reducing soil fertility. As one area of land becomes desert so groups of people move on and clear others. In more developed areas, where land is scarcer, the infrastructure is there to distribute fertilizer to sustain the land artificially and to distribute the results in the form of crops; but impoverished areas do not have this option.

The poorer the population the more likely they are to rely on wood for fuel. Dr. Norman Myers, an independent expert, has estimated that 2.5 million hectares of forest a year goes simply as a result of firewood collection. The poorer the country, also, the more likely the population is to be increasing fast. Brazil is set to double its numbers in the next 30 years. The Amazon is vast, but not big enough to cope with its present demands being doubled that quickly. Many countries regard getting rid of the jungle as a visible sign of progress, but allowing it to be cleared only increases the problems of the poor in the future.

THE POLITICS OF LAND

In under-developed countries governments often trade natural resources with foreign investors in exchange for development of the infrastructure – roads, schools, health posts and so on. However international interests and commercial pressures do not always coincide with benefits to the local people. Forest use will be part of a general plan for 'progress' which some see as inevitable.

In addition to exploiting the natural forest, encouraging the use of plantation timber is important. But the comparatively long growth cycles can make plantations seem an unattractive proposition for private investors. Many companies consider it the government's responsibility to re-plant, paid for by taxes levied on the timber. Frequently the money is not used for this purpose. In the Solomon Islands there is now a regulation that the logging companies must re-plant one third of the areas logged. However the 7½% re-planting tax has not been reduced so the companies are reluctant to comply. Further complications may arise when the traditional land users do not want trees planted, either because they have alternative plans for the land or because they are fearful of losing their rights over it if they allow others to take responsibility for what is grown there.

All too often development is on too large a scale and planned too far away from the reality. Smaller-scale owners become skilled in the utilization of their own land and can care for its future, although they can seldom afford the time and money to experiment with alternatives crops or means of production.

Floods in Bangladesh, (LEFT and ABOVE) are a visible and tragic result of deforestation and erosion in the foothills of the Himalayas.

This logging road in Kimbe, New Britain, Papua New Guinea,(LEFT) is 12 years old; it is cleared three times a year, such is the speed of forest growth. The small village, also in Papua New Guinea,(RIGHT) has only caused limited forest clearance; this is the picture before the roads. After access, (BELOW); 'slash and burn' clearance for cultivation in Java.

However, small-scale methods do not always work when conditions change. Shifting cultivation is carried out by about 200 million people worldwide. In Papua New Guinea this is in the form of 'slash-and-burn' on rotating small 'garden plots'. These work well if 10- to 15- year fallow periods are left between uses, but now larger numbers of people are moving closer together to be near services like medical aid posts, the fallow period is often reduced to around five years. This is not enough time for the soil to recover its fertility, so yields fall and further land must be cleared. Thus indirectly the provision of aid and services can cause a drop in nutritional standards and an increase in forest clearance.

HOPE FOR THE FUTURE

The key to the whole issue of the future of the forest is the possibility of sustainable timber yield through efficient forest management. It is vitally necessary to change current policies and practices from short-term gain to long-term forest management.

Experiments have shown that it is possible to use the forest for sustainable yield of timber and other products in perpetuity without disrupting its function as a natural ecological habitat and environmental protector. It may well become secondary forest, differing in richness from the virgin forest, but still infinitely better than desert.

Management consists of a series of different techniques, including 'enrichment planting' with indigenous species in degraded areas of natural forest, the planting of buffer zones of trees for local use around the forest, selective clearance to encourage particular locally suited species, and careful methods of felling to minimize soil erosion and damage to the canopy. Knowledge is increasing on exactly how much can be extracted before the ability of the forest to regenerate is impaired. It is vitally important that this knowledge is implemented as soon as possible.

It is necessary to encourage such sustainable timber harvesting as an alternative to other cash crops and then to devise a system by which we can identify timber so produced. A decision to patronize

only those producers acting responsibly would be an irrefutable economic argument in favour of conservation.

However, this will mean an increase in the price of the raw material. For far too long timber has been a relatively cheap commodity. Its price should come to reflect not just the cost of extracting and transporting it now, but of replacing it for the future.

TRADE REGULATION

The International Tropical Timber Agreement (ITTA) was negotiated and adopted by 36 producing countries and 33 consuming countries in November 1984 – representing 90% of the world's timber trade. The Agreement, which aims to establish a system of consultation and co-operation between producers and consumers of tropical hardwoods, is administrated by the International Tropical Timber Organisation, which first met in early 1987. It is a focus for optimism because ITTA is the only commodity agreement designed to deal with the rational use and conservation of tropical forests, it is supported by both trade and environmental organisations, and its voting system works in favour of those countries which practise forest conservation. It exists to encourage national policies and to devise means by which sustainably managed timber can become the trade staple.

But there are negatives, of course; like any self-interested system, the timber trade is unlikely to reform itself without a great deal of prompting, and anyway it is generally accepted that trade is not the primary cause of forest destruction. It is very significant, however, and carries with it 'knock-on' effects like migrant populations spreading into the forest on roads built for logging. The FAO has estimated that 70% of annual forest loss is 'directly or indirectly' attributable to logging.

Agro-forestry includes dual cropping with plants like avocado (ABOVE LEFT), here being tended in a Rwandan nursery before going to the plantation. Regeneration (ABOVE); selective felling allows new growth, although the damage looks bad it is the beginning of a new cycle of growth.

One of the crucial questions for impoverished Third-World producer countries is whether their tropical forests can be sustainably managed and still give adequate revenue. A 1988 International Institute for Environment and Development estimate is that only 1% of trade in tropical timber is 'sustainable', which does not paint a rosy future for the producers and traders.

Many countries have now introduced regulations requiring the processing of timber locally, several banning the export of round logs altogether. Converting them into planks on the spot, or better still plywood which involves even more work and thus added value, means that the loggers (usually international companies) have to become processors as well.

In Indonesia in 1978, for example, the government banned log exports by all foreign companies holding forest concessions except those which invested in joint plywood manufacturing projects with Indonesian interests before 1983, thus adding value to its exports. As a result log exports from Indonesia fell from 19.2 million m^3 in 1978 to 3.2 million m^3 in 1982.

Investment like this involves a much greater financial stake in the country concerned, and in order to justify that there must be a medium- to long-term future use of the equipment. This in turn leads to more careful use of the available forest since it must be made to yield indefinitely. Planning for the future is necessary in a way which straight 'grab-and-run' logging did not require. Additional advantages are the much larger labour requirement which improves the local economy, and the provision of permanent roads, schools, and socio-economic infrastructure by the companies.

PLANTING AND PLANTATIONS

Total preservation of the rainforest is not practical where land is scarce, but utilizing rather than destroying it can contribute to conserving it. The answer lies in easing the economic pressures on its owners first – exploitation zones can help pay for reserves, and both will protect the environment.

Conservation must be related to the people using the land. It can be very wasteful to extract one product only from a given area. Integrated use, taking all possible timbers and other products, can be achieved on a sustainable basis. By fully using one small area the economic gain can be sufficient to leave the rest alone.

In the past loggers ranged over wide areas to find particular trees, damaging many others in the process. Changing technology and research have resulted in more species becoming known, so smaller areas can be more intensively exploited.

Successful experiments with agro-forestry point out another hopeful direction. Here tree planting is used in conjunction with other crops. Particular species have been developed which give a

Loading debarked softwood for pulp in Chile (ABOVE). African hardwood logs (RIGHT), handled by Global Wood (GW) and German group Danzer (KD) who controlled half of Zaire's log production in 1983.

A New Guinea teak plantation gets under way (LEFT). This picture was taken when the trees were two years old! For Burma and South-East Asia teak has traditionally been a plantation timber, but often grown and extracted by foreign concerns. Now the Philippines has banned the export of any indigenous tropical hardwood; the nation makes use of its own resource.

wide range of benefits to subsistence farmers. One type of rapidly growing tree – *Morus Alba* – can give fruit, firewood, animal fodder and building poles. Such species contribute to anti-erosion measures, shade crops and may even, like *Zantixollion girata*, have nitrogen-fixing properties to enrich the soil. All this further relieves the pressure on the natural forest.

In many cases if the natural forest will not sustain the required timber output then harvesting is combined with new plantations, involving planning ahead 30 years or more. A coniferous plantation in the tropics, growing a timber such as *Pinus patula*, would expect to take thinnings for the first time after five years and take a mature crop after 30 to 40 years. Cypress, another plantation timber, takes 50 years to mature, but 60% of the original planting would have been removed by the end of that time. A mahogany plantation in Nyungwe, Rwanda, in Africa, boasts treees which reach a height of 18m (60ft) in nine years.

Cash-crop timber kamarere in this case – in New Britain, Papua New Guinea (BELOW). The scene is not one of logging devastation, but of a plantation in its first year; decaying roots, trunks and branches rot back into the soil, adding nutrients, protection to the young saplings and cohesion to the earth. A kamarere plantation after 18 years growth (RIGHT), with trees of a height that would take temperate hardwoods hundreds of years. The tiny figure dwarfed by the soaring trunks is Lucinda Leech, author of this section of the book.

Agro-forestry, here in Rwanda (LEFT)*, optimises crops and the environment and thus the local economy. A forester examines the nitrogen fixing nodules on the roots of a* Leucena leucocephala *sample, which will enrich rather than deplete the soil. Trial planting of* Morus alba*,* (ABOVE) *a fast-growing species which can yield both utility and cash crops – fruit, fodder, and building poles.*

Plantations are labour intensive in the early stages. Rapid growth of the trees is matched by that of the weeds – one type of strangler vine grows 1m (39in) per week in Papua New Guinea!

Plantations can look very untidy at the beginning, typical of what you imagine the worst clear-felled devastation would look like. But this mess of hacked and burnt stumps is not only easier to achieve than neat rows in ploughed earth, it is also sounder agriculture. The crushed vegetation looks terrible at first, but it rots down rapidly and the nutrients are returned to the soil. Ash from burning provides further nutrients, and large stumps protect seedlings from sunlight and anchor the soil.

In selective felling also, the devastation frequently looks much worse than it is. The same huge stumps and mangled crowns litter the ground, but later provide natural fertilizer. Controlled selective felling by man copies the effects of natural tree fall which is common in the 'undisturbed' forest – actually a very unstable environment. The trees have evolved to recover from damage from cyclone, floods, landslides and fire. The structure of the forest is such that the saplings which have been waiting beneath shoot up after a mature tree has been removed, competing for the light flooding in through the hole in the canopy. This environment, and the trees' genetic composition, make them grow tall and straight in crowded conditions, yielding 46m (150ft) of near-perfect timber before the first branches. These re-generation qualities are the basis on which sustainable yield policies are built, but they are only possible if the forest is not over-exploited and the processes of nature can take their course.

RAINFOREST PROJECTS

MALAYSIA-RUBBER WOOD

Solutions to the problems in rainforest areas relate to particular places and are not necessarily transferable. A unique development in Malaysia to reduce the pressure on the natural forest has been the exploitation of the wood from rubber trees (*Hevea brasiliensis*).

Large areas of Peninsular Malaysia are covered by rubber plantations which have to be replaced every 25 years, and in the past the old trees were burnt. Now they are being used for timber as an alternative to ramin which is in short supply in the natural forest. The roads and local labour existing for managing and distributing the rubber are equally useful when utilizing the timber, and the dual-purpose nature of the tree gives welcome added value.

The timber itself needs rapid treatment after felling to prevent rot and discolouration, but this too is practical in a relatively developed country. Although it does not grow to large sizes during the 25-year period it has particular uses within the furniture and construction industries – especially locally – being known for its consistency and ease of working despite a rather bland appearance.

RWANDA – PLANTING AND SAWING

In Rwanda, Africa, foreign aid has helped set up a project to conserve Nyungwe, one of the last remaining forest areas. Most of the rest of the country has been deforested very rapidly in the last 50 years for agricultural use, necessitated by large population increases.

The central part of Nyungwe forest is a reserve where complete protection is offered to the animals and plants. Around this is an exploitation area where selective felling takes place, together with assisted regeneration and enrichment planting of indigenous species. Outside this again there is a buffer zone, approximately 1km deep, of plantation trees. This is available for firewood gathering for local people, many of whom used to use the natural forest. Indeed some of them had to be moved out and re-located when the reserve was introduced. The first plantations are now reaching maturity, providing additional income.

A sawmill was installed to process the timber, and after some years of foreign management it was handed over to the nationals.

Tapping rubber (CENTRE), – a traditional plantation scene in Malaysia. But rubber wood is becoming a traded timber; UK retailer Habitat's kitchen range uses it (TOP LEFT). Pit-sawing in Rwanda (LEFT), an activity that underlines the questionable applicability of technology. Manual labour creates employment and is more reliable and economic than petrol-dependent machinery.

When the generator blew up, rather than wait for expensive spare parts to be imported, local pit-sawyers were employed to convert the logs by hand. Interestingly this proved to be more economical than running the mill even though about 250 extra people were involved. A 4m x 2m (13ft 6in x 6ft 6in) log of Entandiophragma *excelsum* takes three people three weeks to saw.

More timber can be sawn, and the money stays in the country as wages instead of being exchanged for imported fuel, spare parts and transport. Local people are gainfully employed with no more time or need to continue with illegal activities in the forest. Damage to the environment is much reduced – setting up a pit next to a felled tree and then carrying out the planks by hand makes far less disturbance than dragging out huge logs by machine.

BRAZIL – HYDROELECTRICITY

Some attempts to harness the resources of the rainforest, however, in contrast to Rwanda's integrated and appropriate (if fortuitous) development, seem to make little sense. Such are the hydroelectric schemes like Curupira and Itaipu in Brazil, which create employment in the building, and ostensibly will attract industry with their cheap power. But Curupira has flooded a forest, and the vegetation decomposes; the water, turned highly acidic, corrodes the turbines of the dam. In another instance where the same thing is due to happen, building is still going ahead although the acid effect will cut the working life of the scheme to an estimated 20 years. It would have taken too long to cut all the trees down before flooding them, apparently. Even so, if they were cut down and used to fuel a power station on the same site, output at the same level would be guaranteed for 100 years!

The other drastic consequence of such enormous projects – the Itaipu dam, on the Brazil – Paraguay border, is the largest hydroelectric development in the world, one of a chain of dams due to flood huge areas of rainforest – is human. Great numbers of people are displaced by the growing expanses of water, and have to find somewhere else to live. Often it is upriver of the new reservoirs; the forest people clear new plots of land and cause the soil to be washed away – into the river and thus the lake, which silts up and loses some of its working life. The cycle seems self-perpetuating. Such ironic chain reactions will continue to emphasize human shortsightedness; there is nothing, except perhaps our dependence on nuclear defence policies, that shows our capacity to destroy ourselves so poignantly as our destruction of the rainforests.

The Itaipu dam on the Brazil/Paraguay border (LEFT), the largest hydroelectric installation in the world – and it is only one of a number of such developments. Huge areas of forest will be flooded, people will be displaced and settle elsewhere – destroying more forest and silting up the river through soil erosion.

Transport: the biggest single contributor to world timber costs. Here, softwood is being hauled in British Columbia

FROM TREE — TO — WOOD

From the huge coniferous forests of the Pacific North West of the USA to the tropical rain forests of Papua New Guinea there is a huge industry at work transforming the uncut tree into regular, regimented timber. Having been transported from the forest the log is skilfully cut to maximize the yield of planks. The planks are then carefully dried. After these skilled and complex processes have been completed the log has finished its journey from forest to timber merchant.

LOGGING AND EXTRACTION

Mechanization and cost-efficiency are at a premium in commercial logging operations, and nowhere more so than in the Pacific North-West United States, where giant paper companies and sawmills can invest in the most advanced equipment.

In these mountainous areas, self-powered mobile tower yarders, with telescopic steel poles from which lines to haul the felled trunks are rigged, have replaced the old 'spar tree' system, in which a selected centrally located tree would be topped and trimmed, rigged with an elaborate system of lines and pulleys, and used as a point to which the logs were hauled. Smaller mobile yarders with a crane-like jib are also used for 'skylines', cables that float overhead and along which pulleys heave the trunks to be loaded on to trucks.

In tropical forests too, especially in areas of comparatively high accessibility, cutting and extraction can be achieved with a degree of mechanization. In difficult terrain there is little or no alternative to intensive use of human labour, although chainsaws replace the axe, and transporting the enormous trunks, here as everywhere, is a matter of controversial roadbuilding for the heavy trucks.

Although tropical hardwoods reach a high value by the time they arrive in the destination country, third-world economics entails comparatively small investment in logging operations unless foreign companies are involved, and even then the sophisticated equipment seen in the softwood forests of the north-west USA has insignificant presence in Borneo or Papua New Guinea. Scandinavian softwood

Campbell River pulp mill, British Columbia, (LEFT). The huge piles of wood chips, as big as coal-mine waste heaps, await chemical or mechanical processes by which the lignin is stripped and the remaining fibres matted. Oak logs being hauled in Cumbria, northern Britain (BELOW LEFT); not for paper, these trees. They are on their way to conversion and kilning for furniture. The scale of logging and milling operations in the Pacific north-west United States and Canada is suggested by this mill in Chase, British Columbia (BELOW).

forests, on the other hand, are highly mechanized because of expensive labour; here, machinery fells the trees, trims them, debarks and even crosscuts – 'bucks' – them to length on site. Thus transport is made easier and cheaper.

MANAGEMENT AND PROCESSING

For timber as well as pulp and paper production, softwood and even temperate hardwood forests in developed countries are sustainably managed to a degree only dreamed of in the tropics. Timber is not just logged but harvested; as many seedlings are planted as trees are felled each year; the forests are managed to a plan that ensures continuing production. In the plantations, such techniques as multiple cropping – planting agricultural crops to grow with the trees – or 'underplanting', where one species acts as a nurse to encourage growth of another, are brought in to use to maximize profit.

Pulp and paper mills, chipboard and plywood plants and sawmills are often close to the forests themselves. For third-world tropical timber producers, it is of course sound economics to gain as much 'added value' as possible for the indigenous crop before it leaves the country. This is why almost every timber-producing country in the world is now also a chipboard producer, because the technology and investment needed are both comparatively accessible.

Traditionally, major softwood exporters such as Canada and the Scandinavian countries have sawn their timber to size before it is shipped, and stamped the endgrain with shipping marks, which show precisely where the wood came from; hardwoods have been chiefly exported in logs, but now, more and more tropical timber does not leave the producing country until it has been through the sawmill. It travels either in the form of boards – wide planks, one or two inches thick and usually waney-edged, that is with the bark still on them – or 'dimension timber', material that is left in greater thicknesses and lengths for constructional purposes. An average section might be 200 x 100mm (8x4in), and for a structural hardwood such as jarrah the length can reach 8m (25ft).

Temperate hardwoods do not come in such great heights as tropical species, and so board length will usually be less for, say, oak or ash than it will for ramin or mahogany. However, transport can be a limiting factor in tropical timber extraction, and despite the original tree growing to perhaps 60m (200ft), the logs might be disproportionately short by the time the timber is on the market. Tropical timbers tend to come in much wider boards, however.

CONVERSION, GRADING AND SEASONING

'Conversion' is the name given to the process of turning a log into 'dimensioned' timber. Grading distinguishes which pieces are suitable for which jobs, both in terms of appearance and strength. It is important for architects and surveyors, for instance, to achieve

An elephant at work in the teak forests of Chiang Mai, Thailand (TOP). A charming scene, but by no means archaic; the Thai elephant-training schools are still busy. Opportunistic logging in Guadalcanal (ABOVE) – a cyclone-felled log of callophyllum is being checked for damage. Site logging in Kayan River, Borneo (LEFT) ; the logger, literally up a gum tree, is rigging cables to his chosen spar.

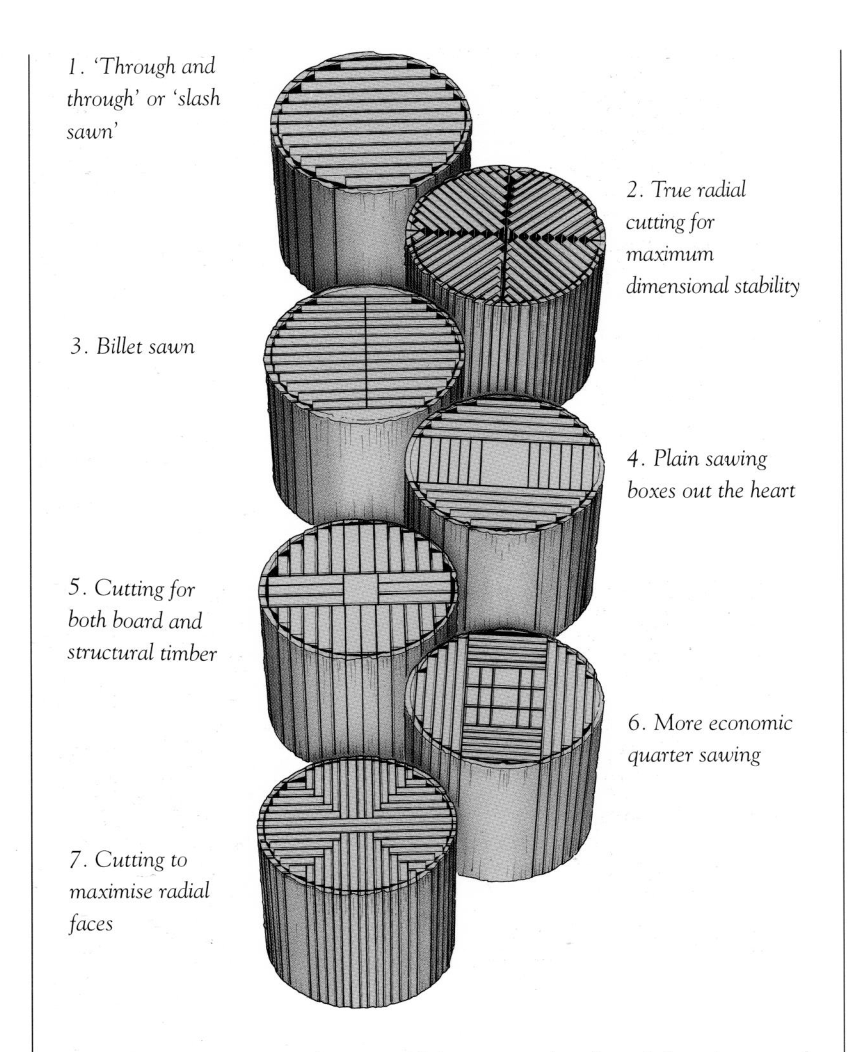

Methods of conversion of the tree to timber, chosen to minimise distortion or knots, or to maximise appearance or board width (LEFT). Piles of hardwood logs show shipper's marks (RIGHT), which identify the grade and other technical information.

maximum economy by specifying exactly the right sizes and thicknesses, the performance of which they can predict with accurate grading.

Unless it is for a sculpture, wood for commercial use will have to be flat and straight, and since this is not how trees grow, it is necessary to understand timber's complex reactions and the implications of producing regular shapes from the organic original.

The way a tree is cut will affect the way the timber absorbs and releases moisture, and shrinks and expands as it does so. Appearance, too, is a major consideration during conversion; the faces of a tangentially cut board will look very different from one cut radially. Further, and in some cases most important, conversion affects strength; it is crucial to ensure that knots do not weaken constructional timber.

There are many variations on the basic methods of log conversion, which amount to whether cuts are made at a tangent to the log or radially. The first – known in different forms as 'through and through', 'flat sawn' or 'slash sawn' – produces the widest but least stable boards, some of which may be severely weakened if knots run across them from edge to edge. It is the most economical conversion method, however; refinements such as billet sawing or plain sawing attempt to limit the disadvantages, perhaps by 'boxing out' the unreliable heart, where rot, splits and shakes are most likely to appear. Growth rings almost always meet the face of flat-sawn boards at less than 45°, so the tendency to curve away from the centre of the tree's original trunk will be greatest. The figure on the face of softwoods, will be strong and attractive.

Radial sawing, or 'quarter sawing', always makes cuts at maximum perpendicularity to the growth rings. It produces the most dimensionally stable boards, and is invariably used in musical instrument soundboards, where stability, great strength and clear tonal quality are needed from wide but thin pieces. It is very uneconomical however, and since the cuts all stop at the centre of the trunk, it limits board width. Timbers with a pronounced medullary ray system, such as oak, benefit greatly in appearance terms from quarter sawing.

The sawyer's skill lies in deciding almost at every cut how to get the best out of a log; with mill equipment which allows a large trunk to be turned and cut again, the optimum combination of strength, appearance and stability can be obtained.

GRADING

Timber is destined for many different uses, some of which set appearance at a premium, like fine furniture making, some of which places the most importance on strength.

Deciding which wood should be used for what purpose is known as grading, the two types of which – visual quality and stress grading – make assessments on the basis of the two criteria.

Systems differ from country to country, but there are accepted standards which form the basis for a general understanding of what a batch of timber will be like. This holds more for softwood than hardwood, and consequently more for the European and North American countries.

Visual grading of a board or batch takes into account the number and direction of knots, the amount of 'wane', or unsquared tree-edge, the amount of surface splits and checks, and the grain direction. Russian and Scandinavian systems, which prevail in Europe, set five or six grades, the top three of which are confusingly known as 'unsorted'.

'Knot area ratio' is one of the standards applied to timber, in other words the proportion of knot as a whole area against clear timber. A ratio of 1:6 for the face and 1:12 for the edge puts timber in the best grades, while fifth grade has a 'KAR' value of 1:3 on the face and 1:6 on the edge.

Stress grading looks at bending strength and stiffness, measuring, in other words, the force required to break a piece of horizontally held timber, and the force to deflect it by a certain amount. Timber so graded is marked so that specifiers and site staff can identify it immediately. This is normally done by machine, but there is also visual stress grading, which includes examination of the slope angles of the grain to face or edge. British Standard 'Special Structural' grade, the top one in this stystem, specifies 1/10, while the American ALS system's top grade – 'select' – requires 1/12. The lowest structural grade in ALS has a grain slope angle of 1/8, while the British Standard General Structural grade accepts 1/6.

There is another stress grading category, the S.C. 1-9 grades classification; softwoods fall into the 3-5 bracket, while extremely strong timbers – in bending resistance terms – such as ekki or greenheart are classified as 8 or 9.

MOVEMENT, MOISTURE AND SEASONING

Timber is hygroscopic; it picks up and releases moisture. While water is evaporating from the centre of the cells, no change in dimension is noticeable, but once the moisture content (MC) goes below about 30% – fibre saturation point – water is leaving the cell walls, which means that they will shrink and come closer together. They do not

Opening up an oak log with an industrial bandsaw,(TOP RIGHT). It is the sawyer's skill to 'read' the log as he cuts and make decisions on the basis of economy or appearance. Sometimes the two are the same thing; wide, knot-free 'clear' boards such as this tree is yielding are at a premium. The first wide cut,(BOTTOM RIGHT) has been made to enable the log to be turned and avoid the heart shake as long as possible.

reduce significantly in length however, which accounts for the fact that timber movement is measurable across the grain, and negligible along it.

Solid wooden furniture will always move according to season and atmosphere – centrally heated homes cause problems for antiques – and construction and finishing should take these vagaries into account. But this distortion is a matter of concern while wood is 'seasoning' – drying – as well as the effect it may have on a wooden structure, or one which incorporates wood. There are differentials in drying rates and amounts within the wood itself; denser latewood with thicker cell walls shrinks proportionately more than thin-walled earlywood, for instance, and pronounced radial cell systems such as the rays in oak serve to inhibit radial shrinkage, while tangential distortion continues apace. A square section of timber will become rhomboid while drying, flat-sawn boards cup away from the tree's heart, and quarter-sawn boards (radially cut) shrink most evenly.

Such differentials cause enormous stresses inside the wood, and it is practically impossible to season a log in the round without at least a radial shake or split across the rings and along the grain. Ring or cup shakes, which follow the rings, and heart shake, which occurs in the centre of the log, are also likely. To minimize these distorting forces, the tree is converted before seasoning.

No-one who uses solid timber in a professional capacity should be without a moisture meter. Simply pressing two needles into the surface of the timber and then pushing a button enables the moisture content to be read directly from a scale. An average of 10% plus or minus 2%, would be appropriate for joinery in a continuous 20-24° centigrade (70-75°F) environment – the warmest likely for a home or office. Even when the timber is to be used in a more humid situation, such as a kitchen or bathroom, it may well be subject to a period of storage or display in warm dry shops or showrooms, and so should be dried to these levels.

(BELOW), 1, wide boards will shrink more in width than length or thickness, and cup away from the tree's original heart; 2, the rings at right angles to the surface cause minimum shrinkage and distortion; 3, again, perpendicular rings mean optimum dimensional stability; 4, square sections with the growth rings diagonally across them will go rhomboid.

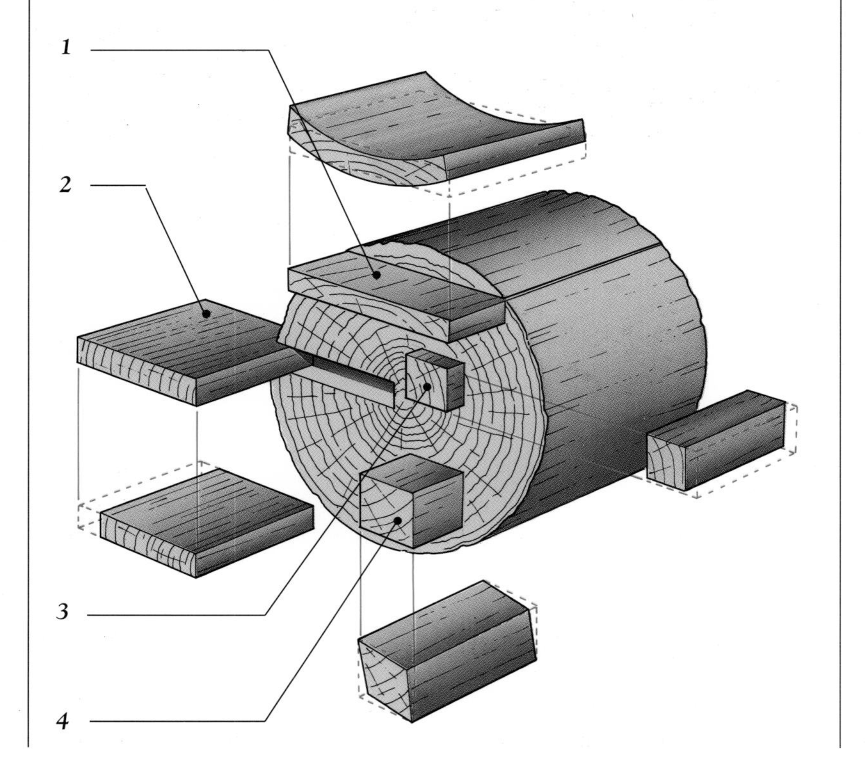

AIR DRYING

Air drying is the traditional method of reducing the moisture content of freshly felled – or 'green' – timber. It seeks to maximize the drying effect of wind, while reducing the effect of fog and wind-blown rain. Clearly, this is simpler in some climates than in others and, as a result, many countries in the northern hemisphere now use air drying mainly for pre-drying sawn timber before it is treated with preservative and kilned, and for drying items such as fence posts where low moisture contents are not essential.

For maximum efficiency, an air-drying yard should be arranged to benefit from the prevailing, drying winds and to allow access for handling. Roundwood, or poles, should have bark removed to prevent insect attack. The timber stacks should be perfectly level and clear of the ground to avoid stagnant, humid air from adversely affecting the lower layers; for the same reason, the base of the stack should be open and the area kept free of weeds and rubbish. This also helps prevent fungus and insect attack.

While maximizing the rate of drying and achieving the lowest practicable moisture content is usually vital in commercial yards, this may not be quite so important for timber which is to be used outdoors, for example. A final moisture content in the mid to high 20s is adequate for gates or fencing.

KILN DRYING

While most kiln drying of timber is carried out in bulk by professional timber dryers, since the method involves the use of expensive and sophisticated equipment, there are smaller workshop-sized methods which involve temperature-controlled de-humidification. The timber is stacked, each layer separated by smaller pieces of dry, clean softwood placed at regular intervals to support the timber and help prevent distortion. This is known as 'sticking' and enables air to flow all around the timber.

Stacks of timber in a large commercial kiln, (RIGHT)*. The operator is selecting a sample from the middle of the pile to check it for moisture content; these are small-section boards which are quite likely to have been dried once before conversion to this size. Note the separate species and section of waney-edged boards on top of the pile to the left.*

'Close piled' air drying keeps the gaps between the boards small, (BELOW LEFT) *to avoid distorting differentials in drying rates between centre and surface. Note the corrugated fasteners that prevent splits in the ends from widening. A 'stickered' pile for air drying* (BELOW)*; timber for kilning must also be stacked like this.*

The stacks are then placed in the kilns, which are sealed units within which both temperature and humidity levels can be controlled. These may vary according to the species and size of stack being dried. The choice of drying schedule – the combination of temperature, humidity and duration of kilning – is an important factor in ensuring the quality and cost-effectiveness of each batch. Too fast can be as bad as too slow.

Throughout the drying process, the timber in the kiln is checked at regular intervals to ensure that all is going smoothly and to avoid any possible degrade, such as splitting and case-hardening, where the outer surface of the timber dries too quickly in relation to its centre.

PRESERVATIVES AND FINISHES

Wood preservation extends the life of timber components by providing a protective 'shell' to guard against stain, decay, mould growth and insect attack. Before deciding to use preservative treatment, however, the natural resistance of the timber species to decay and insect attack should be considered. A timber that is naturally durable can, without preservative treatment, give an effective service life of anything between a few years and infinity, depending on the degree of exposure. But, regardless of how durable the heartwood is, the sapwood of any species is in the lowest of the five categories of natural durability. Even a naturally durable species of timber will require preservative treatment if it has substantial amounts of sapwood.

The three main types of preservative in common use today are tar oils, water-borne preservatives and solvent-borne preservatives. Traditional creosote belongs to the first group; used mainly on fences and outbuildings, it gives excellent durability but has environmental disadvantages. Its oily nature also slows down weathering.

Most water-borne preservatives are applied by vacuum/pressure impregnation. This fixes the most common copper-chrome-arsenic (CCA) types as insoluble in the timber, giving high levels of protection. Some softwoods take on a greenish colour after treatment, although this tones down with weathering. No maintenance is required, but timber treated in this way will tend to become grey after prolonged exposure to the elements.

Organic solvent-borne preservatives can be applied by double-vacuum, by vacuum-pressure impregnation, or by dip. Brushing, provides only minimal protection. It is generally preferable to cover them with some sort of finish – a wood stain, paint or varnish – and this is essential if they are to perform well in exterior situations.

Before deciding on a preservative it is important to consider how long the component must last. Wood does not decay or disintegrate immediately, even when it is in contact with the ground – the worst situation. If only a limited life is required, and the component is considered expendable, treatment may be unnecessary.

The recent development of microporous or 'breather' paints, stains and varnishes has given such finishes a protective dimension they have not had before; instead of cracking and peeling, they allow moisture in the wood to evaporate, but do not allow the passage of free water. Moisture can leave the timber but not enter it, so the risk of paint failure and decay is dramatically reduced.

VENEERS AND HIGH-TECHNOLOGY TIMBER PRODUCTS

Veneering has been a method of maximizing the use of rare, exotic and expensive timber from as early as pre-Egyptian times. Until comparatively recently, saw-cutting was the only method of producing the leaves, which would usually be as thick as 3mm (1/8in). With sophisticated modern production methods, thickness can be brought down to as little as 1 millimetre (1/24in), which obviously achieves great efficiency in terms of area covered.

For straightforward decorative purposes, veneers also have a conservation side-effect in that timber species under threat of extinction can be made to last far longer. For the exporting country, veneer production means work for local labour, added value and a higher return on an indigenous resource. For the individual furniture maker, the relative cheapness of the material makes it attractive, an advantage with especially exotic timbers, but veneering is a labour-intensive craft. Such economies make huge sense in large-scale manufacturing, where mechanized application processes virtually eliminate the need for highly skilled craftspeople.

Constructional veneers go into the production of plywood, blockboard and other laminated timber products, but appearance is not the main criterion in such uses, and manufactured boards do not depend on valuable woods, other than visual selection for the facing plys, which are better quality than the internal ones. There is, however, a high demand for pre-veneered plys and blockboards, which are faced with high-value and exotic timbers. Used in furniture and panelling, they combine the advantages of beautiful hardwoods with the structural stability and comparative cheapness of sheet material.

Some veneers are still saw-cut, usually where the timber is so hard that mechanized knife systems would not work. Ebony and lignum vitae veneers, if available at all, are likely to be saw-cut, but the kerf

An English mahogany-veneered bureau cabinet (RIGHT), made about 1755, while the timber was still at the beginning of its popularity. Eight foot high, the piece is remarkable for its concept and veneering rather than the workmanship; the saw-cut veneers, at least 1in thick, have split on the face of the writing surface. The mahogany has been laid over oak, visible on the inside surfaces of the cupboards.

(the gap left by the saw-blade) is practically as thick as the veneer itself, and at 50% wastage, production is prohibitively uneconomic.

The wonderful figuring of truly decorative veneers is a quality for which the best logs of a batch are often earmarked. Other characteristics, which would be considered defects in a log for conversion into timber, can make a veneer yet more highly prized; colour variation, growth irregularities, even diseases, can be turned to advantage. Burrs, bird's-eye and crotch timber – from the junction between branch and trunk – would make no sense in a structural component.

Such effects are best exploited on flat-cut veneer, which is made by slicing a square-cut baulk, first softened with steam or hot water. The block is mounted on a carriage which carries it against a large blade and pressure plate, much like the mouth of a giant plane, thickness by thickness. This flat-cut method produces a figure exactly like that of an ordinary board, with the visual advantage that the pattern repeats from sheet to sheet. Thence comes the traditional 'book matching', where consecutive leaves can be laid on, say, a pair of doors to give a mirror-image effect.

The knife and pressure plate is also used in the various forms of rotary veneer production; a log mounted eccentrically and turned against the blade will produce an unusual decorative face pattern unobtainable in any form of board, while constructional veneers for ply are rotary cut from a centrally mounted log. This is more economic because the sheet is continuous. These veneers can be anything from 1-4mm (1/24-5/32in) thick. Builder's constructional plywood ('shuttering'), made from veneers cut in this way, exhibits the strange characteristic of repeating flaws, like a wallpaper pattern.

HIGH-TECH TIMBER

Wood-based panel products have been available commercially since the beginning of this century. The development of technological products and processes, such as added adhesives and the application of systematic and intense heat and pressure, has turned wood waste into an enormous variety of boards used for purposes as different as formwork and furniture.

Man-made or manufactured boards, sheet materials, or wood-based panel products, to quote a few of the names by which they are known, all contain a significant amount of wood in the form of strips, veneers, chips, flakes or fibres. The three major categories –

Production of the de Havilland Mosquito aircraft in a factory 'somewhere in England', 1943 (ABOVE and RIGHT). The design forced the development of resin glues for product manufacture in plywood; synthetic formaldehyde cement, introduced during the machine's production run, was found to be an improvement over casein glues for surface jointing. The Mosquito used ash, spruce, birch and balsa in its sophisticated construction.

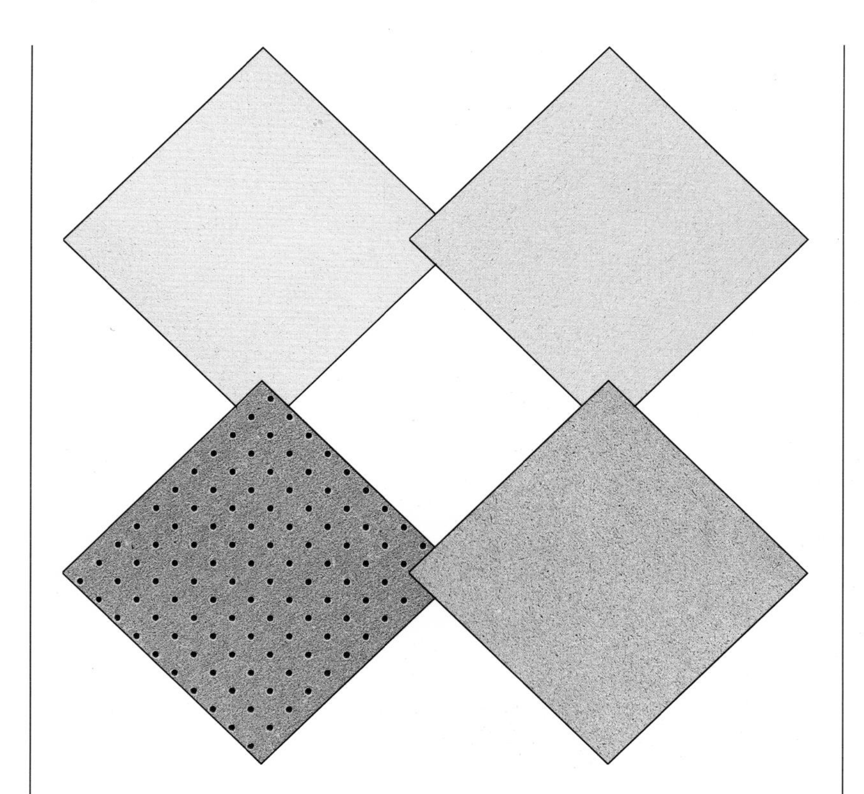

Manufactured boards (ABOVE): clockwise from top left, medium-density fibreboard (MDF), hardboard (Masonite US), standard softboard and perforated hardboard or 'pegboard'. MDF has revolutionized design in sections of the furniture industry. A chair in MDF and steel (BELOW) from Rycotewood College, Oxfordshire.

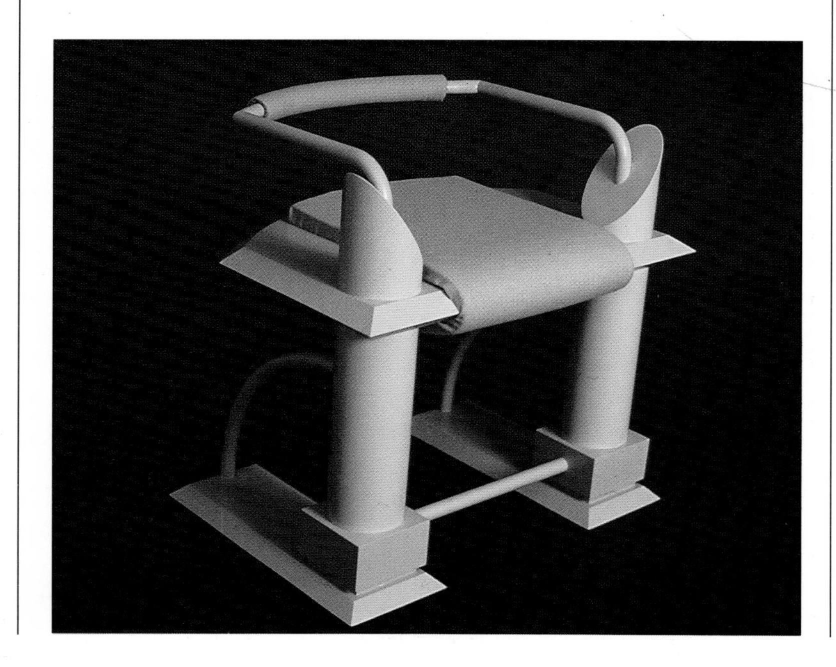

plywood, particleboard and fibre building board – include other general-purpose or utility boards and special products such as fire-resistant or super-light construction materials.

Plywood was developed to provide panels with dimensional stability and good strength properties both along and across the sheet. Wood chipboard, other particleboards and fibre building boards were developed, using forest thinnings and sawmill waste, to provide stable, predictable sheet materials with uniform properties. From these have developed whole families of panel products, amongst which can be listed, from the traditional established varieties: veneer plywood, that is to say ply made from veneers; blockboard and laminboard, known as core plywood because they consist of wood strips covered with face veneers; wood chipboard; other particleboards such as flaxboard and bagasse board; fibre building boards like hardboard, medium board, and fibre insulating board. More recent developments add a new generation to this list, including waferboard, orientated strand board (OSB), medium-density fibreboard (MDF) and wood cement particleboard.

PLYWOOD

The practice of cross-laminating veneers for special uses can be traced back to the pre-Christian Egyptian Empire. Crude forms of plywood were bonded with natural adhesives such as animal glue and blood albumen. Techniques changed little until the 19th century with the advent of the rotary peeling machine, which peels a log section for ply veneers rather as the blade of a pencil sharpener sharpens pencils.

Plywood, made up of successive veneers glued and assembled at right angles to one another, is a comparatively light, strong sheet material, whose size and thickness can be varied easily during manufacture. Plywood combines an attractive surface appearance with excellent performance under strenuous conditions such as marine applications and building construction.

PARTICLEBOARDS

Wood chipboard has been developed since the Second World War, with the advent of thermosetting adhesives. Its manufacture uses wood residues such as forest thinnings, planer shavings and other joinery ship residues. The product is not as demanding as plywood in terms of raw materials and skilled labour and wood chipboard mills are now found in most countries.

Particleboards like wood chipboard are made from small particles of wood, mixed with adhesives and formed into a mat. They come in thicknesses 3-50mm (1/8-2in), and vary in quality; they can be of uniform construction through their thickness, of graded density, or of distinct three-or five layer construction, to give enhanced properties without excessive weight. The differences in constructional quality

Modern furniture in MDF (LEFT and BELOW). The material has the stability and predictability of chipboard, but is far easier to work and takes built-up and moulded shapes and detail outstandingly well. It also provides an excellent surface for flat-colour spray finishes. The machining dust, however, can present health hazards; good factory air extraction is vital.

allow chipboard to be used in various situations, from furniture to flooring, and some boards are designed to give a certain amount of moisture resistance. Generally, chipboard has a sponge-like reputation when it comes to water contact.

FIBRE BUILDING BOARDS

Fibre building boards usually exceed 1.5mm (1/16in) in thickness, and are manufactured from ligno-cellulosic material. The primary bond is usually derived from the inherent adhesive properties of the material, when the fibres are formed into a mat under pressure – a process known as 'felting'.

The earliest fibre building boards, produced in the late 19th century, contained large amounts of repulped newsprint and were of relatively low density. Later, insulating boards were produced from ground wood pulp. During the 1920s and early 1930s further techniques were developed to break solid wood down into fibres and reconstitute them under heat and pressure as a strong and durable panel – hardboard.

Medium density fibreboard (MDF) is the latest development in 'dry process' fibre building board technology. It is a furniture grade board with superior characteristics in terms of surface texture, smoothness and machinability. Fibre building boards are used in a very wide range of applications, from pin boards, sheathing panels and insulating boards through to high-class joinery and shopfitting.

GLU-LAM

The latest technological advance in the use of timber is glue lamination of timber into members such as beams. This involves small cross-section boards in commercial sizes, layered up and glued together under pressure, so the grain of the adjacent boards is parallel. In this way, large cross-sections and very long lengths can be obtained. Standard straight glued-laminated ('glulam') beams in a fixed range of sizes are now available and can be bought from normal trade outlets.

Specialist manufacturers can produce laminated beams to specific lengths, widths and shapes. Glulam beams can be curved or constructed with a varying section. This means that tapering beams, columns, arches or portal frames can be used to create interesting shapes, while strength is retained. Even in large section sizes, glulam is not particularly heavy; its strength-to-weight ratio is therefore very high, making handling relatively easy and the construction process less labour-intensive.

A less well-known glulam member is the laminated timber slab, where the laminations are aligned vertically; slabs are used as roof or floor decking. They have a high load-carrying capacity coupled with good thermal and sound insulation, and as a bonus are esthetically pleasing when left exposed in floors or ceilings. They have often been used with a concrete top surface where heavy wheel loads are expected, for instance on bridges, wharves and warehouses.

A forest of Radiata pine in Canterbury, New Zealand. A tree with a wide variety of commercial uses from paper pulp to broom handles.

THE DIRECTORY — OF — WOOD

From the many different species of trees comes a huge variety of timbers for numerous different purposes; delicate turnery wood to timber for railway sleepers. Each of the world's 150 top commercial timbers is described in detail and illustrated to show its typical grain. Where it grows, its special properties, its resistance to insect attack and its commercial and creative uses are all covered in the authoritative text.

CONTENTS

HOW TO USE THE DIRECTORY

Each entry has a map which shows the major areas of the world in which the timber is grown commercially.

At the bottom of the right hand column of each entry is a visual summary of the most important characteristics of each wood.

◆ A diamond placed in the top of the grid shows that the wood is more dense, for example, than if the diamond were to be placed at the bottom of the grid. The values relating to the position of the diamond should be taken as an approximate guide .

Impact bending: is the timber's resistance to suddenly applied loads; a measure of the toughness of the timber. It is tested by dropping a constant weight from increasing heights on to a beam supported near each end until it fractures.

Stiffness: a measure of the elasticity of the wood. It is considered in conjunction with the bending strength. An important test for wood that is going to be used as a long column or strut.

Density: is measured as specific gravity, the ratio of the density of a substance to that of water.

Workability: how easily a wood is worked and whether it has a significant blunting effect on tools.

Bending Strength: also known as the maximum bending strength. Pressure is applied to each end of a board until it cracks.

Crushing Strength: the ability of wood to withstand loads applied to the end grain. A critical test for wood that is going to be used as short columns or props.

Each wood has been colour coded to give an at-a-glance guide to whether it is a hardwood or a softwood.

HARDWOOD

SOFTWOOD

NOTE: (S) denotes softwood

Abies spp. **Family**: *Pinaceae*
SILVER FIR (WHITEWOOD) (S)

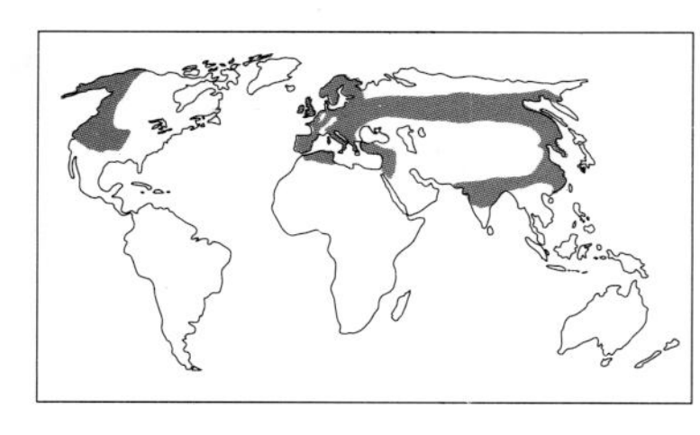

Where it grows
Only the *Abies spp.* produces true fir trees. *A.lasiocarpa* produces alpine fir, which occurs from Alaska to New Mexico, and grows to 40m (131ft). *A.procera* provides noble fir from the western USA, which can reach up to 70m (230ft) in height. The average height of *A.alba*, the common 'silver fir', ranges from 38-45m (125-148ft). *Abies grandis*, a closely related species, has topped 56m(183ft) and is the tallest tree in Britain. The silver fir grows extensively in Britain and mid-Europe, from Corsica and the Balkans and through Poland to the Carpathians and western Russia. It is included with European spruce in shipments of whitewood from Central and southern Europe, and sold as whitewood in Europe and with hemlock in supplies from western Canada. It is known by the trade name of 'European silver pine', as white deal in the UK, and as Baltic fir, Finnish fir and so on, according to its port of shipment.

Appearance
The colour varies from creamy-white to pale yellow-brown, closely resembling European spruce, but is slightly less lustrous. The timber is straight grained with a fine texture.

Properties
Silver fir is slightly resinous, and weighs 480 kg/m^3 (30 lb/ft^3) when dry. It should be kiln dried for best results. It will air dry very rapidly with little tendency to warp, but may split or check, and knots may loosen and split. There is medium movement in service; it has low stiffness and resistance to shock loads, with medium bending and crushing strengths, but a very poor steam-bending classification. Fir works well with both hand or machine tools and has little dulling effect on cutters if they are kept reasonably sharp. It nails satisfactorily, can be glued without difficulty, and brought to a smooth finish; it takes stain, paint or varnish well. There is often damage by pinhole borer, longhorn and Buprestid beetles, and sometimes by *Sirex*. The wood is non-durable and moderately resistant to preservative treatment, but the sapwood is permeable.

Uses
Fir is excellent for building work, carcassing, interior construction, carpentry, boxes, pallets and crates. Small trees are used in the round for scaffolding, poles and masts; it is used in conjunction with *Pinus sylvestris* for plywood manufacture.

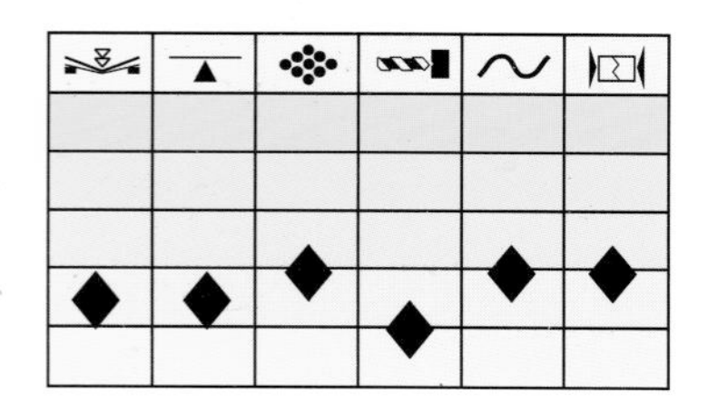

Acacia melanoxylon **Family**: *Leguminosae*
AUSTRALIAN BLACKWOOD

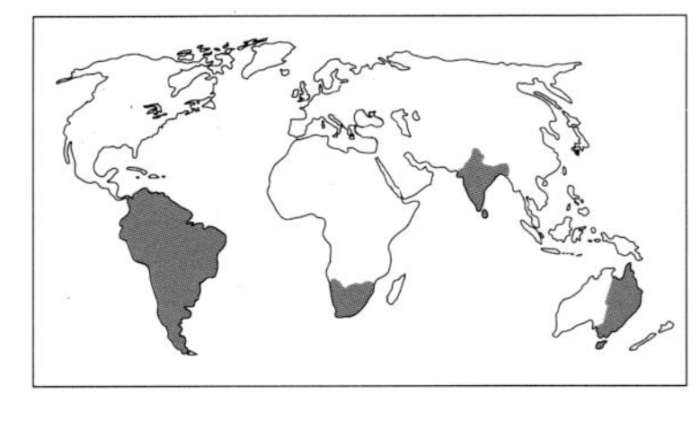

Where it grows
There are hundreds of species of wattle belonging to the *Acacia* genus found in India, South Africa and South America, but this is one of the most attractive. Australian blackwood, one of the largest of the wattles, grows as an understorey tree in forests of giant mountain ash in New South Wales, Queensland, south eastern Australia and Victoria, and is also found in Tasmania. It reaches a height of 25-30m (80-100ft); the diameter of the bole, which is known as black wattle, is about lm (3ft).

Appearance
The sapwood is straw coloured. The heartwood timber is not black, despite its name, but reddish-brown to almost black, with attractive bands of golden to dark brown with a reddish tint. Regular dark brown zones mark the growth rings. It is usually straight grained, but with a handsome fiddleback figure when the grain is interlocked or wavy. It is medium and even textured with a lustrous appearance.

Properties
Blackwood weighs about 660 kg/m^3 (41 lb/ft^3) when seasoned. It is a fairly heavy, dense timber with medium bending strength and stiffness, and a high crushing strength. It has good resistance to impact loads and a very good steam-bending classification. The wood dries fairly easily and is stable in service. It works satisfactorily with both hand and power tools, and offers only a moderate blunting effect on cutting edges.The grain tends to pick up when planing or moulding interlocked or wavy grain on quartered stock, and a reduction of the cutting angle is recommended. Wattle can be nailed and screwed satisfactorily, and takes stain and polish for an excellent finish. The heartwood is durable, but liable to attack by the common furniture beetle and termites, while the sapwood is liable to attack by powder post beetle. It is extremely resistant to preservative treatment.

Uses
This is a highly decorative timber, in great demand for high-quality furniture, cabinets and panelling. It is also used for shop, office and bank fitting and interior joinery. Billiard tables, tool handles and gun stocks are made of wattle; ornamental turnery, bent work for cooperage, coach and boat building and wood block flooring are other applications. Selected logs are sliced for beautiful decorative veneers used in plywood faces and flush doors, cabinets, and architectural panelling.

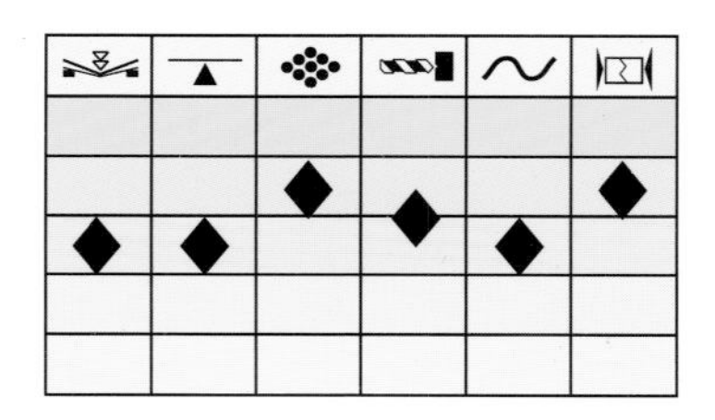

Acer pseudoplatanus **Family**: *Aceraceae*
SYCAMORE

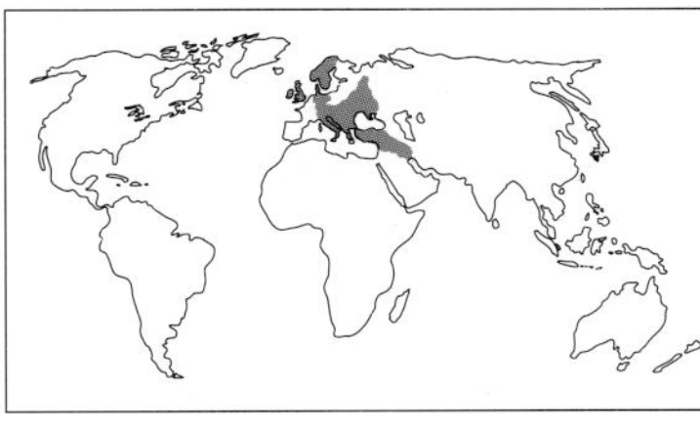

Where it grows
This medium height tree grows in widely differing soil and exposure conditions in the UK, but is a native of central and southern Europe and western Asia. It reaches about 35m (105ft) in height with a large broad-domed crown, and the bole is 1.5m (5ft) in diameter. It is known as sycamore plane, great maple (England) or plane (Scotland). Field maple (*Acer campestre*) and Norway maple (*A.platanoides*) are similar and both grow in Europe.
Note: For American sycamore see *Platanus hybrida* and related spp., also pp.174-5.

Appearance
There is little difference between the sapwood and heartwood of sycamore; a creamy-white colour with a natural lustre. Slowly dried timber changes to a light tan colour and is known as 'weathered sycamore'. It is usually straight grained, but curly or wavy grain produces a very attractive 'fiddleback' or lace ray figure on quartered surfaces. The texture is fine and even.

Properties
The wood is of medium density and weighs about 610 kg/m^3 (38 lb/ft^3) when seasoned. It air dries fairly rapidly and well, but is inclined to stain unless end stacked; it kiln dries well. There is medium movement in service. It has medium bending and crushing strength, low shock resistance and very low stiffness, giving it very good steam-bending properties. The wood has a moderate blunting effect on tools and cutting edges; the grain tends to pick up when planing or moulding interlocked or wavy grain on quartered stock, so a reduction of the cutting angle is recommended. It has good nailing and gluing properties and can be given an excellent finish. Sycamore is perishable and the sapwood is liable to attack by the common furniture beetle and by *Ptilinus pectinicornis*, but it is permeable for preservation treatment.

Uses
Sycamore is the traditional wood for fingerboards and ribs of lutes – the chief instrument in 15th- and 16th-century Court music. Fiddleback sycamore is still used for violin backs today. It is an excellent turnery wood for textile rollers and bobbins, brush handles, domestic and dairy utensils, laundry appliances and food containers. It is also the traditional wood for chemical treatment into various shades of silver grey, sold as harewood. Selected logs are sliced to produce attractive figure for cabinets and panelling, and it is dyed for marquetry and inlays.

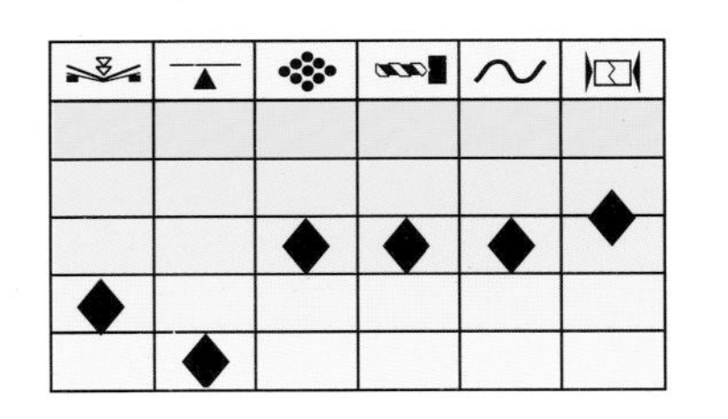

Acer spp. **Family:** *Aceraceae*
SOFT MAPLE

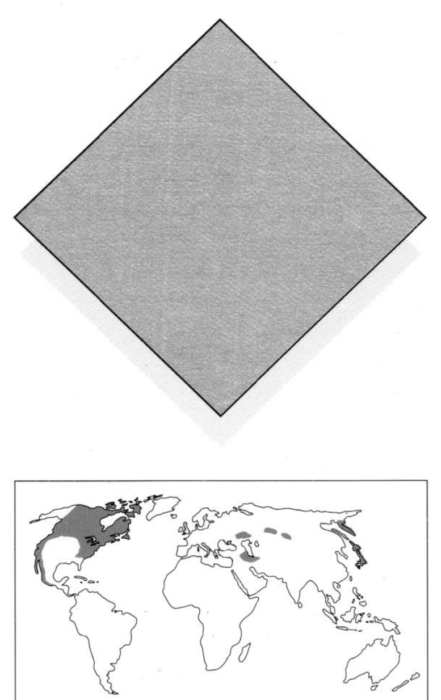

Where it grows
More than 10 species of this genus grow in north temperate regions of Canada and the eastern USA and on the Pacific coast, but only about five are important sources of timber. Maples are famed for their brilliant spectrum of multi-coloured autumn leaves of yellow, golden-orange and red. Soft maple is provided by both *Acer rubrum*, which provides red maple, and *A.saccharinum*, silver maple, which is one of the largest and fastest growing of all the maples. It reaches a height of 40m (131ft), but red maple is smaller, reaching only 30m (98ft) with a diameter 0.6-1.2m (2-4ft). *A. macrophyllum* produces Pacific maple.

Appearance
The sapwood is indistinguishable from the heartwood, which is creamy-white in colour with a close, straight grain and indistinct growth rings on plain-sawn surfaces. The texture is even, fine and slightly less lustrous than rock maple and lighter in weight.

Properties
The weight of *A.rubrum* is about 610 kg/m^3 (38 lb/ft^3), and *A.saccharinum* and *A.macrophyllum* are about 540 kg/m^3(34 lb/ft^3) when seasoned. The wood dries rather slowly with little degrade and there is medium movement in service. Soft maple is of medium density, with good bending and crushing strengths and low stiffness and shock resistance. It has a good steam-bending classification and works well with both hand and machine tools in all operations, as it offers only a moderate blunting effect on tools. Nailing and screwing are satisfactory with care. Gluing is variable but it can be brought to a good finish. The wood is non-durable and moderately resistant to preservation treatment; the sapwood is liable to insect attack but permeable.

Uses
This attractive timber is softer and lower in strength than rock maple but is eminently suitable for furniture making, interior joinery work, turnery and domestic woodware. Numerous specialized uses include the manufacture of shoe lasts, dairy and laundry equipment, sports goods, musical instruments and piano actions. It also makes an excellent light domestic flooring. Selected logs are peeled for plywood manufacture and sliced to produce a range of excellently figured veneers for cabinets, flush doors and architectural panelling.

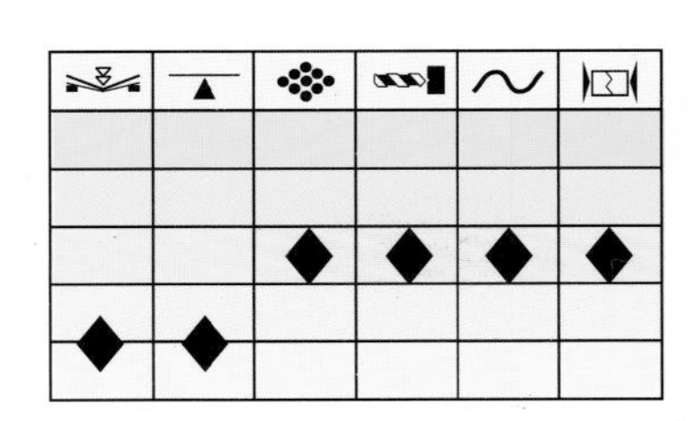

Acer saccharum **Family:** *Aceraceae*
ROCK MAPLE

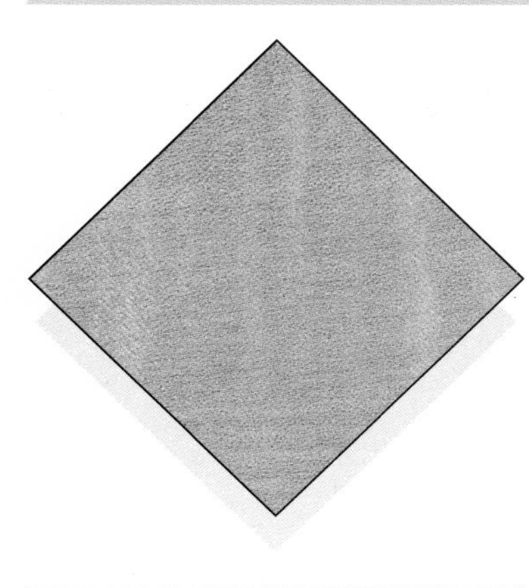

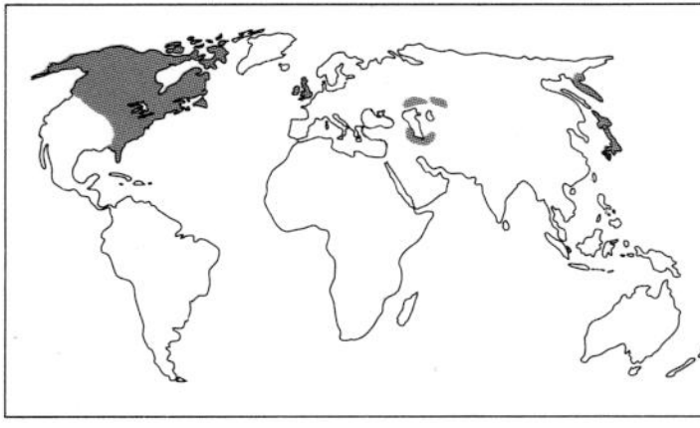

Where it grows
A.saccharum and *A.nigrum*, jointly sold as rock maple are known as hard maple (UK, Canada and USA); white maple (sapwood) (USA); and black maple. Rock maple is one of the most valuable timbers growing east of the Rocky Mountains in Canada and in the northern and eastern states of America. Canada displays the red leaf of the *A.saccharum* as its national flag. Sugar maple is also the source of maple sugar and maple syrup, obtained by tapping the sap in spring.
Rock maple grows to 40m (130ft) tall with a diameter of 0.6-1m (2-3ft).

Appearance
The wood is creamy-white with a reddish tinge, sometimes with a dark brown heart. It is usually straight grained but often curly or wavy, with fine brown lines marking the growth rings on plain-sawn surfaces. The texture is even, fine and lustrous. Pith flecks are sometimes present.

Properties
Rock maple weighs about 720 kg/m^3 (45 lb/ft^3) seasoned. It dries fairly slowly with little degrade, and there is medium movement in service. The wood is of medium density, has good bending and crushing strengths, with low stiffness and shock resistance and a good steam-bending classification. It has a moderate blunting effect on tools, with a tendency to create tooth vibration when sawing. Irregular grain tends to pick up when planing or moulding on quartered surfaces, and a reduced cutter angle is recommended. The wood has a tendency to ride on cutters and burn during endgrain working. Rock maple requires pre-boring for nailing, but it glues very well and polishes to an excellent finish. The wood is non-durable, liable to beetle attack, and subject to growth defects, known as pith flecks, caused by insects. The heartwood is resistant to preservation treatment but the sapwood is permeable.

Uses
Rock maple makes excellent heavy industrial flooring, for roller skating rinks, dance halls, squash courts and bowling alleys. It is used for textile rollers, dairy and laundry equipment, butchers' blocks, piano actions and musical instruments and sports goods. It is also a valuable turnery wood. Selected logs are peeled for 'bird's eye' figure, or sliced to produce fiddleback, curly or blistered and mottled maple veneers for cabinets and architectural panelling.

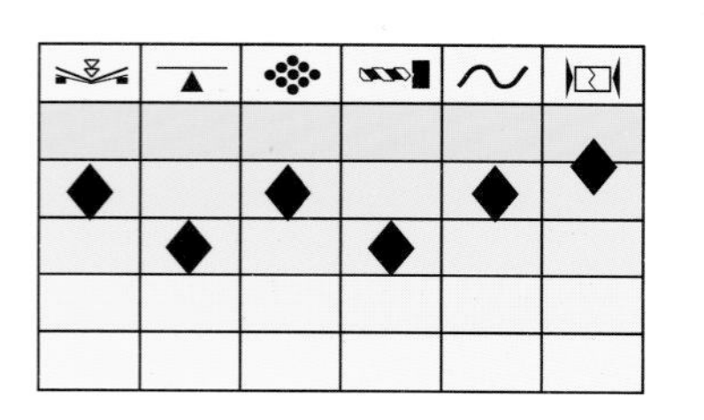

Aesculus hippocastanum **Family:** *Hippocastanaceae*
HORSE CHESTNUT

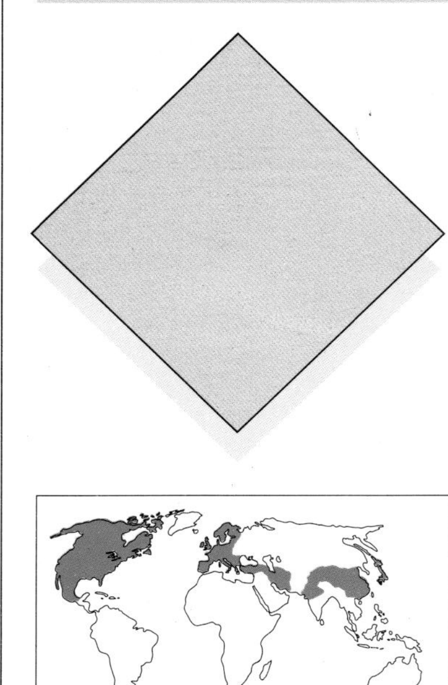

Where it grows
Native to Albania, this wood thrives in the mountain regions of northern Greece, Bulgaria, Iran and northern India. It is now widespread throughout Europe and north America. More than 20 species occur in America, where it is called *buckeye* because the base of the nuts resemble the eyes of a deer. The tree is also grown in China, Japan and the Himalayas, and especially in Europe, as an ornamental parkland tree. In cultivated positions it attains a grand stature of 40m (131ft) tall, but in parklands it has a short bole rarely exceeding about 6m (20ft), with its low branches hung with 'lanterns'. The diameter is 1.5-1.8m (5-6ft).

Appearance
If the tree is felled in early winter it is extremely white like holly, but timber felled later in the year is a pale yellow-brown colour. Spiral grain is usually present, and the wood is inclined to be cross or wavy grained. It has a very fine, close, uniform texture, however, caused by minute pores and fine storied rays which give the surface a lustrous sheen. Longitudinal surfaces sometimes show a subdued mottle.

Properties
This medium density wood, weighs about 510 kg/m^3 (32 lb/ft^3) when seasoned. It dries well with little degrade, and with small movement in service. It has a low bending strength, very low stiffness, and low-to-medium crushing strength, with a good steam-bending classification. It can be worked easily with both hand and machine tools, with only a slight blunting effect on cutting edges – which must be kept very sharp. Nailing and screwing are satisfactory, it glues well and gives a good finish when stained and polished. The sapwood is liable to attack by the common furniture beetle. The wood is perishable, but permeable for preservation treatment.

Uses
Horse chestnut is extensively used as a substitute for holly for cabinet making, furniture and carving, and is popular for general turnery for brushbacks and handles, dairy and kitchen utensils, fruit storage trays and racks, engineering and moulders' patterns. It is also used for the hand pieces of tennis, badminton and squash racquets. Selected logs are sliced for handsome decorative veneers, and dyed for marquetry veneers as harewood.

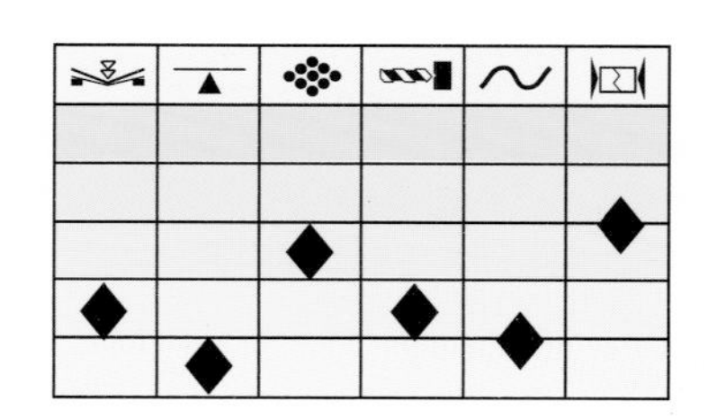

Afzelia spp. **Family:** *Leguminosae*
AFZELIA

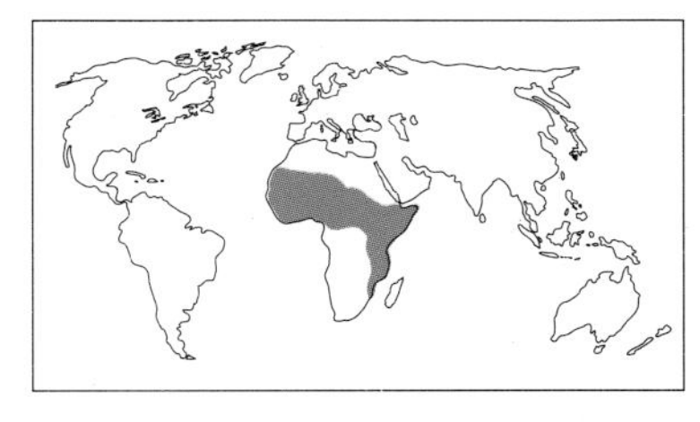

Where it grows,
Afzelia grows between the savannah forests of the dry areas of eastern Africa and the dense forests of the more humid regions of the west. Sold as a single commercial timber, *A.africana* (Smith); *A.bipindenis* (Harms); *A.pachyloba* (Harms), from West Africa, are known as doussié. *A.quanzensis* (Welw.) known as chamfuta, from Mozambique and Tanzania, is marketed separately. Other names include apa, aligna (Nigeria); mkora, mbembakofi (Tanzania); mussacossa (Mozambique). Trees from the moist deciduous forests reach a maximum height of about 30m (100ft) and East African trees grow to 21-25m (70-80ft). The diameter averages 1.2m (4ft).

Appearance
The sapwood is a pale straw colour and quite sharply demarcated from the light brown heartwood, which matures to a rich red-brown mahogany colour on exposure. Yellow or white deposits in the grain may cause staining. The grain is often irregular and interlocked, with a rather coarse but even texture.

Properties
This dense timber weighs 620-950 kg/m³ (39-59 lb/ft³) averaging 820 kg/m³ (51 lb/ft³) when seasoned. Afzelia can be kiln dried satisfactorily but very slowly from green, and there may be some distortion in the extension of existing shakes and fine checking. It is an exceptionally stable timber, comparable to teak. The wood has high strength, outstanding durability and stability. It is fairly hard to work and has a moderate blunting effect on cutting edges. Gluing can be difficult. Afzelia has a moderate bending classification because it distorts and exudes resin during steaming, but *A.quanzensis* can be bent to a small radius. A satisfactory finish can be obtained when the grain is filled. The sapwood is liable to attack by powder post beetle; the heartwood has outstanding durability, with extreme resistance to preservation treatment.

Uses
Afzelia is highly valued for interior and exterior joinery, window frames, doors and door-frames, staircase work, bank and shop counters and ships' rails. It is used for heavy construction, dock and harbour work, and bridge building. It is popular for school, office and garden furniture, and is especially useful for laboratory benches and vats and presses for acids and chemicals. It is used for flooring in public buildings.

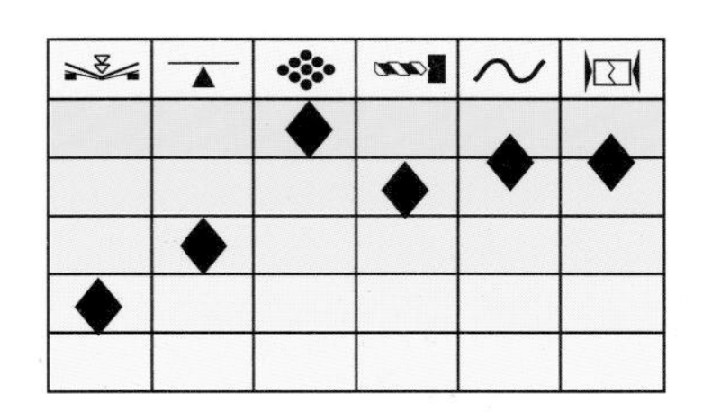

Agathis spp. **Family:** *Araucariaceae*
'KAURI PINE' (S)

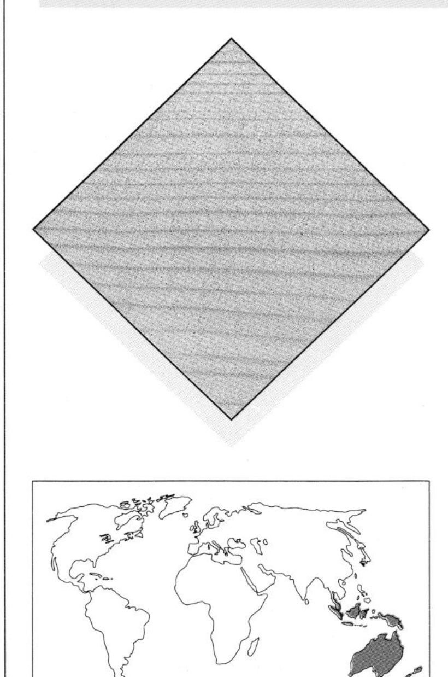

Where it grows
The cone-bearing softwood kauris occur singly or in small groves intermingled with broadleaved trees; and are distributed from Malayasia to Australia, from New Guinea to New Zealand and Fiji. The Maoris from the North Island of New Zealand called them *Tanemahuta* – 'King of the Forest' – as the trees soared above the forest canopy up to 45m (147ft), with a diameter of 1.5-4m (5-13ft) *A.australis* produces New Zealand kauri; *A.robusta*, *A.palmerstonii* and *A.microstachya* produce Queensland kauri; *A.dammara* East Indian kauri; *A.vitiensis* Fijian kauri.

Appearance
These valuable straight-grained timbers are not true pines. They resemble the botanically related 'parana pine' in appearance, but are darker in colour and coarser in texture. The heartwood colour varies from pale biscuit to pink or even dark red-brown. The darker wood contains the most resin, though kauri does not contain either resin cells or resin canals. It comes from the ray cells, and in vertical tracheids near the rays, in the form of hard resin plugs, which do not affect its finishing properties. Kauri has a fine, even silky texture and a lustrous surface.

Properties
The weight of New Zealand kauri is 580 kg/m³ (36 lb/ft³), and Queensland kauri is lighter at 480 kg/m³ (30 lb/ft³) when seasoned. It dries at a moderate rate with a tendency to warp but is stable in use. The wood has high stiffness, medium bending and crushing strength and resistance to shock loads, but it is not suitable for steam bending. It works easily with both hand and machine tools, and has only a very slight dulling effect on cutters. The wood planes or moulds to a smooth finish but in boring or mortising it needs to be properly supported at the tool exit. The wood holds nails and screws well. It glues easily and can be brought to an excellent finish. Kauri is subject to attack by the common furniture beetle, but is moderately durable and resistant to preservation treatment.

Uses
Kauri top grades are used for vats, wooden machinery and boat building, and lower grades for building construction. Queensland and Fiji kauri are used for high-class joinery and cabinet work, battery separators, pattern making and butter boxes and churns. Lower grades are used for cheap plywood manufacture, boxes and crates.

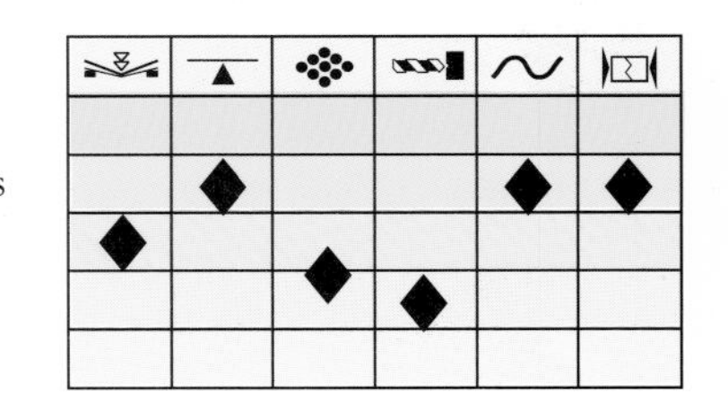

Albizia spp. **Family:** *Leguminosae*
ALBIZIA

Where it grows
The genus *Albizia* includes more than 30 species in Africa, many from the savannah forests. Commercial timbers come from the high forests and occur from Sierra Leone through central and east Africa down to Zimbabwe. West African albizia comprises *A.adianthifolia* (Schum) W.F.Wright; *A.ferruginea* (Guill. & Perr.) Benth., and *A.zygia* (D.C.) McBride. *A.ferruginea* produces heavy albizia. East African albizia is sold as red or white nongo according to colour and comprises *A.grandibracteata*,Taub, and *A.zygia* (D.C.) McBride.The trees grow to an average height of about 37m (120ft) or more, with a diameter of about 1m (3ft). The wood is known as okuro in Ghana, ayinre in Nigeria and sifou in Zaire.

Appearance
The sapwood is clearly demarcated from the heartwood and is pale yellow to straw in colour and about 2in (50mm) wide. The valuable heartwood colour varies from red-brown to chocolate-brown, often with a purplish tinge. The grain is irregular and often interlocked and variable in direction, with a coarse texture.

Properties
This dense wood weighs on average 700 kg/m^3 (44 lb/ft^3) when seasoned. Kiln drying must be carried out very slowly to avoid checking or twisting, but there is small movement in service. It has a medium bending strength, low stiffness and very low shock resistance, but a high crushing strength, and a moderate steam-bending classification. The wood requires care in machining, and fine dust can cause nasal irritation. It has a moderate blunting effect on tools, and irregular grain tends to pick up when planing or moulding on quartered surfaces. A reduced cutter angle is recommended in this case. It also tends to break out when machining across the grain, when recessing, or on arrises. Pre-boring is required for nailing. The grain requires filling before the surface can be brought to a good finish. The sapwood is liable to powder post beetle attack. The heartwood is very durable and extremely resistant to preservative treatment, but the sapwood is permeable.

Uses
Heavy albizia is used for marine construction and piling. The lighter species are used for utility and general joinery, domestic flooring, general carpentry and vehicle bodywork. Selected logs are sliced for veneers.

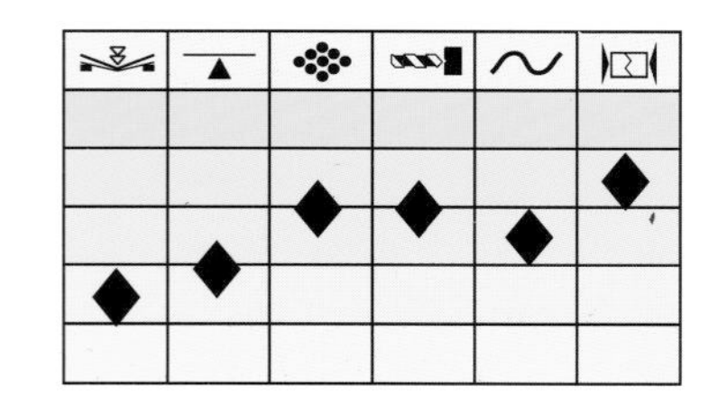

Alnus glutinosa **Family:** *Betulaceae*
COMMON ALDER

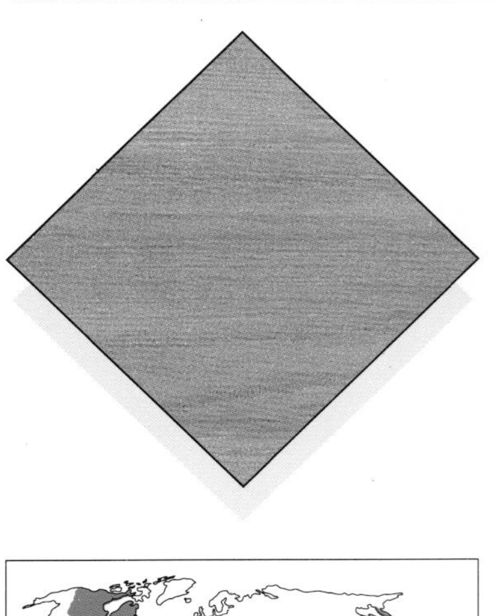

Where it grows
The tree is native to Europe and North Africa but has a wide distribution throughout the northern hemisphere in Russia, western Asia and Japan. Black alder (UK) and grey alder (*A.incana* Moench) are found in northern Europe and western Siberia, red alder (*A.rubra,* Bong.) is one of the most common commercial hardwoods and is widely distributed on the Pacific coast of Canada and the USA. The trees grow on streamside damp and moist sites to a modest height of 15-27m (50-90ft), with a diameter of 0.3-1.2m (1-4ft).

Appearance
There is little difference between the sapwood and heartwood colour, which is a dull, lustreless bright orange-brown when freshly cut, maturing to light reddish-brown, with darker lines or streaks formed by the broad rays. It is straight grained except near the butts, and has a fine texture.

Properties
The medium density heartwood dries fairly rapidly and well, and weighs about 530 kg/m^3 (33 lb/ft^3) when seasoned. It can be machined easily if the cutting edges are kept sharp, with only a slight blunting effect on tools. It possesses a low bending strength and shock resistance, a medium crushing strength and very low stiffness, which earns it a moderate steam-bending classification. It is inferior to many other hardwoods such as beech. Nailing and screw holding are satisfactory, and it can be glued without difficulty. The timber can be stained and polished to a good finish. The sapwood is liable to attack by the common furniture beetle. The wood is perishable but permeable for preservation treatment.

Uses
Alder is widely used for wood carving and turnery for domestic woodware, broom handles, brush backs, hat blocks, textile rollers and wooden toys. It is the traditional wood for clog making and is highly regarded as ideal for the manufacture of artificial limbs. It is also a source of charcoal for making gunpowder. Gnarled pieces of the tree are highly prized in Japan for decorative sculpture and carving. It is rotary cut for utility plywood and used for packing crates, while selected logs are sliced to provide an attractive figure suitable for decorative veneers for furniture and panelling.

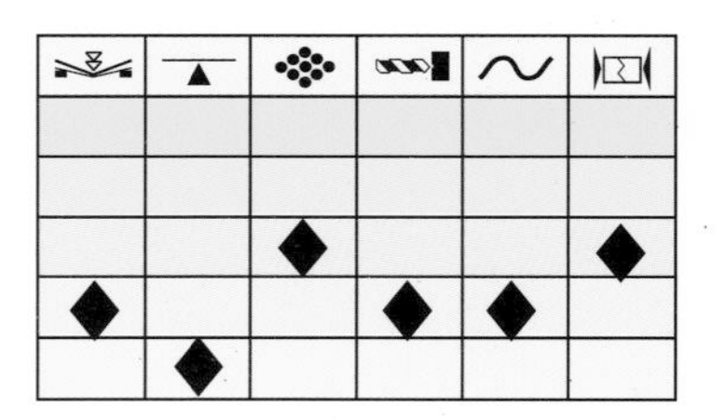

Anisoptera spp. **Family:** *Dipterocarpaceae*

MERSAWA and KRABAK

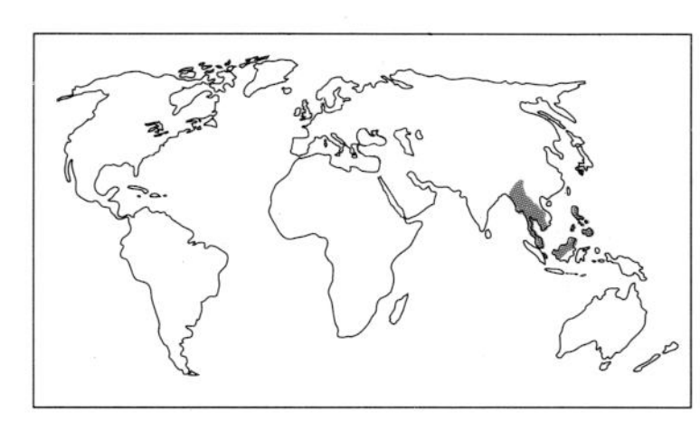

Where it grows

There are many species of the genus *Anisoptera*, which are usually mixed together in the countries of origin and exported in a group under the trade names of mersawa or krabak. The timber is known as mersawa in Malaya, Sabah, Brunei and Sarawak, krabak in Thailand, kaunghmu in Burma and palopsapis in the Philippines.

Mersawa comprises principally A.*costata*, Korth., A.*laevis*, Ridl., A.*scaphula*, Pierre, A.*curtisii*, Dyer, and A.*marginata*, Korth. It is also known as pengiran in Sabah. Krabak from Thailand comprises A.*curtisii*, Dyer, A.*oblonga*, Dyer, and A.*scaphula*, Pierre. Kaunghmu from Burma is A.*scaphula*, Pierre, and palosapis from the Philippines is A.*thurifera*. The trees vary in size but generally reach a height of 45m (150ft) with a diameter of l-1.5m (3-5ft).

Appearance

The sapwood, not clearly different from the heartwood, is usually attacked by a blue fungus which stains it. The heartwood varies from pale yellow to yellow-brown with a pinkish tinge, and is moderately coarse but even in texture. The timber is rather plain, but with a slight silver flecked figure and ribbon stripe on quartered surfaces from prominent rays. Generally the grain varies from straight to interlocked and it has a fairly coarse, even texture.

Properties

Weight averages 640 kg/m³ (40 lb/ft³) when seasoned. The timber dries very slowly from green, and it is difficult to extract moisture from the centre of thick stock. The interlocked grain and silica content affects machining, causing severe blunting of cutting edges. Mersawa possesses a low bending strength and shock resistance, with medium crushing strength; it has very low stiffness and a poor steam-bending rating. It can be glued and nailed satisfactorily and brought to a good finish. The sapwood is liable to attack by the powder post beetle. The wood is moderately durable and resistant to preservation treatment.

Uses

Mersawa/krabak is used for furniture making and general construction, for interior joinery, domestic flooring, vehicle bodies, and in boat building for planking. It is rotary cut for utility plywood, and sliced for decorative veneers.

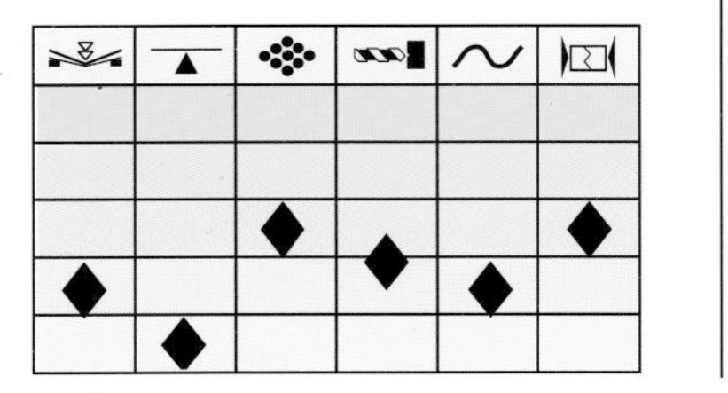

Araucaria angustifolia **Family:** *Araucariaceae*

'PARANA PINE'(S)

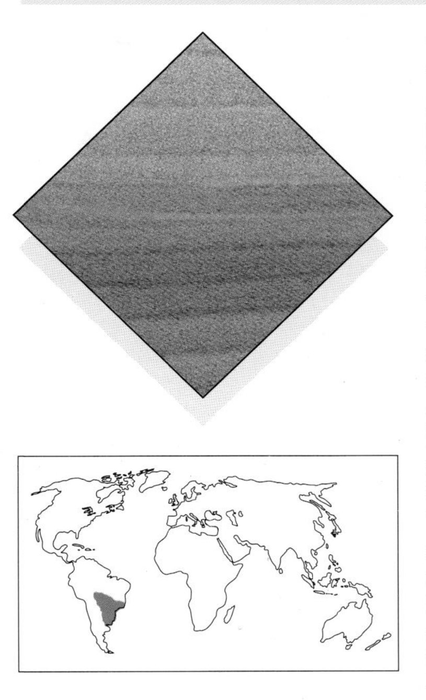

Where it grows

The tree is not a true pine of the *Pinaceae* family. It grows mainly in the Brazilian state of Parana and is also found in Paraguay and northern Argentina. It is also known as 'Brazilian pine' (USA) and is closely related to A.*araucana* 'Chile pine' or Monkeypuzzle tree, which is of no commercial importance. It grows up to a height of 40m (131ft) with a flat-topped crown and has a diameter of about 1.2m (4ft), with a clear straight bole.

Appearances

'Parana pine's' very close density, almost complete absence of growth rings, and unusual colouring make it a very attractive wood. It is mainly straight grained and honey coloured, although dark grey patches appear at the inner core of the heartwood, along with (sometimes vivid) red streaks, which fade in time. The texture is fine and uniform.

Properties

The weight varies widely, between about 480 kg/m³ and 640 kg/m³ (30-40 lb/ft³) seasoned. You can get a very light or a very heavy piece of parana pine. The timber is not durable, and has only medium bending and crushing strengths, with a very low resistance to shock loads. It is extremely difficult to dry, showing a marked tendency to split in the darker areas, and needs to be monitored constantly. If it is not well dried it can distort alarmingly in service, particularly if the commonly available wide boards are used. It works extremely well with both hand and machine tools and planes and moulds cleanly to a very smooth finish. It glues and finishes nicely. The sapwood is liable to attack by the pinhole borer beetle and the common furniture beetle. The wood is non-durable and moderately resistant to preservation treatment, but the sapwood is permeable for preservation treatment.

Uses

This is Brazil's major export timber and only the higher grades are shipped. They are used for internal joinery, especially staircases because of the sizes available and freedom from knots. It is not tough enough, however, for applications like long ladder strings. It appears in cabinet framing, drawer sides, shop fitting and vehicle building. Locally it is used for joinery, furniture, turnery, sleepers and general constructional work. The sizes and ease of working make 'parana pine' an attractive DIY timber, but the moisture needs checking carefully. It is also used for plywood manufacture and sliced for decorative veneers.

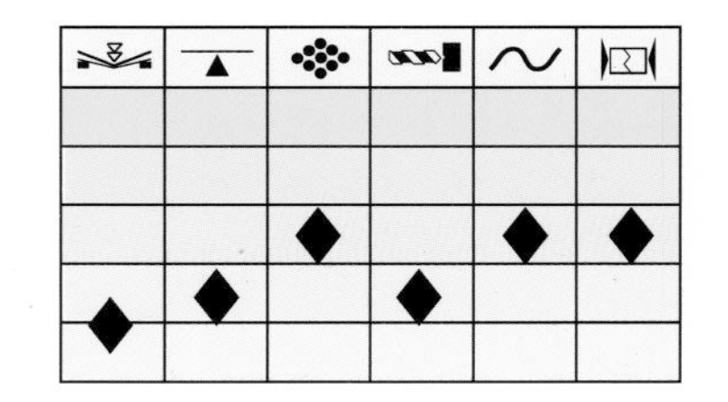

Aspidosperma spp. **Family:** *Apocynaceae*

ROSA PEROBA

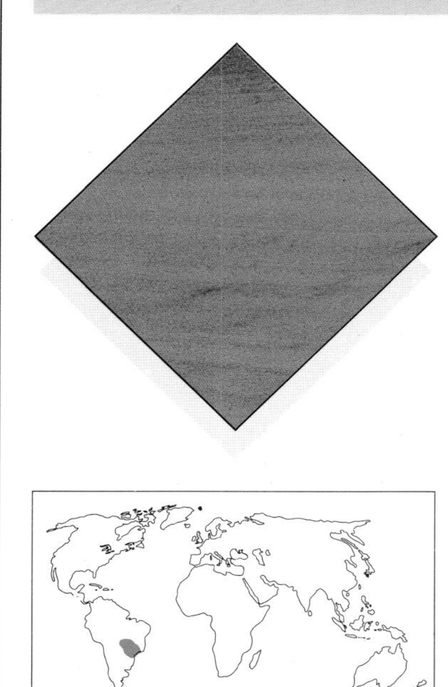

Where it grows

This tree occurs in the south east regions of Brazil, chiefly from Goias, Minais Gerais and San Paulo. It is derived principally from *A.peroba*, Fr.All. (*A.polyneuron*, Muell.Arg.), and known as rosa peroba and red peroba (UK). It grows to a height which varies from 15m (50ft) up to about 38m (125ft), averaging around 27m (90ft), with a well-formed straight bole of average diameter up to 1.2-1.5m (4-5ft).

Appearance

The cream-yellow sapwood blends gradually into the rose-red heartwood, which varies considerably from tree to tree. Sometimes is has purplish-brown streaks and patches, which turn orange-brown on exposure. Although the grain is very variable, from straight to very irregular, the texture is fine and uniform but without lustre. Peroba preta is rose-red with black streaks, peroba revesa has bird's eye figuring, peroba muida red with darker patches, peroba poca is almost white, peroba rajada is pink-red with large black patches, peroba tremida is yellow with golden patches.

Properties

This wood is hard, heavy and very dense. It varies in weight from about 700 to 850 kg/m^3 (44-53 lb/ft^3), averaging 750 kg/m^3 (47 lb/ft^3) when seasoned. Special care is needed in drying to avoid distortion and splitting. Because of its irregular grain, there is considerable variation in strength properties. On average the timber has medium to high bending strength, medium resistance to shock loads, low stiffness and a high crushing strength, but it is not normally used for steam bending. This durable wood is fairly easy to work, with a slight blunting effect on cutting edges. A reduced cutter angle is recommended. The wood requires pre-boring for screwing and nailing. It can be glued easily and takes stain and polish finishes excellently. The sapwood is liable to insect attack. The heartwood is durable and extremely resistant to preservative treatment, but the sapwood is permeable.

Uses

In Brazil rosa peroba is used externally for construction work, joinery and ship building, also for superior furniture and cabinet making, panelling, strip and parquet flooring and turnery. Selected logs are sliced to produce a really beautiful range of decorative veneers for architectural panelling and marquetry.

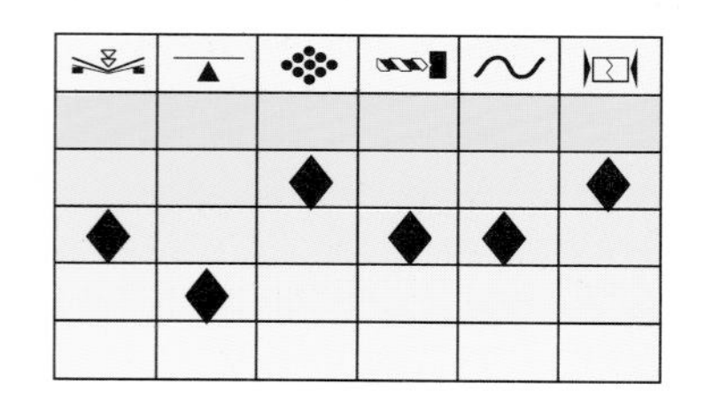

Aucoumea klaineana **Family:** *Burseraceae*

GABOON

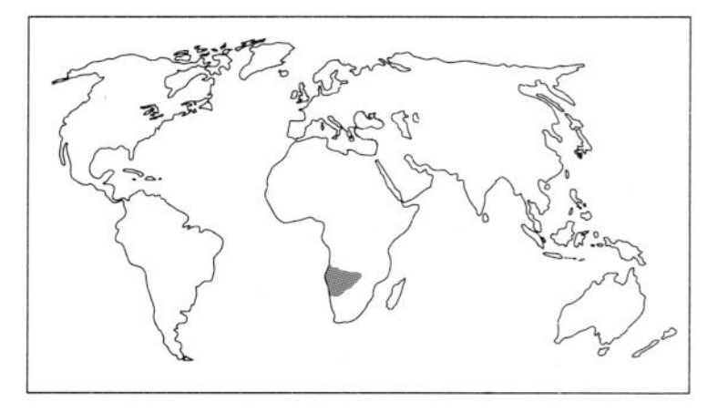

Where it grows

More gaboon is exported than any other African wood. This large tree grows mostly in Equatorial Guinea, Gabon and the Congo Republic. It is known as okoum – and angouma (Gabon); gaboon (UK); combogala (Congo); and mofoumou, n'goumi (Equatorial Guinea). It grows up to a height of around 60m (200ft) and has a slightly curved but clear cylindrical bole with a diameter of 1-2m (3-6½ ft).

Appearance

The heartwood is light pink, toning to pink-brown on exposure. It is commonly straight grained and sometimes shallowly interlocked or slightly wavy, producing an attractive stripe on quartered surfaces. It has a medium, uniform, even texture with a natural lustre.

Properties

Gaboon is a weak timber of light density, the weight varying from 370 to 560 kg/m^3 (23-35 lb/ft^3), it averages around 430 kg/m^3 (27 lb/ft^3) when seasoned. It dries fairly rapidly, without difficulty with little degrade, and there is medium movement in service. It has low bending strength, very low stiffness, a medium crushing strength, and a poor steam-bending classification. It is generally a weak wood that is not durable or resistant to decay, but as it is mostly used internally these faults are not critically important. Although the timber works fairly well with both hand and machine tools, it is rather woolly and as it contains silica it can result in moderately severe blunting of cutting edges. The grain tends to pick up when planing or moulding irregular grain on quartered surfaces. It nails without difficulty and glues easily. The wood can be polished to a lustrous finish if the surface is scraped and sanded well. The sapwood is liable to attack by the powder post beetle. The wood is non-durable and resistant to preservation treatment.

Uses

This very important timber is usually rotary cut into constructional veneers for the manufacture of plywood, blockboard and laminboard, which are used extensively for a wide variety of purposes such as flush doors, cabinet making and panelling. The solid timber is used for edge lippings, facings and mouldings as a substitute for mahogany, and interior frames for furniture construction. It is also used for cigar boxes and sports goods. Selected logs are sliced for mottled and striped decorative veneers for cabinet work, architectural wall panelling and so on.

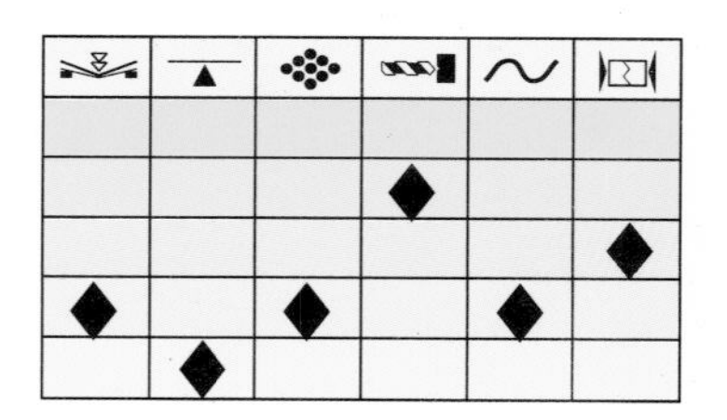

Baikiaea plurijuga **Family:** *Leguminosae*
'RHODESIAN TEAK'

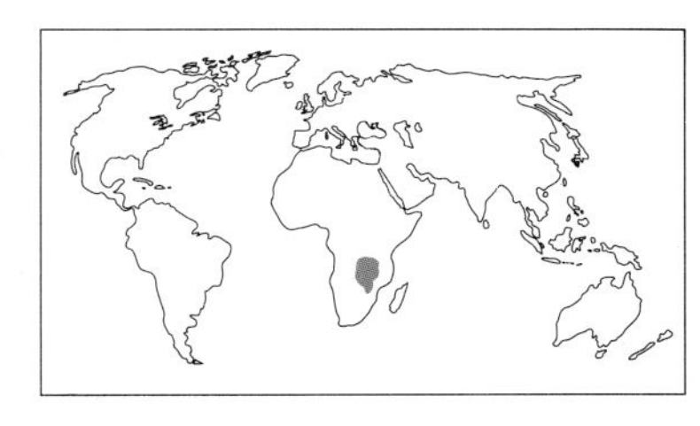

Where it grows
This attractive tree grows in Zambia and Zimbabwe, usually scattered in open forests or in small stands. It is small to medium size with a great many branches and grows to a height of 15-18m (50-60ft) with a diameter of about 0.8m (2½ft). It is also known as Zambesi redwood, umgusi, mukushi and mukusi. It is not a true teak (*Tectona grandis)*.

Appearance
The heartwood is a handsome red-brown, often marked with irregular black lines or flecks, and clearly defined from the narrow and much paler sapwood. The grain is usually straight or sometimes slightly interlocked, with a fine even texture which provides a smooth lustrous surface. The presence of tannin in the wood makes it liable to stain in contact with iron or iron compounds in damp conditions.

Properties
The average weight is about 900 kg/m^3 (56 lb/ft^3) when seasoned. 'Rhodesian teak' seasons well but rather slowly, and care is needed to avoid warping or splitting of knots or the extension of initial shakes. Once dry it is stable, with very small movement in service. It is a very heavy, hard timber with a high resistance to abrasion, and therefore difficult to machine. It offers a high resistance in cutting with a severe blunting effect on cutting edges of tools, and there is a considerable build-up of resin on tipped saw teeth. It tends to char in some operations. The wood has high bending and crushing strengths, but low stiffness and shock resistance and a moderate steam-bending rating. It tends to buckle and cannot be bent at all if small knots are present. It requires pre-boring for nails and screws, but glues and stains well, and provides an excellent finish. The sapwood is liable to attack by powder post beetle.The heartwood is very durable and extremely resistant to preservation treatment and the sapwood is moderately resistant.

Uses
'Rhodesian teak' is an excellent turnery wood. It has a very high resistance to wear and the outer heartwood offcuts are exported to make excellent decorative or heavy-duty flooring. Large-sectioned timber is used locally for furniture making, wagon building and railway sleepers.

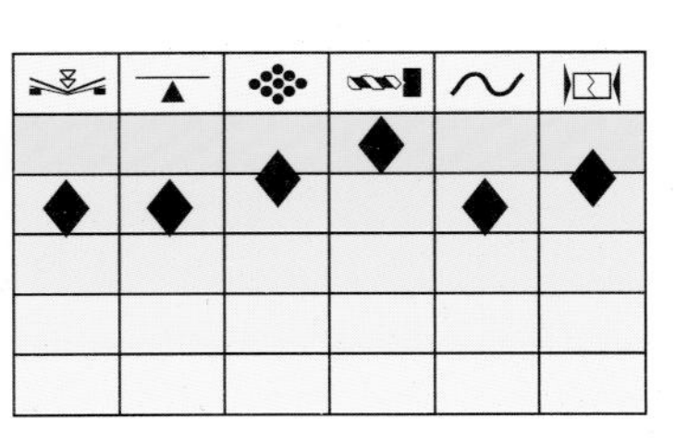

Balfourodendron riedelianum **Family:** *Rutaceae*
PAU MARFIM

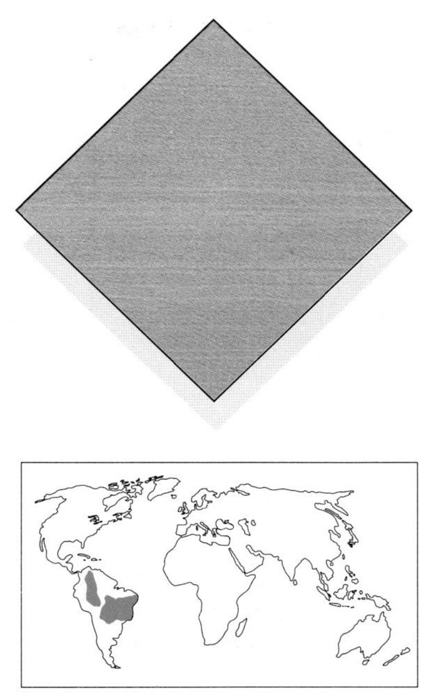

Where it grows
This moderately sized tree grows chiefly in southern Brazil,around Rio Paranapamena, São Paulo and Rio de Sul. It also occurs in Paraguay and northern Argentina. It may reach a height of 25m (80ft), and a rather slender diameter 0.8m (2½ft). It is usually shipped as sawn lumber and squares. It is also known as moroti, guatambu moroti (Brazil and Argentina); quatamba, farinha seca, pau liso (Brazil); kyrandy (Paraguay); quillo bordon (Peru); yomo de huero (Colombia); and ivorywood (USA).

Appearance
Pau marfim has an almost featureless appearance and a pale creamy-lemon yellow colour. There is very little difference between the sapwood and heartwood, which sometimes has darker streaks and a very fine, even texture. The growth rings are visible on flat-sawn surfaces. The grain is mostly straight, but sometimes irregular or occasionally interlocked, with a medium natural lustre.

Properties
This heavy, dense wood weighs around 800 kg/m^3 (50 lb/ft^3) when seasoned. It dries without difficulty and is stable in use and easy to work, although it may quickly blunt the cutting edges of tools. It can be brought to a smooth, fine finish. It is a really tough, strong timber and has high strength properties in all categories, especially resistance to shock loads. It is considered too tough for steam bending. Irregular grain tends to pick up when planing or moulding on quartered surfaces, and a reduced cutter angle is recommended. The wood can be nailed and screwed without difficulty and glues readily. It can be stained and polished to a fine, smooth finish. The sapwood is permeable and liable to insect attack, while the heartwood is both non-durable and resistant to preservative treatment.

Uses
This tough wood is ideal for uses such as striking tool handles and oars. It is used in the countries of origin for construction, furniture and cabinet work. It is also an excellent turnery wood, being fine textured and compact; it is used for shoe lasts, textile rollers, drawing instruments and rulers, and is also an ideal substitute for maple for hard-wearing floors. Selected logs are sliced for highly decorative veneers suitable for cabinets, panelling and marquetry. Other pale, fine-textured woods in South America, especially species of *Aspidosperma*, are also called pau marfim.

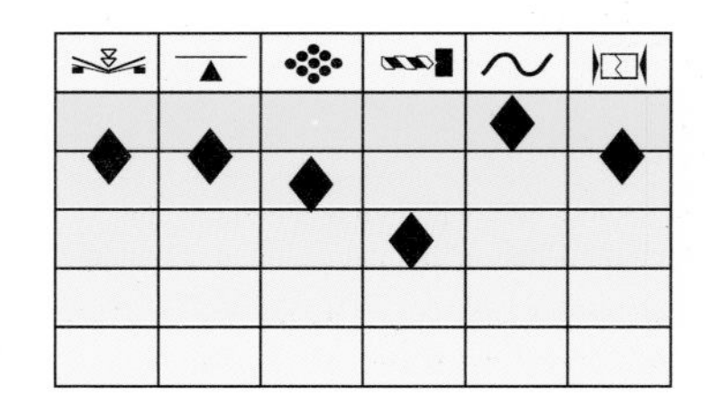

Betula alleghaniensis **Family:** *Betulaceae*

YELLOW BIRCH

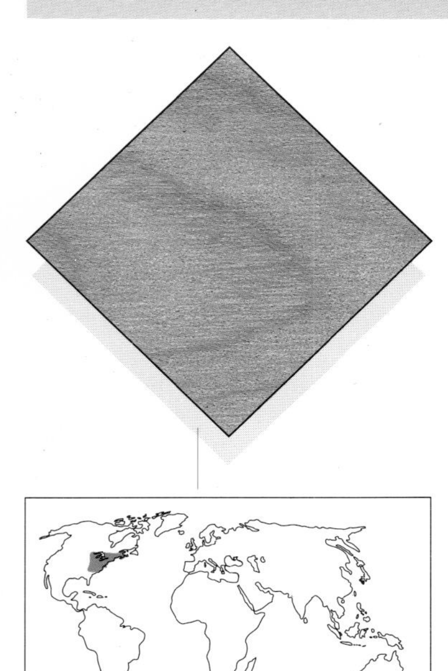

Note: syn. *B.lutea*, Michx.f. (principally) and *B.lenta*, L.

Where it grows

Yellow birch is a very important and common tree in Canada and the eastern USA. It is also known as Quebec birch and American birch (UK), and hard birch or betula wood (Canada). The wood is graded by uniformity of colour; in the UK, strongly figured woods are called 'Canadian silky wood', sapwood is sometimes called 'white birch' (occasionally confused with paper birch), and heartwood known as 'red birch'. It grows in the Canadian Maritime Provinces from Lake Superior down into the USA as far as Tennessee. This tree grows straight and upright to a maximum height of 30m (100ft), with a diameter of 1-1.2m (3-4ft).

Appearance

The light yellow sapwood is distinct from the heartwood, which is reddish-brown, with growth rings marked with dark reddish-brown lines. This is a straight-grained wood with a fine, even texture. The sapwood and heartwood often contain growth defects known as pith flecks caused by insect attack.

Properties

This dense, heavy wood weighs about 690 kg/m³ (43 lb/ft³) when seasoned, and dries rather slowly with little degrade, but suffers from considerable movement in use. It is as tough as European ash, with high bending, shock resistance and crushing strengths, medium stiffness, and a very good steam-bending rating. It works well with both hand and machine tools, with only a moderate blunting effect on tools. Pre-boring is required for screwing or nailing. It glues well. This timber is liable to attack by the common furniture beetle. It is perishable and moderately resistant to preservative treatment, but the sapwood is permeable.

Uses

Yellow birch is used for high-grade best-quality plywood. It is highly valued for furniture and upholstery frames, and is excellent for turnery, bobbins, shuttles and spools. It is also used for cooperage and parts of agricultural instruments. Its high resistance to wear makes it ideal for light-duty flooring in factories, schools, dance halls or gymnasia. Selected logs are sliced for highly decorative veneers for high-quality cabinets and panelling.

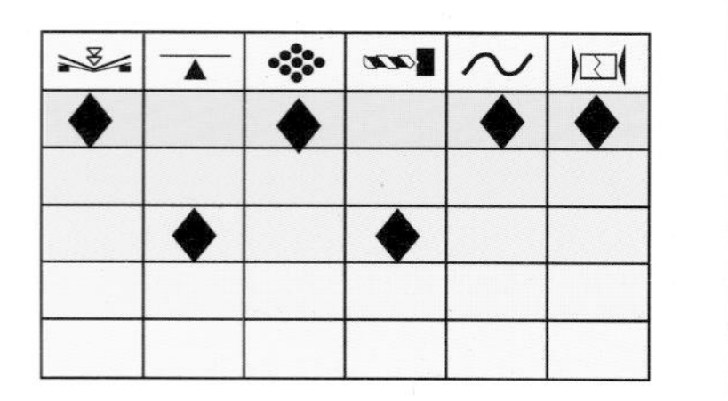

Betula papyrifera **Family:** *Betulaceae*

PAPER BIRCH

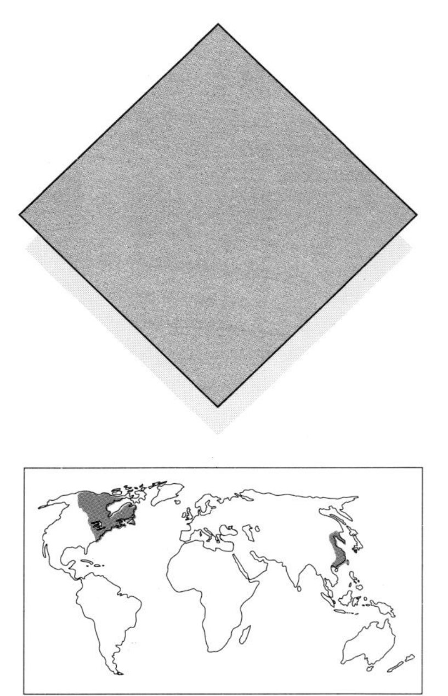

Where it grows

The American Indians made their canoes from the bark of this tree. Today, it is an important commercial timber growing from the Yukon to Hudson Bay, and Newfoundland down to the eastern USA. It is known as American birch in the UK and white birch in Canada. The tree grows to a height of 18-21m (60-70ft) with a nice long, clear, cylindrical bole, and an average diameter of 0.5m (1½ft).

Appearance

Paper birch has a wide, creamy-white sapwood with a pale brown heartwood, normally with a straight grain and fine, uniform texture. A good proportion of the logs produce a lovely curly figure known as 'flame' birch. Growth defects known as pith flecks caused by insects sometimes form a decorative feature.

Properties

This medium density tree weighs about 620 kg/m³ (39 lb/ft³) when seasoned. It dries rather slowly with little degrade. Paper birch is closer to European birch in characteristics than yellow birch, and is slightly inferior in strength properties but shrinks less. It has medium strength in all categories except stiffness, which is rated low, and has only a moderately good steam-bending rating. It works reasonably well with both hand and machine tools, with a moderate blunting effect on the cutting edges of tools; sometimes curly grain may pick up when planing or moulding on quartered surfaces, and a reduced cutter angle is recommended. The wood glues well, takes stain and polish easily, and can be brought to an excellent finish. It is non-durable and moderately resistant to preservative treatment, but the sapwood is permeable.

Uses

The bulk of the timber is rotary cut for plywood manufacture. It is also an excellent turnery wood, and used for spools, bobbins, dowels, domestic woodware, hoops and crates, toys and parts of agricultural machinery. Selected logs are sliced for attractive decorative veneers for cabinets and architectural panelling. It is also valuable as a pulp wood for writing paper, and used for fruit basket making, ice cream spoons and medical spatulas. The World War II Mosquito fighter bomber aircraft was built from Canadian birch plywood made principally from paper and yellow birch.

Betula pendula **Family:** *Betulaceae*
EUROPEAN BIRCH

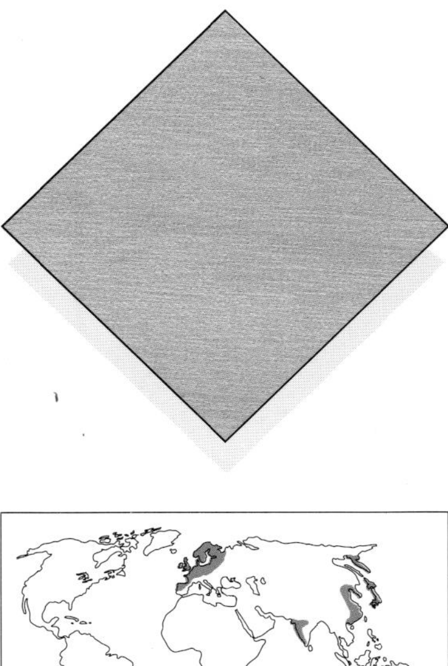

Where it grows
The birch, which will endure extremes of cold and heat, can be found further north than any other broadleaved tree, as far as Lapland. It also grows throughout Europe, from Scandinavia down to central Spain. It is sold according to its country of origin, thus English, Finnish, Swedish birch and so on. Two forms are valuable commercially, *B.pendula* (*B.alba*, L., partly) and *B.pubescens*, Ehrh. European birch attains a height of 18-21m (60-70ft), with a diameter of 0.6-lm (2-3ft).

Appearance
There is no distinction between sapwood and heartwood in colour, a featureless creamy-white to pale brown. The wood is straight grained and fine textured. The larvae of burrowing insects create pith flecks which show as irregular dark markings and local grain disturbance. When this happens the timber is called masur birch when peeled into veneer. Grain deviation causes 'flame' and 'curly' figure.

Properties
Birch is a heavy dense wood and weighs about 660 kg/m^3(41 lb/ft^3) when seasoned. It dries fairly rapidly with a slight tendency to warp. The wood is moderately stable in use and possesses high bending and crushing strengths with medium stiffness and shock resistance. Knots and irregular grain commonly present limit its use for steam bending. The wood works easily with both hand and machine tools, but is inclined to be woolly. To prevent tearing of cross or irregular-grained material when planing or moulding, use a reduced cutter angle. The wood requires pre-boring for nailing or screwing, and can be glued, stained and polished to a good finish. The sapwood is liable to attack by the common furniture beetle. The heartwood is moderately resistant to preservative treatment but the sapwood is permeable.

Uses
This is the major material for birch plywood in Finland and the USSR. In the solid it is used for upholstery framing, interior joinery and furniture making. It is an excellent wood for turnery and also suitable for brushes, brooms, bobbins and dowels. Selected logs are peeled or sliced for highly decorative veneers for doors and panelling.

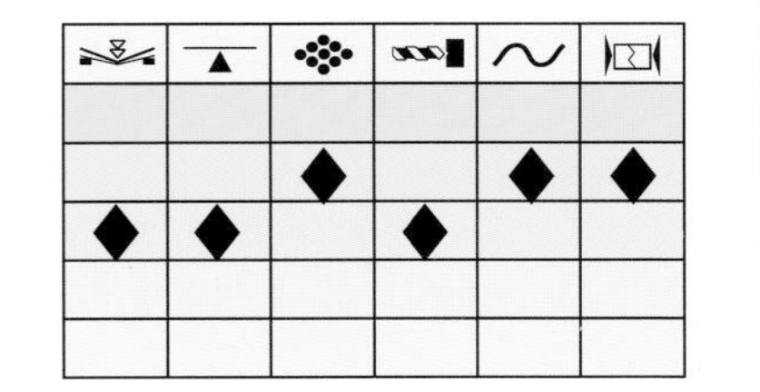

Brachylaena hutchinsii **Family:** *Compositae*
MUHUHU

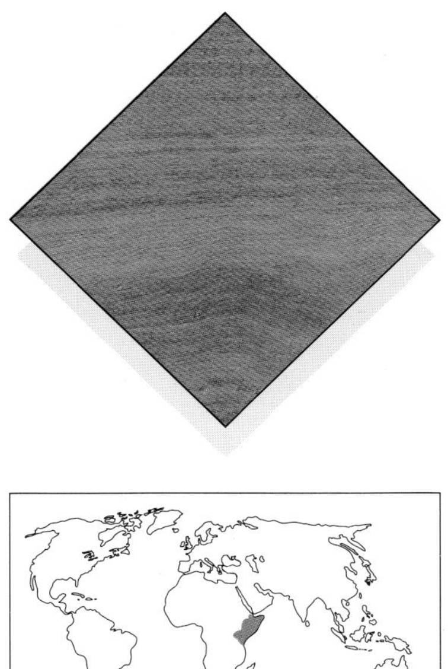

Where it grows
The tree grows in the semi-evergreen and lowland dry forests of the East African coastal belt and also occurs in the highland forests of Tanzania and Kenya. It is medium sized, growing to a height of about 25m (80ft), with a diameter of 0.5-0.6m ($1\frac{1}{2}$-2ft). The tree is often twisted or fluted and difficult to get in large sizes.

Appearance
This very heavy, dense wood has a narrow grey-white sapwood, quite distinct from the heartwood, which is orange-brown in colour, sometimes with a greenish tinge. It matures to a medium brown on exposure and although usually straight grained, can also contain closely interlocked or wavy grain. It has a very fine, even texture and a pleasant odour when cut.

Properties
The weight varies from 830 to 1,000 kg/m^3 (52-62 lb/ft^3), but averages around 910-960 kg/m^3(57-60 lb/ft^3) when dry. The timber needs to be dried slowly and carefully to avoid a tendency for hair checking and end splitting, but when dry it is stable in use. It has medium bending strength, low stiffness, very low shock resistance and high crushing strength. Unless pin knots are present, it has a medium steam-bending rating. It also has a high resistance to indentation and abrasion. The wood is difficult to machine, moderately blunting cutting edges, on which gum tends to build up. This very durable timber requires pre-boring for nailing, but it glues well and a very good finish is obtained when stained and polished. It is extremely resistant to preservative treatment.

Uses
This attractive, very hard-wearing timber is available only in short lengths, which tends to limit its usefulness. It is used for flooring blocks and strips for high-quality floors in public buildings like hotels, where there is heavy pedestrian traffic, and in factories and warehouses where it is exposed to industrial traffic such as fork lift trucks. In East Africa it is used for carving animals, and is a very fine turnery wood. When longer lengths are available it is used for heavy construction, bridge decking, girders and railway sleepers. It has a spicy aromatic oil which is distilled and sold as a substitute for sandalwood oil, and the timber has been exported to India for use in crematoria instead of sandalwood.

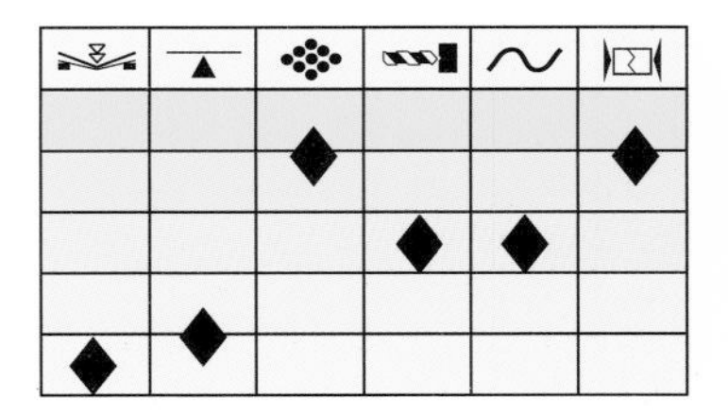

Buxus sempervirens **Family:** *Buxaceae*
EUROPEAN BOXWOOD

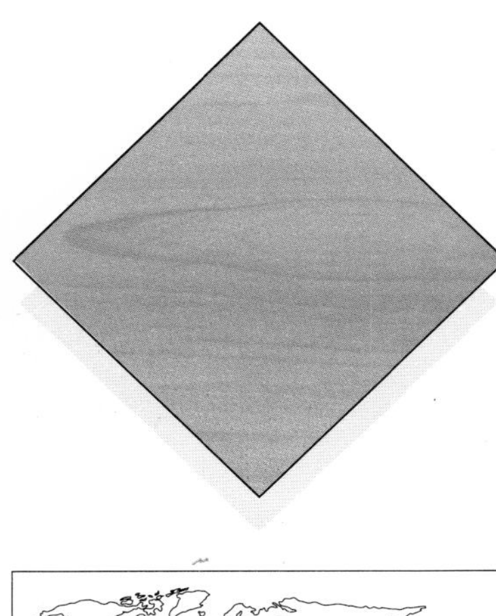

Where it grows
This is one of the very few evergreen broadleaved trees which occur in mild temperate climates. It grows in the UK, Europe, North Africa, Asia Minor and western Asia. It is known as Abassian, Iranian, Persian or Turkish boxwood, according to the country of origin. It is a small tree growing to a height of 6-9m (20-30ft). Imported billets are only 0.9-1.2m (3-4ft) long and only 0.1-0.2m (4-8 in) in diameter.

Appearance
The wood is a pale bright orange-yellow colour, occasionally straight grained but often irregular, especially from small trees grown in Britain. It has a very compact, very fine and even texture.

Properties
The weight varies from 830 to 1140 kg/m^3 (52-71 lb/ft^3) averaging about 910 kg/m^3 (57 lb/ft^3) when seasoned. Box dries very slowly with a strong tendency to develop surface checks or split badly if dried in the round. The billets are usually soaked in a solution of common salt or urea before drying and end coatings applied. Billets should be converted into half rounds and dried under covered storage conditions. It is a very dense, hard, durable wood with a high resistance to cutters, which should be kept very sharp to prevent the wood from burning when boring or recessing. The irregular grain tends to tear in planing. It is an excellent turnery wood and has a good steam-bending rating, with high stiffness, good crushing strength and resistance to shock loads. Pre-boring is required for nailing. It glues very well, and when stained and polished gives an excellent finish. The sapwood is liable to attack by the common furniture beetle. The heartwood is durable and resistant to preservative treatment.

Uses
Boxwood has an outstanding reputation for its fine, smooth texture and excellent turning properties and is used for textile rollers, shuttles, pulley blocks, skittles, croquet mallets and especially tool handles. It is also in demand for wood sculpture and carving. Specialized uses include measuring instruments, rulers and engraving blocks, parts of musical instruments, chessmen, corkscrews and so on. It is also used for inlay lines and bandings in marquetry and reproduction of period furniture and the repair of antiques.

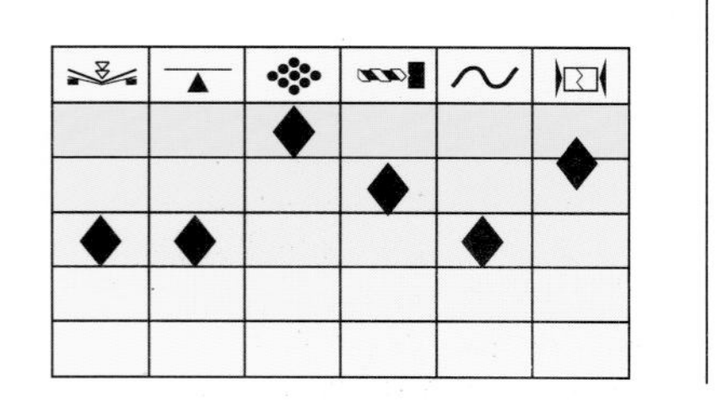

Caesalpinia echinata **Family:** Leguminosae
BRAZILWOOD

Where it grows
The name 'brazil' (*brasa* in Portuguese) had been used since the Middle Ages to describe a red dye with the colour of glowing fire embers which came from the East Indian wood *C.sappan*. The Portuguese gave the name of the wood to the country when they came to Brazil. This timber is also known as Bahia wood, para wood and pernambuco wood (UK), Brazil ironwood (USA), hypernic, brasilette, brasilete or brasiletto (Brazil). The small tree grows along the coastal forests of eastern Brazil, from Bahia towards the south. It grows to a height of 6-9m (20-30 ft), yielding short billets lm (3ft) long and up to 0.2m (8 in) in diameter.

Appearance
The almost white sapwood is sharply distinct from the heartwood, which matures from bright orange to a rich deep dark red. It has a marble-like figure of dark red-brown variegated stripes, sometimes dotted with pin knots. It can be selected for straight grain, but is often interlocked, and has a fine, compact, smooth, even texture with a natural lustre.

Properties
Brazilwood is a very hard, heavy wood weighing about 1200-1280 kg/m^3 (75-80 lb/ft^3) when seasoned. It needs to be dried very slowly to avoid degrade but is very stable in service. It has high strength properties in all categories, but is unsuitable for steam bending. It is sometimes difficult to work, with a severe blunting effect on cutting edges, which must be kept very sharp, but it finishes well. It needs pre-boring for nailing but has good screw-holding properties. The very durable wood can be glued easily and brought to a very smooth, lustrous, brilliant finish, often showing a snake-like ripple. It is resistant to both insect attack and to fungal decay.

Uses
Brazilwood has a richly deserved worldwide reputation as a dye wood, but it is also regarded as the world's finest timber for violin bows, for its weight, flexibility, strength and resilience. It is also highly prized for ornamental turnery and for high-quality gun butts and rifle stocks. It is used locally for exterior joinery, and it makes excellent heavy-duty parquetry flooring. Selected logs are sliced for decorative veneers for panelling and sawn for inlays in reproduction antique furniture.

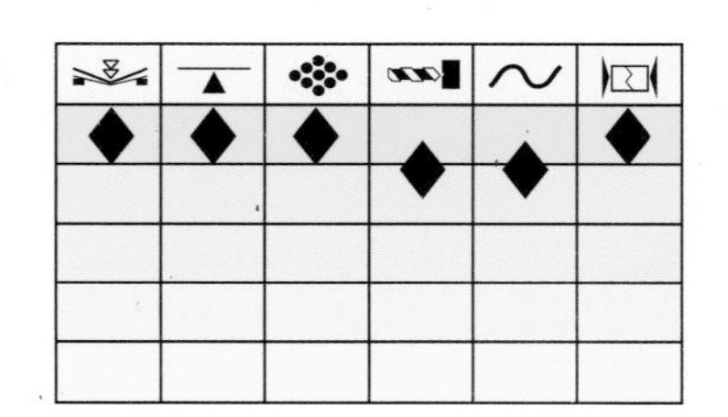

Calophyllum brasiliense **Family:** *Guttiferae*
JACAREUBA (SANTA MARIA)

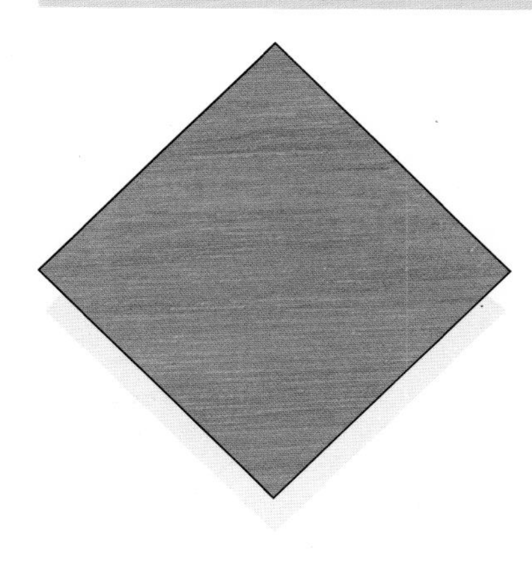

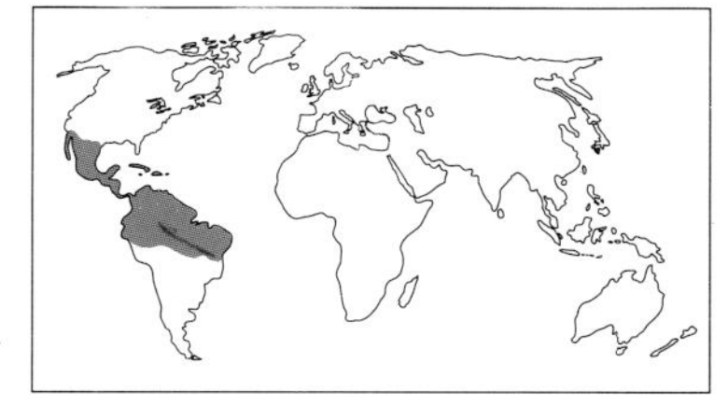

Where it grows
This genus is widely distributed in the tropics. It grows throughout the West Indies, and from southern Mexico down into the northern parts of South America. *C.brasiliense*, Camb. (var. *rekoi*, Standl.) produces santa maria from Central America and the West Indies, and is known as jacareuba in Brazil. It is also known as guanandi, pau de Maria (Brazil); krassa (Nicaragua); and koerahara (Surinam). The tree grows to a height of 30-45m (100-150ft) and has a diameter of 1-1.5m (3-5ft), with a clear cylindrical bole of around 20m (65ft).

Appearance
The heartwood varies from yellow-pink to rich red-brown. Its grain is commonly interlocked, sometimes straight, often with fine dark red parenchyma stripes which show on flat-sawn surfaces, and an attractive ribbon striped figure on quartered surfaces. The indistinct band of narrow sapwood is lighter in colour. The wood has a fairly uniform, medium texture.

Properties
Weight varies from about 540-700 kg/m^3 (34-44 lb/ft^3), averaging about 590 kg/m^3 (37 lb/ft^3) when seasoned. The timber is difficult to air dry because it dries slowly with a considerable amount of warping and splitting, and it is especially difficult to extract moisture from the centre of thick material. There is medium movement in use. It has medium bending strength and shock resistance, with low stiffness and high crushing strength. It has a moderate steam-bending rating. The timber is moderately easy to machine but the soft parenchyma tissue tends to pick up badly, and brown gum streaks can cause rapid blunting of cutting edges when planing or moulding quarter-sawn surfaces. A reduction of cutting angle is advised here. It requires pre-boring for screwing or nailing. This very durable wood glues well and stains and polishes to a good finish. The heartwood is extremely resistant to preservation treatment but the sapwood is permeable.

Uses
Jacareuba is used locally for exterior joinery, general construction, bridge building, shingles, ship building and so on. It is also used for interior construction and joinery, shop fitting and furniture making. Selected logs are sliced for decorative veneers for cabinets and panelling.

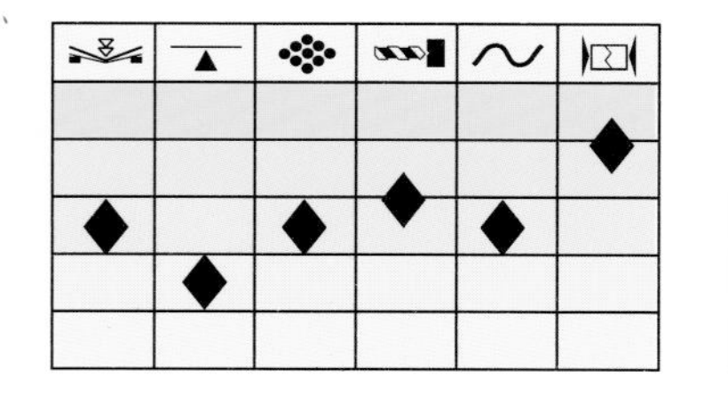

Calycophyllum candidissimum **Family:** *Rubiaceae*
DEGAME

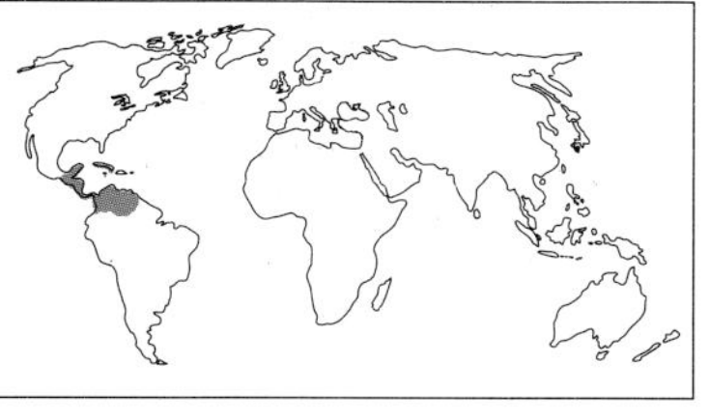

Where it grows
This small to medium sized tree occurs in Cuba, and from southern Mexico through Central America to Colombia and Venezeula in tropical South America. It grows to a height of 12-20m (40-65ft), and an average diameter of 0.4m (16in). Exported in small round, unbarked logs, it is also known as degame, lancewood in the UK and lancewood or lemonwood in the USA.

Appearance
The wood has a wide thick sapwood of white to very pale brown, not clearly defined from the heartwood, which is a variegated light olive-brown. The grain is generally straight, occasionally irregular, with an exceedingly fine and even texture.

Properties
Degame's average weight is about 820 kg/m^3 (51 lb/ft^3) when seasoned. It dries slowly with little degrade, and is stable in use. Degame is a hard, heavy, tough and resilient timber, not easy to split. It has high bending and crushing strengths, medium stiffness and shock resistance, and a very good steam-bending rating. The wood is not difficult to work with either hand or machine tools if the cutting edges are kept sharp. It holds screws and nails without difficulty, glues well, takes stain and polish easily and can be brought to an excellent finish. The wood is not durable or resistant to decay.

Uses
Degame is an excellent wood for sculpture and wood carving and has a first-class reputation for turnery. It is extensively used as a good alternative for true lancewood (*Oxandra lanceolata*) for tool handles of exceptional hardness and the top joints of fishing rods; also for billiard cues, shuttles, shoe lasts, pulleys and measuring instruments. It is valued for archery bows, as it will bend and not break, and it is also used for parts in organ building and textile machinery. It is a tough flooring timber, superior for interior joinery and cabinet work. Locally it is used for agricultural implements, and for internal frames for buildings.

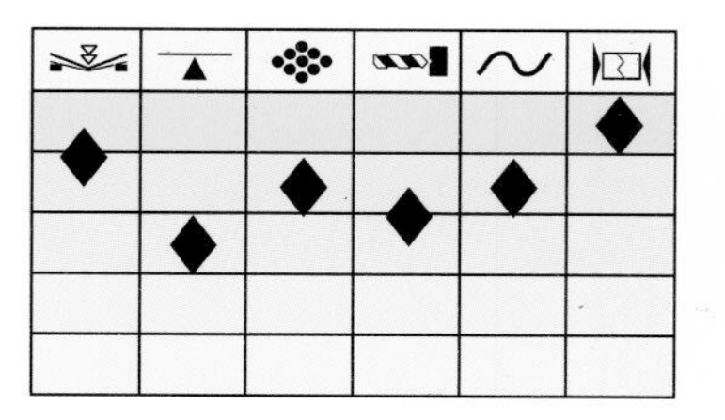

Cardwellia sublimis **Family:** *Proteaceae*
AUSTRALIAN SILKY OAK

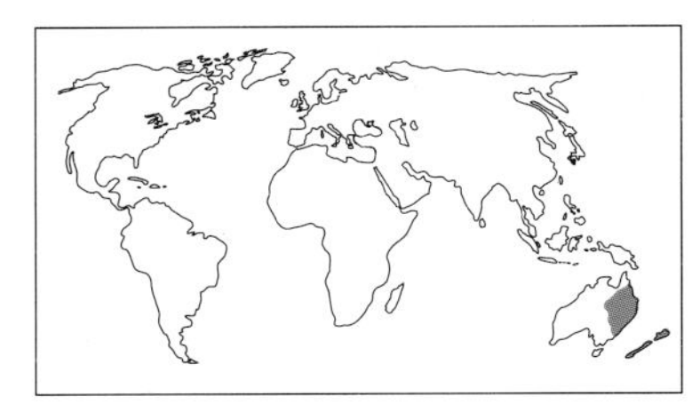

Where it grows
The name 'silky oak' is given to a number of different genera and species in Australia and New Zealand. The name originally referred to *Grevillea robusta*, a native of southern Queensland which also occurs in New South Wales. *C.sublimis*, from north Queensland, has similar characteristics, and became known as silky oak and northern silky oak. *G.robusta* has also been planted as a shade tree in coffee and tea plantations in East Africa. It is a large tree, about 37m (120ft) high and 1.2m (4ft) in diameter.

Appearance
This very attractive timber has large rays and a very well-marked silver grain ray figure showing on quartered surfaces, but it is not a true oak (*Quercus spp*). The heartwood of both main species is pink to reddish-brown, maturing into a darker red-brown. But *Cardwellia* is generally darker than that of *Grevillea*, which is a paler pink-brown. It is usually straight grained, except where the wood fibres distort around the rays, and has a coarse but even texture. Narrow gum lines are sometimes present and the wood has an attractive golden sheen.

Properties
The weight of *Grevillea* is about 580 kg/m³(36lb/ft³), and *Cardwellia* is about 550kg/m³ (34lb/ft³) when seasoned. It is a difficult timber to dry, and severe cupping of wide boards may take place. It has medium movement in use. The wood works readily with both hand or machine tools with only a slight blunting effect on cutting edges, but quarter-sawn material tends to pick up when planing or moulding and a reduction of the cutting angle is advised. It has good nail holding properties. It has below average strength in all categories relative to its density, especially bending and compression, but a good steam-bending classification. The wood glues well, takes stain readily and provides an excellent finish. The sapwood is liable to attack by the powder post beetle. The heartwood is moderately durable and resistant to preservation treatment.

Uses
Silky oak is used in Australia instead of softwood for building and shuttering. Top grades are used for furniture, interior joinery and panelling, office, bank and shop fitting, coachwork and cask staves, also for hard-wearing floors. It is rotary cut for plywood manufacture and sliced for decorative veneers for panelling.

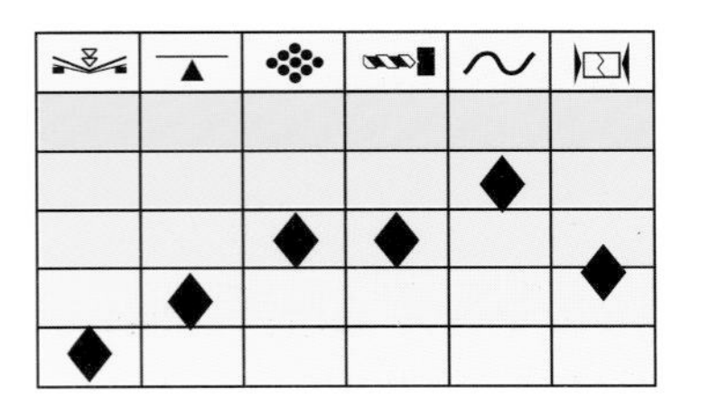

Carpinus betulus **Family:** *Betulaceae*
HORNBEAM

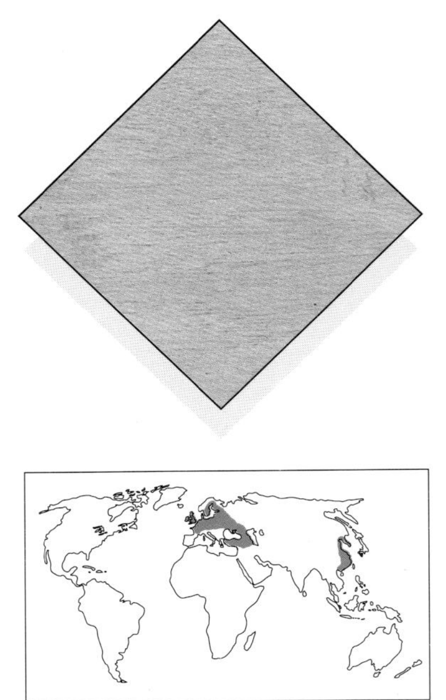

Where it grows
About 20 species of hornbeam grow in European northern temperate zones in rich soil on low ground. It occurs from Sweden south to Asia Minor, but the species of commercial interest grow in France, Turkey and Iran. In Britain it is a woodland and hedgerow tree. It grows to a height of 15-25m (50-80ft), with a fluted bole which is seldom clear of branches. The trunk is usually elliptical instead of round. The diameter is 0.9-1.2m (3-4ft).

Appearance
There is no distinction between sapwood and heartwood – a dull white colour marked with grey streaks and flecks caused by the broad ray structure, which produces a flecked figure on quartered surfaces. It is usually irregular or cross-grained, but has a fine, even texture.

Properties
The weight averages 750 kg/m³ (47 lb/ft³) when seasoned. Hornbeam dries fairly rapidly and well with little degrade, but with considerable movement in service. This heavy, dense wood has high bending and crushing strength, medium stiffness and resistance to shock loads and excellent shear strength and resistance to splitting. Similar to ash in toughness, it has a very good steam-bending classification. It is fairly difficult to work, as there is high resistance in cutting with a moderate blunting effect on tools. The wood finishes very smoothly, glues well, takes stain and polish easily and can provide an excellent finish. The sapwood is liable to attack by the common furniture beetle and the heartwood is perishable but permeable for preservation treatment.

Uses
Hornbeam is an excellent turnery wood and is used especially for brushbacks, drum sticks, the shafts of billiard cues, skittles and Indian clubs. It is also used for piano actions, clavichords, harpsichords and other musical instrument parts such as violin bridges; wooden cog wheels, pulleys, millwright's work, dead-eyes, mallets and wooden pegs. Its high resistance to wear makes it a flooring timber for light industrial use. Selected logs are sliced for decorative veneers.

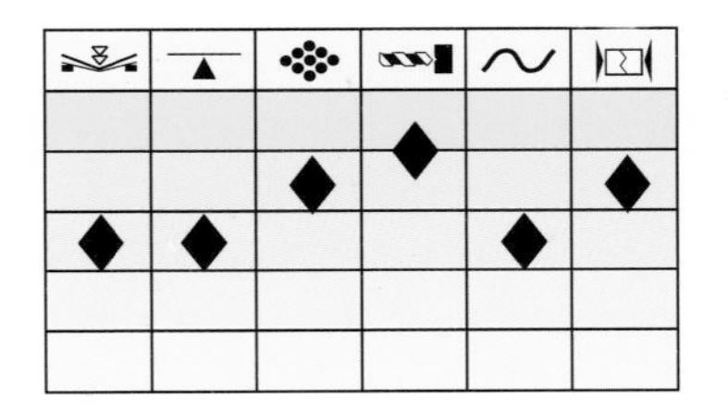

Carya spp. **Family:** *Juglandaceae*
HICKORY

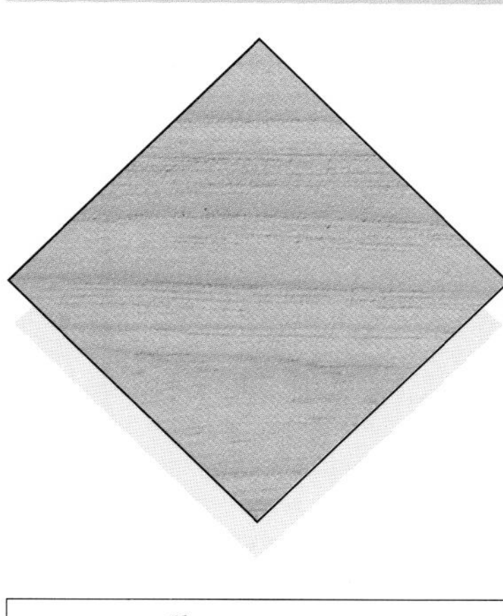

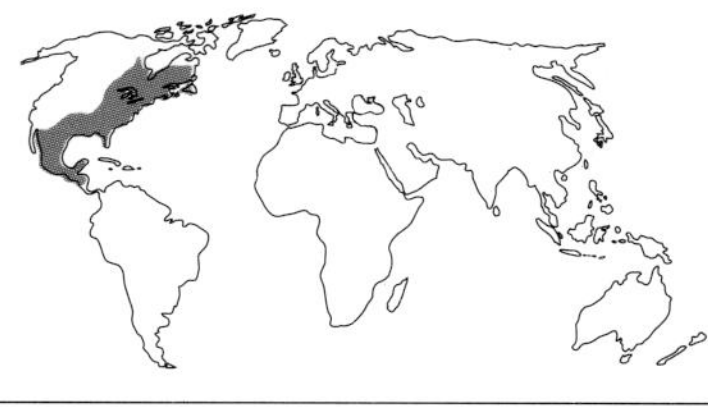

Where it grows

Although more than 20 species of hickory grow in the large forests of eastern Canada and eastern states of America, there are only four commercial species: C.*glabra* (Mill.) Sweet produces pignut hickory; C.*tomentosa*, mockernut hickory; C.*laciniosa*, shellbark hickory and C.*ovata*, shagbark hickory. These four true hickories occur from Ontario in Canada to Minnesota, Florida and Mexico in the deciduous forests. The trees vary according to species and grow from 18-36m (60-120 ft) high. They have a straight cylindrical bole, with a diameter 0.6-0.9m (2-3ft). C. *illinoensis* is known as sweet pecan and pecan hickory; C.*aquatica* is sold as bitter pecan or water hickory.

Appearance

The very pale grey wide sapwood, sold as 'white hickory', is generally preferred to the heartwood, which is red to reddish-brown – 'red hickory'. It is usually straight grained but occasionally wavy or irregular, with a rather coarse texture.

Properties

The weight ranges from about 700 to 900 kg/m³ (45-56 lb/ft³) but averages 820 kg/m³ (51 lb/ft³) when seasoned. Hickory needs very careful drying but is stable in service. It is very dense and has high toughness, bending, stiffness, and crushing strengths, with exceptional shock resistance. It has excellent steam-bending properties. The wood is difficult to machine, and has a moderate blunting effect on tools. It is also difficult to glue, but stains and polishes very well. Hickory is non-durable. The sapwood is liable to attack by the powder post beetle, and the heartwood is moderately resistant to preservation treatment.

Uses

Hickory is ideal for the handles of striking tools, such as hammer, pick and axe handles; also for wheel spokes, chairs and ladder rungs. It is a valuable sculpture and carving wood; it is extensively used for sports equipment such as golf clubs, lacrosse sticks, baseball bats, the backs of longbows, laminae in tennis racquets and skis. It appears in the tops of heavy sea-fishing rods, drumsticks, picking sticks in the textile industry and vehicle building. It is rotary cut for plywood faces and sliced for decorative veneers.

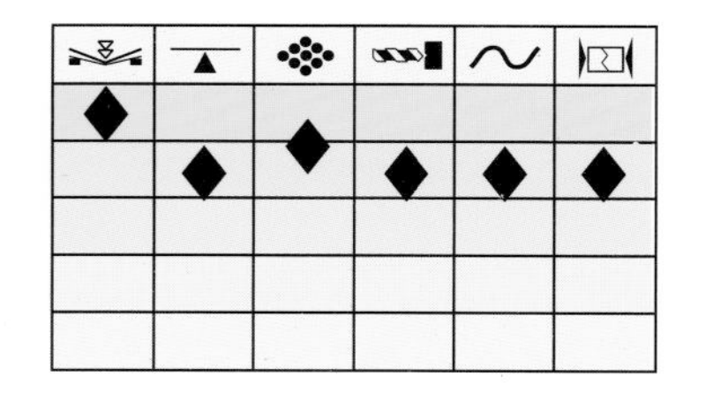

Castanea sativa **Family:** *Fagaceae*
SWEET CHESTNUT

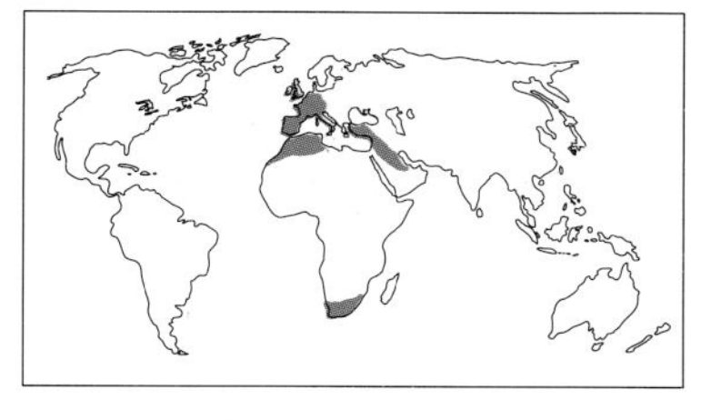

Where it grows

This stately tree is a native of south-west Europe, and grows in Britain, France, Germany, North Africa and western Asia. In favourable conditions of growth it reaches a height of about 30m (100ft) or more, and has a diameter of 1.5m (5ft). In less favourable sites the short bole branches into several limbs. It is known as Spanish chestnut or European chestnut.

Appearance

The heartwood is a straw-to-brown colour resembling oak, distinct from the narrow sapwood. The wood has prominent growth rings but finer rays, so it does not have the silver grain figure of oak on quartered surfaces. The grain is straight but often spiral, especially from mature trees, and has a coarse texture. Many logs are subject to cupping or ring shakes.

Properties

The weight of sweet chestnut averages about 540 kg/m³ (34 lb/ft³) when seasoned. It is rather difficult to air dry as it retains moisture in patches and tends to collapse or honeycomb; it does not respond well to kiln reconditioning. Its acidic character tends to corrode iron fastenings in damp conditions, and tannin in the wood causes blue-black stains to appear in the wood when it is in contact with iron or iron compounds. It is a medium density wood, possessing low bending strength, very low stiffness and shock resistance, and medium crushing strength. Air-dried wood has a good bending rating if free from knots. It works satisfactorily with both hand and machine tools, to which it offers only a slight blunting effect. It has good screw and nail holding properties, glues well and stains and polishes to an excellent finish. The sapwood is liable to attack by the powder post beetle and the common furniture beetle. The heartwood is durable and extremely resistant to preservation·treatment.

Uses

Sweet chestnut is used as a substitute for oak in furniture, also for coffin boards, domestic woodware, kitchen utensils and turnery for walking-sticks, umbrella handles, bowls and so on. It is cleft for fencing, gates and hop poles. Staves are used for casks for oils, fats, fruit juices and wines. Selected logs are sliced for decorative veneers.

		◆	◆		◆
				◆	
◆	◆				

Castanospermum australe **Family:** *Leguminosae*
BLACKBEAN

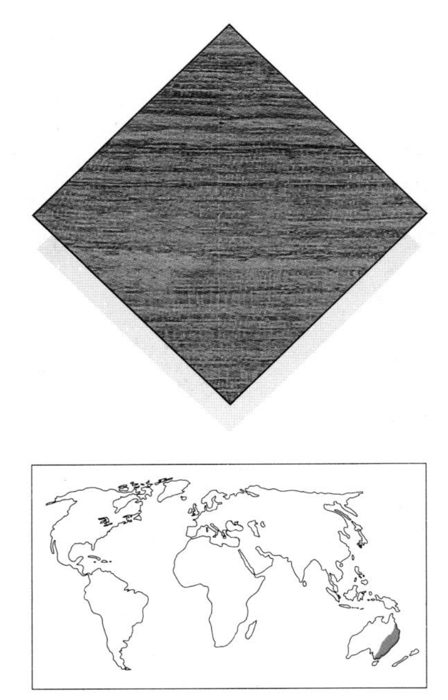

Where it grows
This rather small to medium sized tree grows in the moist scrub forests of Australia, in New South Wales and eastern Queensland. It is also known as Moreton Bay bean, Moreton Bay chestnut and beantree. It reaches a height of 40m (130ft) and the diameter of the bole averages 1.2m (4ft).

Appearance
The fairly narrow heartwood has a chocolate-brown colour, with narrow grey-brown streaks of parenchyma tissue surrounding large pores. Grain is generally straight but may be interlocked, producing a decorative striped effect on quartered surfaces. The wood is coarse textured.

Properties
Blackbean weighs about 700 kg/m³(44 lb/ft³) when seasoned. It is difficult to dry and very slow air drying is required before kilning to avoid splitting. It has a marked tendency to degrade and does not respond well to reconditioning treatment. There is medium movement in service. This dense, tough wood has medium strength in most categories but is fairly brittle and will not withstand shock loads. By its nature is also has a very poor steam-bending classification. It is fairly difficult to machine, the soft patches of lighter tissue crumble during planing or moulding unless cutters are kept very sharp. There is a moderate blunting effect on cutting edges. It screws and nails satisfactorily, but the greasy nature of the wood gives varying gluing results. It stains and polishes to an excellent finish. The sapwood is permeable, and liable to attack by the powder post beetle. The heartwood is extremely resistant to preservation treatment.

Uses
This very attractive wood is valued for high-class furniture, cabinet making and interior joinery. It is also an excellent turnery wood, used for fancy turnery, brushbacks, tool handles, mallet heads and so on. Selected pieces are in great demand for wood sculpture and wood carving. It is used for measuring instruments, and in electrical appliances because of its insulating qualities. It is used locally in Australia for heavy construction work and interior joinery. Selected logs are sliced for highly decorative veneers used as plywood faces for flush doors, cabinets and architectural panelling. The veneers are also popular for marquetry and inlay work.

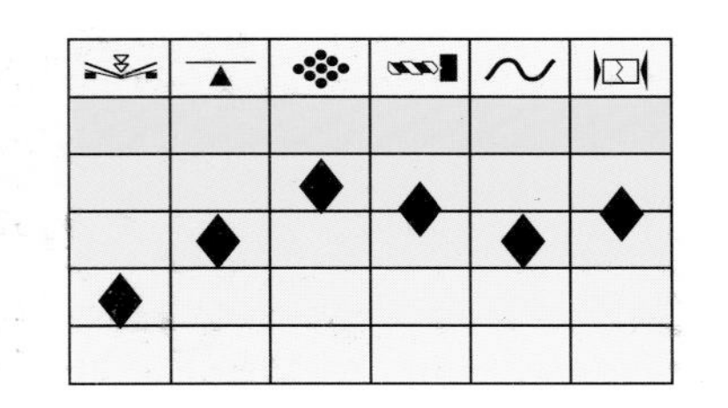

Cedrela spp. **Family:** *Meliaceae*
SOUTH AMERICAN CEDAR

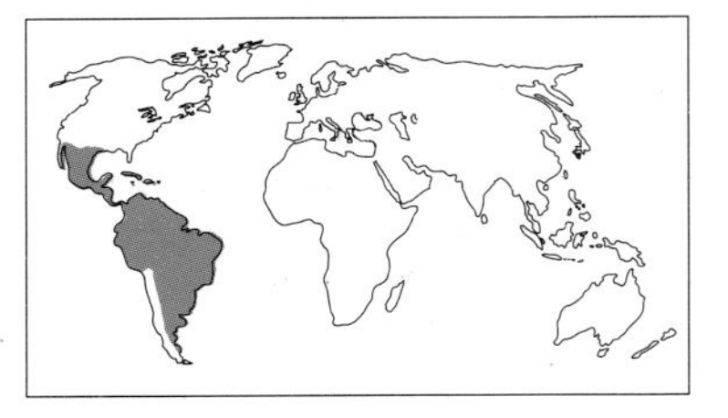

Where it grows
This is not a true cedar *(Cedrus spp.)*, which is a softwood. The name cedar has been given to many hardwoods of similar fragrance which occur in practically every country south of the United States, with the exception of Chile. The species comprise: *Cedrela fissilis*, Vell. – South American cedar, also known as Brazilian cedar, British Guiana cedar, Peruvian cedar (UK) and cedro (South America); also C.*odorata*, L. (C.mexicana, Roem) producing Central American cedar, known as Honduras cedar, Mexican cedar, Nicaraguan cedar, Tabasco cedar and Spanish cedar according to country of origin; also cigar-box cedar. The growth varies according to species and locality from 21-30m (70-100ft) tall, with a diameter of 1-2m (3-6ft).

Appearance
The pale pink-brown coloured young trees are not so straight grained, but have a fine texture; the darker reddish-brown types, occasionally with a purplish tinge, are from mature, slow-grown trees, and are more resinous, straight grained, but often interlocked and with a coarser texture. The wood contains gum which gives it a characteristic odour similar to genuine softwood cedar.

Properties
The weight varies from 370 to 750 kg/m³ (23-47 lb/ft³) with a fair average of about 480 kg/m³ (30 lb/ft³) when seasoned, and is very stable in use. It dries fairly rapidly and without difficulty. It is an easy wood to work with hand or machine tools, with low resistance to cutters, which should be kept sharp to avoid a tendency to woolliness. There is only slight blunting of cutting edges. The wood has a medium density and strength in all categories, and a moderately good steam-bending rating despite gum exudation. It nails and glues well, and can be brought to an excellent finish. The wood is durable, but the sapwood is liable to attack by the powder post beetle. The heartwood is extremely resistant to preservation treatment but the sapwood is permeable.

Uses
Cedar is used for high-quality cabinets and furniture, clothing chests, interior joinery and panelling, boat building, racing boat skins and canoe decks, organ sound boards and cigar boxes. It is also rotary cut for plywood and sliced for decorative veneers for cabinets and panelling.

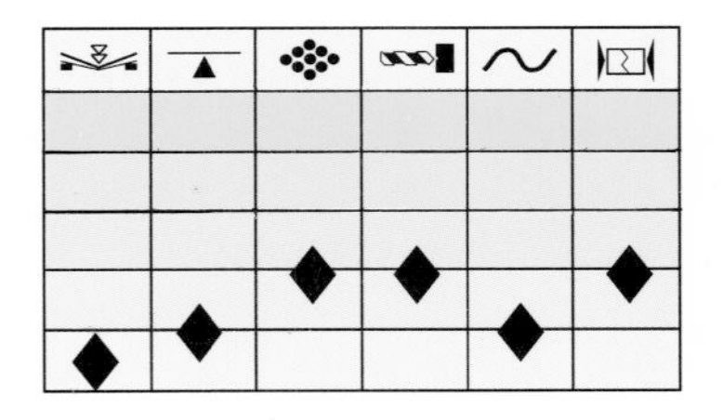

Cedrus spp. **Family:** *Pinaceae*
CEDAR (S)

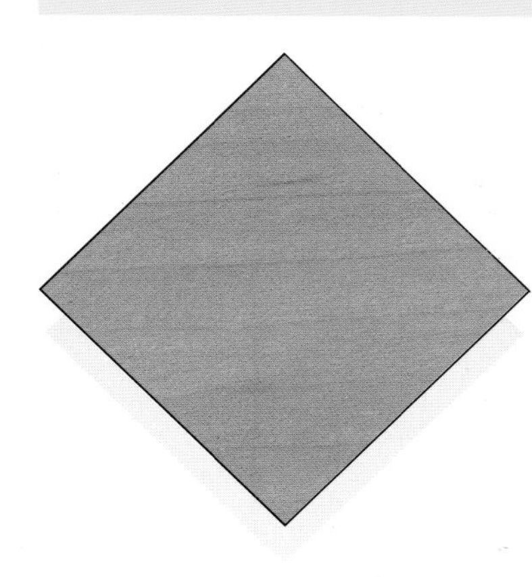

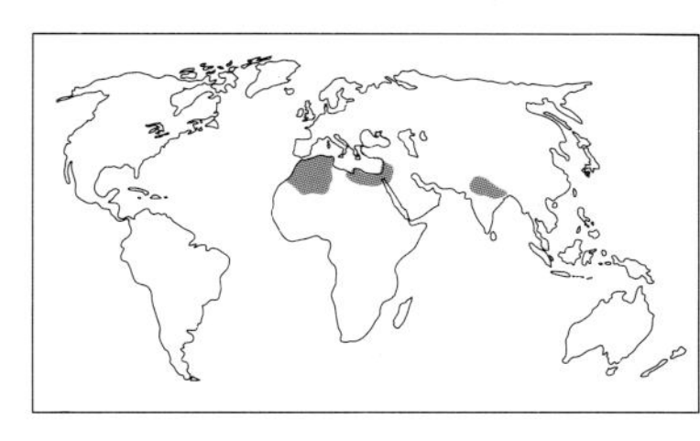

Where it grows

King Solomon built his temple with cedar from Mount Lebanon. The *Cedrus* species are the true cedars and should not be confused with other softwoods and hardwoods called 'cedar' because of their fragrant scent.

C.atlantica (Endl.) Carr., Atlas cedar, Atlantic cedar occurs in Algeria and Morocco; *C.deodara* (Roxb.) G.Don. produces deodar cedar from the western Himalayas, and is an important tree in India; *C.libani*, A.Rich. produces cedar of Lebanon.

The deodar grows to a height of about 61m (200ft) with a diameter of about 2.1m (7ft); Atlantic and Lebanon cedars vary from 36.5-45.5m (120-150ft) tall, and up to 1.5m (5ft) in diameter. Parkland timbers have a low flattened crown with large spreading branches near the ground, often from several stems.

Appearance

The heartwood is resinous, and has a very strong cedar odour. It is light brown in colour with a prominent growth ring figure and quite distinct from the narrow lighter coloured sapwood. Deodar cedar is straight grained, but Atlantic and Lebanon cedars are usually knotty, with much grain disturbance. These also tend to produce pockets of in-growing bark in the wood. The texture is fairly fine.

Properties

Weight averages about 560 kg/m^3 (35 lb/ft^3) when seasoned. The wood dries easily, with medium movement in use. It has medium bending strength but is low in other strength properties, and has a very poor steam-bending classification due to resin exudation. Cedar works easily with both hand and machine tools, with only slight blunting effect on cutters. The large knots and in-growing bark may cause problems in machining and cutters should be kept sharp. This durable wood has good holding properties, and stains, varnishes, paints or polishes to a good finish. The sapwood is liable to attack by the pinhole borer and longhorn beetle; the heartwood is resistant to preservative treatment. The sapwood varies from moderately resistant to permeable.

Uses

The best grades are used for furniture, interior joinery and doors. Lower grades are used for paving blocks, sleepers, bridge building and house construction. Knotty material is used for garden furniture. Selected logs are sliced for decorative veneers for cabinets and panelling.

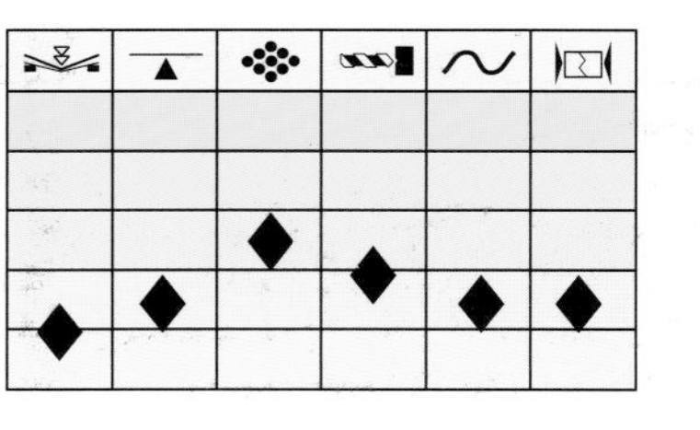

Ceratopetalum apetalem **Family:** *Cunonaceae*
COACHWOOD

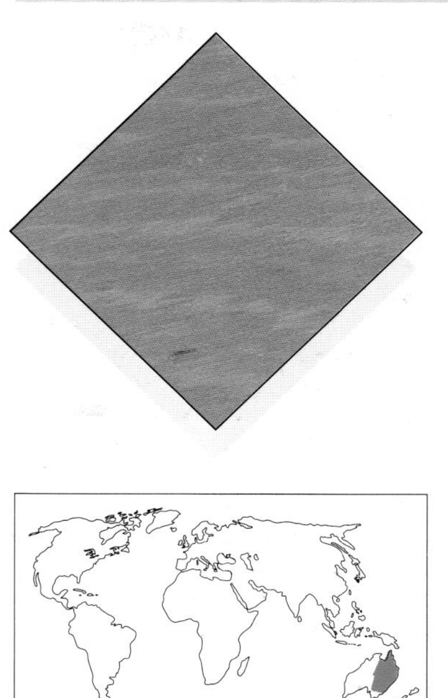

Where it grows

This medium-sized Australian tree grows in New South Wales and Queensland to a height of 18-24m (60-78ft), and a diameter of 0.5-0.75m (1½-2½ft). It is also known as scented satinwood.

Appearance

The tree has a light brown to pink-brown coloured sapwood, not clearly demarcated from the only slightly darker coloured heartwood. It has a close, straight grain with fine rays which produce a flecked figure on quarter-sawn surfaces. This is an attractive timber with a fine, even texture and a fragrant odour.

Properties

The wood weighs about 630 kg/m^3 (39 lb/ft^3) when seasoned. It dries fairly rapidly with some tendency to split and warp, and requires slow and careful seasoning to avoid shrinkage. It has medium movement in use. The timber is of medium density, bending strength, stiffness and shock resistance, but with a high crushing strength and a good steam-bending classification. The wood works fairly easily with both hand and machine tools, and gives a smooth finish. It does tend to chip out at tool exits when drilling or mortising, and nailing requires pre-boring. The wood glues easily, takes stain well and provides an excellent finish. The sapwood is liable to attack by the powder post beetle. The wood is non-durable, but permeable for preservation treatment.

Uses

Coachwood, with its decorative appearance, delicate flecked figure and fine texture, is used for furniture and cabinet making, and is often used for wall panelling, where a uniform effect is desired without too much contrast. In Australia it is used extensively for interior joinery and mouldings, and favoured for bent work for boat building and sporting goods. It is also used for gunstocks and parts of musical instruments, appears widely in light domestic flooring and skirtings and is ideal for turnery for a wide range of uses including bobbins, shoe heels, domestic woodware and so on. It is also rotary cut for plywood corestock. Selected logs are sliced to produce decorative veneers for cabinets and architectural panelling.

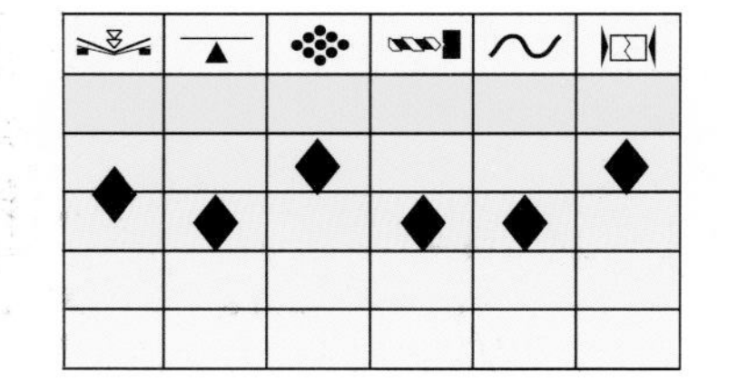

Cercidiphyllum japonica **Family:** *Trochodendraceae*

KATSURA

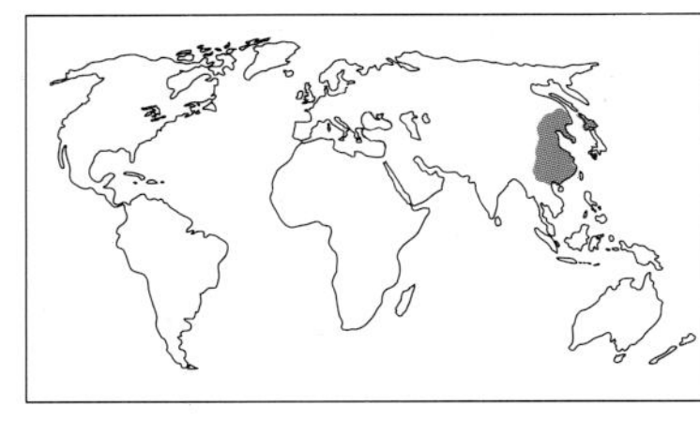

Where it grows
This valuable tree grows mainly in the northern temperate forests of Japan, but also occurs in China and Korea. It is a large deciduous broadleaved tree which forms several furrowed trunks, and is often spirally twisted. It grows to a height of about 30m (98ft) and a diameter of 1.2m (4ft).

Appearance
The heartwood is a light medium nut-brown colour. Logs selected for timber are straight grained, with a very fine, compact and even texture and a high lustre. The plain appearance of the background colour is relieved by light narrow lines of the growth rings, with a distinct narrow band of parenchyma being visible on flat-sawn surfaces. Although darker and brown in colour, it is otherwise similar to American whitewood.

Properties
This soft and compact wood weighs about 470 kg/m^3 (29 lb/ft^3) when seasoned. It dries easily and without any problems of distortion or degrade and is very stable in use. It is of low density and medium bending and crushing strengths, with low stiffness and resistance to shock loads. It has a moderately good steam-bending classification. The wood is a delight to work with both hand and machine tools, and provides a very smooth surface. Because it is compact, it is ideal for moulding and carving where there is a need for sharply detailed arrises to remain intact and not chip out. It does not hold nails or fastenings well, but it glues, takes stain and polishes with ease and can be brought to an excellent finish. The sapwood is liable to insect attack; the wood is non-durable but permeable for preservation treatment.

Uses
This excellent timber is mainly used in Japan, but is also exported in the form of square-edged boards. It is highly valued for sculpture and wood carving, pattern making in foundries, delicate mouldings, engravings, lacquered work and drawing boards. It is widely used in cabinet making, furniture and high-class interior joinery. It is also used for pencil manufacture and cigar boxes and for the Japanese shoes, geta. Selected logs are rotary cut for plywood and sliced for decorative veneers.

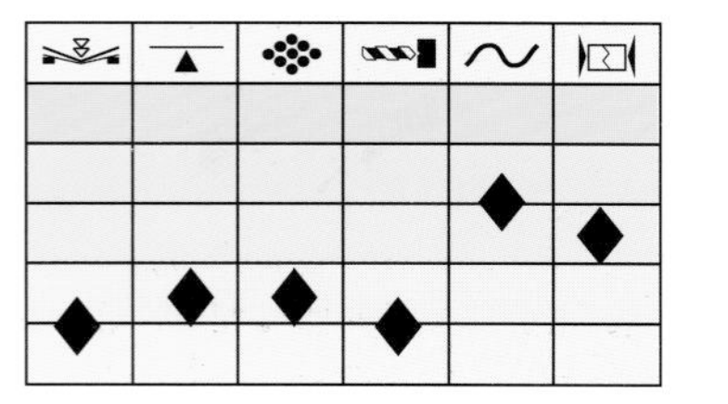

Chamaecyparis lawsoniana **Family:** *Cupressaceae*

PORT ORFORD CEDAR (S)

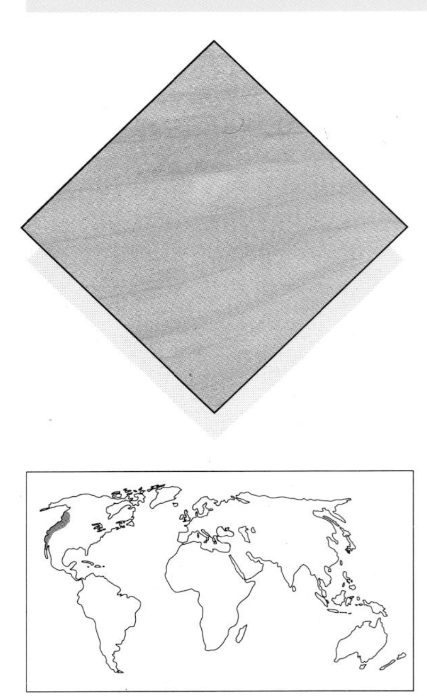

Where it grows
The tree grows over a limited range, scattered along the edge of the coastal forests of south western Oregon and north west California. It is also being planted in many other countries. Other names are Lawson's cypress (UK) and Oregon cedar (USA).The tree attains a height of about 61m (200ft), with a diameter of 3.7m (12ft).

Appearance
The heartwood is a pale pink-brown, and not clearly defined from the sapwood. It has a straight grain and a fine, even texture with a pungent spicy but fragrant odour. Although typically non-resinous, the timber does occasionally have an orange-yellow resin exudation.

Properties
Weight is about 485 kg/m^3 (30 lb/ft^3) when seasoned. The timber dries readily with little degrade and is stable in use. It has medium bending and crushing strength, low stiffness but good resistance to shock loads, and a very poor steam-bending classification. The timber works easily with both hand and machine tools, and has only a slight blunting effect on cutting edges. The wood holds screws and nails without difficulty. It can be glued satisfactorily, gives good results with stain, paint, varnishes and polishes, and provides an excellent finish. Damage by longhorn beetle is often present, and sometimes damage by *Sirex*. The wood is moderately durable and moderately resistant to preservative treatment, but the sapwood is permeable with oil-based preservatives by pressure or boron salts by the diffusion process.

Uses
In America Port Orford cedar is extensively used for ship and boat building, for oars, canoe paddles and arrows. It is also used for interior joinery and furniture, especially for chests and closet linings, where the fragrant scent is taken to act as a moth repellent. Special uses include organ pipes, acid battery separators and match splints. It is used externally for cladding and shingles, and in the round for posts, poles and piling. Selected knotty logs are sliced for decorative veneers for furniture and panels.

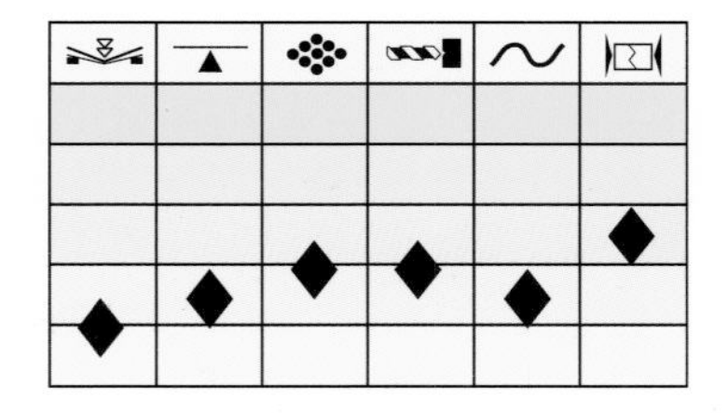

Chamaecyparis nootkatensis **Family:** *Cupressaceae*
YELLOW CEDAR (S)

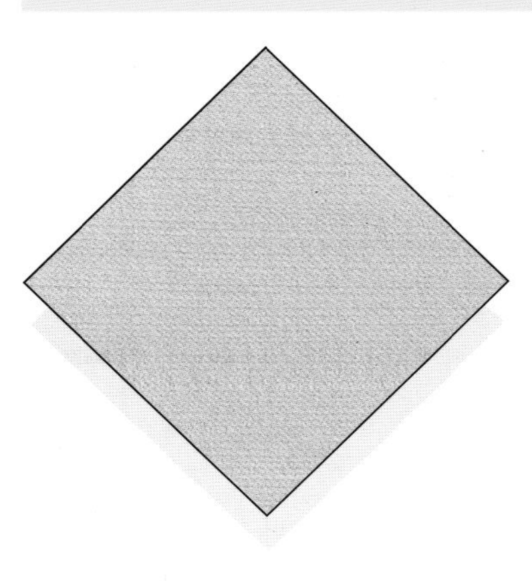

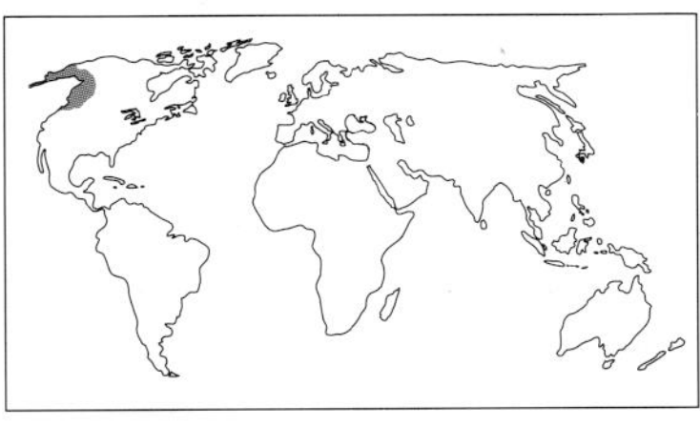

Where it grows
This large forest tree is native to the narrow coastal belt from Alaska down to northern Oregon, but the largest sizes grow in Alaska and north British Columbia. It is also known as 'Alaska yellow cedar' and nootka false cypress in the USA, and yellow cypress and 'Pacific coast yellow cedar' in Canada. It grows to a towering 53m (174ft) and 2m (6ft) diameter, but more usually about 24.5m (80ft) with a diameter of 0.6-0.9m (2-3ft), and a sharply tapering bole. It is not a true cedar (*Cedrus*).

Appearance
This pale yellow wood is straight grained with a fine, even texture. It has a strong spicy scent when freshly cut, but no appreciable odour when dry.

Properties
Weight is about 500 kg/m^3 (31 lb/ft^3) when seasoned. The wood should dry slowly to avoid surface checking in thick stock and end splitting if the drying is hurried. But once dry, it is noted for its durability and stability in use. The timber is moderately strong, with medium bending and crushing strengths, low stiffness and resistance to shock loads. It has a very poor steam-bending rating. It works easily with both hand and machine tools, with only a slight dulling effect on cutting edges. It holds screws and nails without difficulty, glues, takes stain, paint and varnish satisfactorily, and can be brought to an excellent finish. The wood is durable, and resistant to preservative treatment.

Uses
Yellow cedar is extremely stable in use and has a natural durability, especially when exposed to fluctuating atmospheric conditions, and is therefore highly valued locally for high-class joinery, window frames and finishing in houses. It is also used for furniture and cabinet work. The timber from Alaska trees is heavier and stronger than from those grown in Oregon. It is extensively used for boat and ship building, for exterior joinery and for cladding and shingles. It is resistant to acids and is considered to be the best wood for battery separators. It is also used for engineers' patterns and surveyors' poles. In the round, the wood is used for posts, poles and marine piling. Selected logs are sliced for decorative veneers for cabinets, doors and panelling.

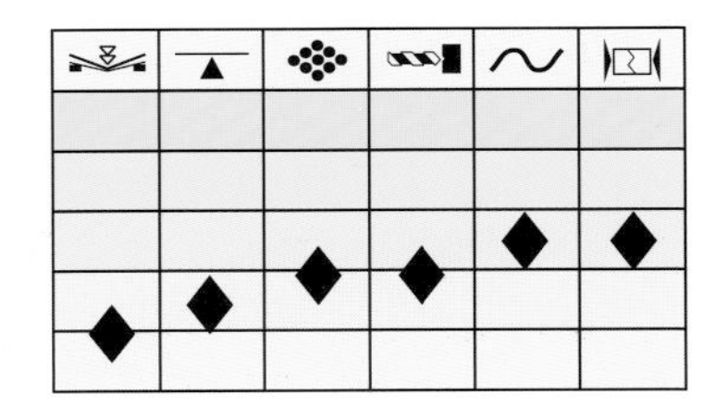

Clorophora excelsa **Family:** *Moraceae*
IROKO

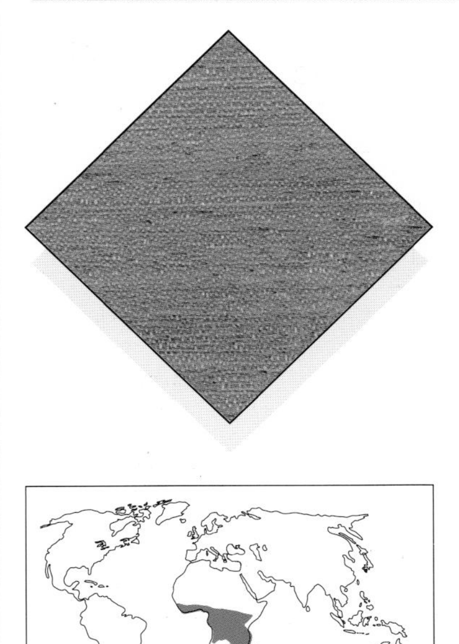

Where it grows
This species grows in the moist semi-deciduous forests of tropical Africa, from Sierra Leone in the west to Tanzania in the east. *C.regia*, A. Chév. grows in West Africa, and occurs from Senegal to Ghana. It is also known as mvule (East Africa), odum (Ghana and Ivory Coast), kambala (Zaire), tule or intule (Mozambique), and moreira (Angola). *C.excelsa* reaches 50m (160ft) in height with a diameter of about 2.5m (8-9ft); the bole is clear and cylindrical up to 21m (70ft) or more. *C.regia* is not quite so tall.

Appearance
The pale sapwood is clearly defined from the yellow-brown coloured heartwood, which matures to a deeper brown with lighter vessel lines conspicuous on flat-sawn surfaces. The grain is typically interlocked and sometimes irregular.

Properties
When seasoned, iroko weighs about 640 kg/m^3 (40 lb/ft^3). The wood dries well and fairly rapidly, with some degrade. This very durable wood of medium density is stable in use, has medium bending and crushing strengths, very low stiffness and shock resistance, and a moderate bending rating. Hard deposits of calcium carbonate known as 'stone' are often completely hidden in the grain and are only detectable from the darker wood surrounding them. They can severely damage the cutting edges of tools, and tipped or hardened saw teeth are required. The fine machining dust can cause nasal and skin irritation. Nailing or screwing and gluing are satisfactory. When filled, it provides an excellent finish. The sapwood is liable to attack by the powder post beetle. The wood is very durable and extremely resistant to preservation treatment, but the sapwood is classed as permeable.

Uses
Iroko is used for similar purposes to, and as a substitute for, teak, but lacks the greasy feel of teak. It is valued for high-class interior and exterior joinery, counter and laboratory bench tops and draining boards. It is a favourite wood for sculpture and wood carving, and is also ideal for turnery. It is widely used in ship, boat and vehicle building (but not for bent work), and a structural timber for piling, marine work, and garden and park bench seats. Another popular use is for domestic parquet flooring where underfloor heating is used. Logs are rotary cut for plywood manufacture and sliced for decorative veneers.

			◆		
		◆		◆	◆
◆	◆				

Chloroxylon swietenia **Family:** *Rutaceae*
CEYLON SATINWOOD

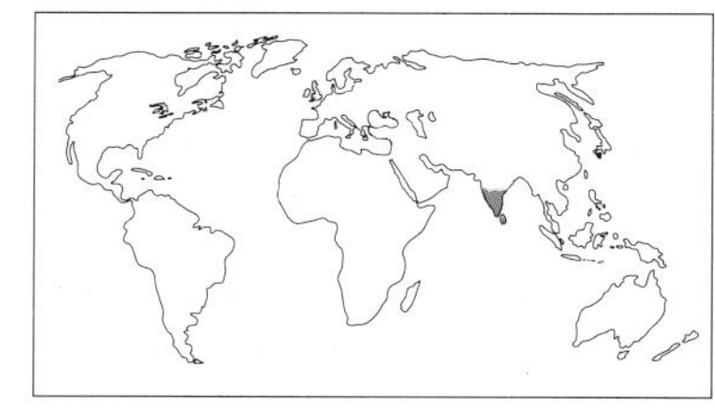

Where it grows
This is a small to medium sized tree which reaches its best development in Sri Lanka, but also occurs in central and southern India. It is also known as East Indian satinwood (USA, UK); burutu (Sri Lanka); and bhera, behra or mutirai (India). It grows to 14-15m (45-50ft) tall and a diameter of about 0.3m (1ft) or more, with a cylindrical bole up to around 3m (10ft).

Appearance
There is very little distinction between the sapwood colour and the heartwood, which is a beautiful golden yellow. The inner heartwood matures into a slightly darker golden brown. The grain is narrowly interlocked, sometimes wavy or variegated, producing roe or narrow ribbon striped figure on quartered surfaces. It is often broken striped, or with 'bee's wing cross mottled' figure. Gum rings can develop thin dark veins on flat-sawn surfaces. The wood is lustrous and fragrant; the texture is fine and very even.

Properties
The average weight is about 980 kg/m^3 (61 lb/ft^3) seasoned. The wood should be allowed to air dry slowly to avoid a tendency to surface cracking and distortion, but it kiln dries well with little degrade and is stable in service. This very heavy, dense wood has high bending and crushing strengths, medium stiffness and low shock resistance, but strength is of little importance for the end-uses of this timber. It is fairly difficult to work with machinery and has a moderate blunting effect on cutting edges. Nailing requires pre-boring. It is a difficult wood to glue, but it takes stain and polishes to an excellent finish when filled. The wood is durable, and extremely resistant to preservative treatment.

Uses
Ever since the 'Golden Age of Satinwood' this timber has been highly valued and in great demand for luxury cabinets, fine furniture making and interior joinery. It is excellent for turnery, for the backs and handles of hair-brushes, and is also used for jute bobbins. It appears widely in panelling in office, shop and bank fitting, and also in the manufacture of traditional inlay motifs, lines and bandings. Selected logs are sliced to produce extremely attractive decorative veneers with a range of ribbon striped or mottled surface figure.

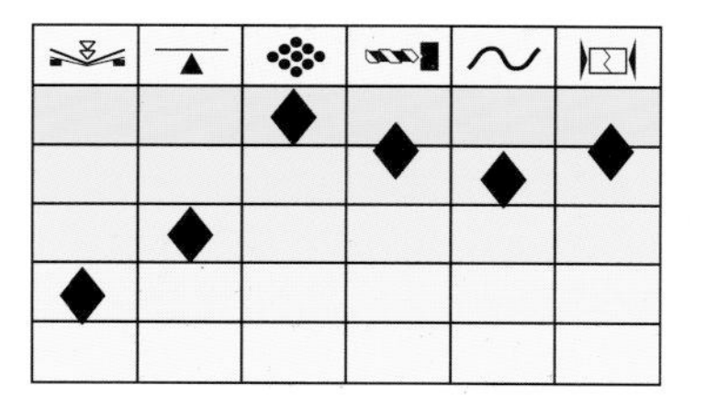

Cordia spp. **Family:** *Boraginaceae*
AFRICAN CORDIA

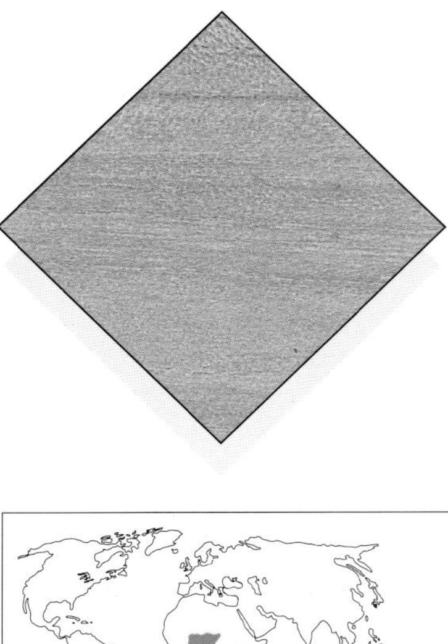

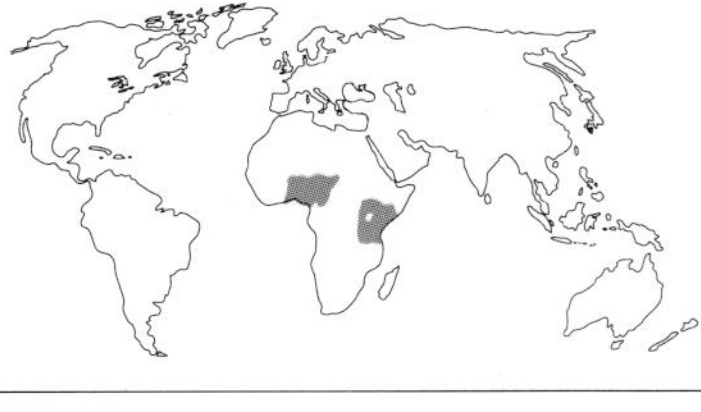

Where it grows
The African species of *Cordia* is a deciduous tree which occurs in the semi-tropical rain forests of Kenya, Tanzania and also Nigeria. The species include: *C.abyssinica*, R.Br., known as mukumari (Kenya) and mringaringa (Tanzania); *C.millenii*, Bak., known as mugoma (Kenya) and mukeba (Uganda); and *C.platythyrsa*, Bak. Both *C.millenii* and *C. platythyrsa* are known as omo in Nigeria. The trees are medium sized, averaging 18-30m (60-100ft) in height, with an irregularly shaped bole about lm (3ft) in diameter.

Appearance
The cream-coloured sapwood is distinct from the heartwood, which varies in colour from golden to medium brown, sometimes with dark streaks and a pink tinge. It usually darkens on maturity to a light red-brown. The grain varies from fairly straight to interlocked or irregular, and the medium-sized rays give an attractive striped and mottled figure on quartered surfaces. The texture is coarse.

Properties
The weight is about 430 kg/m^3 (27 lb/ft^3) seasoned. The wood dries well and rapidly, without too much splitting or warping, and is stable in service. It has low density, low bending strength, very low stiffness, medium crushing strength, very low shock resistance, and a poor steam-bending classification. Brittleheart is fairly common in this species. The wood works easily with both hand and machine tools, but cutters should be kept very sharp to prevent the surface from becoming woolly. Nailing and gluing are satisfactory, and when filled the timber can be brought to a good finish. The outer heartwood is very durable but the inner heartwood is only moderately durable and resistant to preservative treatment.

Uses
This rather weak and soft timber is mainly used for decorative parts of furniture where strength is unimportant, such as interior framing and fitments, library fittings, edge lippings and so on. It is used locally for making traditional drums and sound boards because of its resonance properties, and is the traditional wood for canoe making and boat building in West Africa. Logs are rotary cut for plywood and laminated corestock, and selected logs are sliced for decorative veneers for cabinets and panelling.

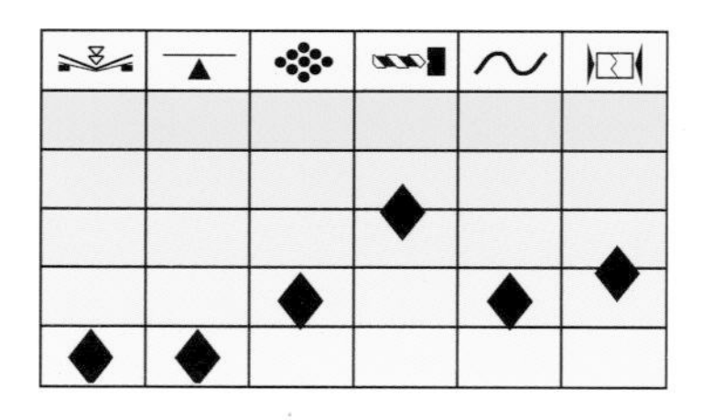

Cordia goeldiana **Family:** *Boraginaceae*
FREIJO

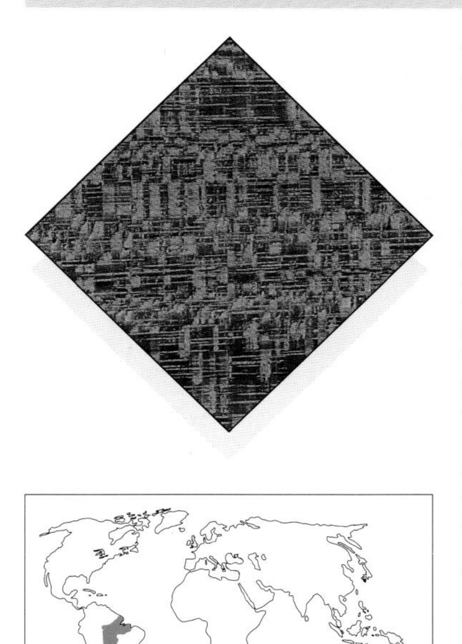

Where it grows

The tree grows in the Amazon basin of Brazil, where it is known as *frei jorge*; in the USA it is cordia wood or jenny wood. It reaches a height of about 30m (100ft) with a diameter of 0.6-1m (2-3ft).

Appearance

The heartwood is a golden-brown colour maturing to dark brown, and is similar to teak in appearance. It is usually straight grained. Quartered surfaces often produce a very attractive ray figure and the rays appear lighter than the background colour. The wood has a medium but uniform texture.

Properties

The weight varies from about 400 to 700 kg/m^3 (25-44 lb/ft^3), averaging about 590 kg/m^3 (37 lb/ft^3) when seasoned. It air dries well with little distortion, but with a slight tendency for end splits to develop. There is small movement and the wood is stable in use. The timber has a medium density and strength in all categories, but with low shock resistance and a poor steam-bending rating – although bends of a moderate radius are possible with selected pieces. However, freijo is a tougher wood than teak. It works easily with hand or machine tools, and has a moderate blunting effect on cutting edges. Cutters should be kept very sharp to avoid tearing the grain. The wood requires support in end grain working, mortising, boring and so on to prevent breaking out at tool exits. Nailing is satisfactory, and it glues, stains and polishes well if the grain is filled. The sapwood is liable to attack by powder post beetle. The wood is durable and resistant to preservation treatment.

Uses

Freijo is used for panelling, interior joinery, furniture and fitments, and in cabinet making. It is also used in boat building – including decking – as a substitute for teak, and for vehicle bodies and exterior joinery. It is used in Brazil for dry cooperage. Selected logs are sliced for a range of highly decorative veneers for cabinets, flush doors and architectural panelling. Note: related spp: *C.alliodora*, Cham. (West Indies, Honduras, Eucuador) and *C.trichotoma*, Allab.(tropical America, Brazil and Argentina) produce American light cordia, weight 550 kg/m^3 (34 lb/ft^3), which is used for purposes similar to freijo.

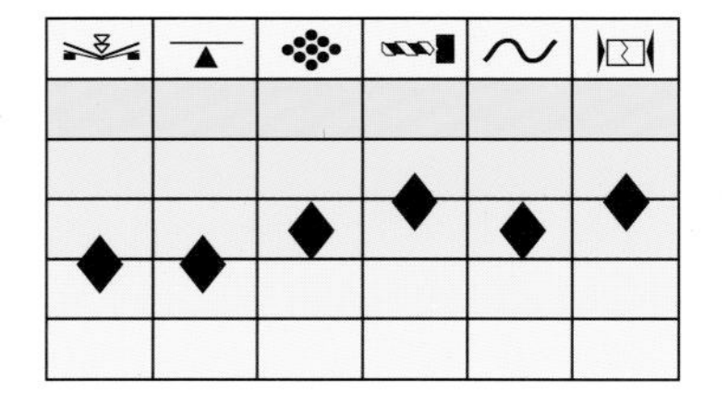

Cupressus spp. **Family:** *Cupressaceae*
EAST AFRICAN CYPRESS (S)

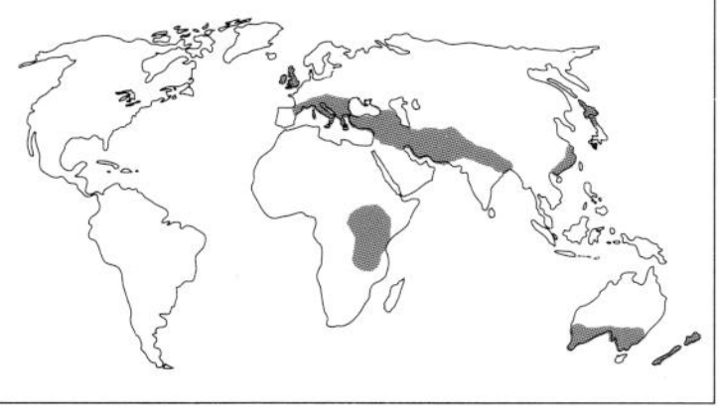

Where it grows

There are very few indigenous conifers in Africa, less than one per cent of the total forest land. They are restricted mostly to the Mediterranean region or the high montane forests of central and eastern Africa, and have been introduced into South Africa. The main species is *C.lusitanica*, Mill. (*C.lindeyi*), which grows in East Africa, and *C.sempervirens*, which is a native of the Mediterranean countries, Asia Minor and the Himalayas. Also *C.macrocarpa*, a native of California, has been planted in Australia and New Zealand, East and South Africa, and in the UK. The trees attain a height of 18-21.5m (60-70ft), with a diameter 0.6-0.9m (2-3ft), but in good conditions up to 30.5m (100ft) with a diameter sometimes up to 1.5m (5ft).

Appearance

The heartwood colour is yellow to pinkish-brown, distinct from the paler-coloured sapwood, which is up to 100mm (4in) wide. The timber is straight grained and the texture fine and fairly even. The inconspicuous growth rings are defined by a narrow band of latewood. Although the timber is classed as non-resinous, resin cells are present and often appear as brown streaks or flecks. It has a cedar-like odour.

Properties

Seasoned timber weighs about 450 kg/m^3 (28 lb/ft^3). The timber is strong in relation to its weight, but it has a very poor steam-bending classification because of knots. The wood dries fairly easily with little degrade, and is stable in service. Although the timber works readily with both hand and machine tools with little dulling effect on cutters, frequent knots may be troublesome. Extra care is needed when working end-grain to prevent break-out at tool exits. The material screws, nails and glues well, and accepts all finishing treatments satisfactorily. This durable softwood is highly resistant to both insect and fungal attack and is resistant to preservation treatment.

Uses

This is a very useful softwood for external constructional work where the timber is in contact with the ground. It is also used in exterior joinery, ship and boat building, and the manufacture of farmhouse-style furniture, closet linings and so on.

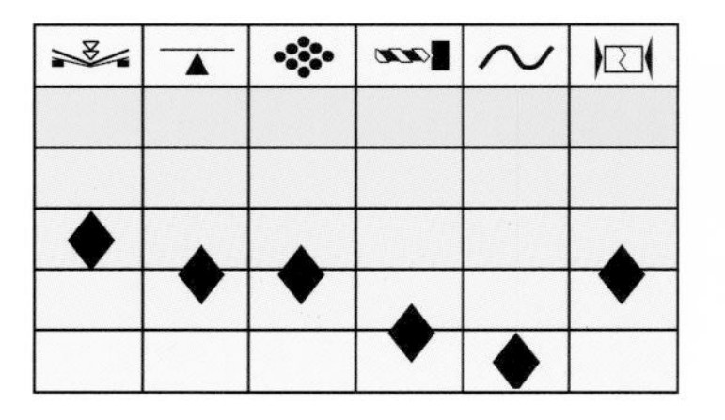

Dalbergia spp. **Family:** *Leguminosae*
ROSEWOOD (BRAZILIAN and HONDURAS)

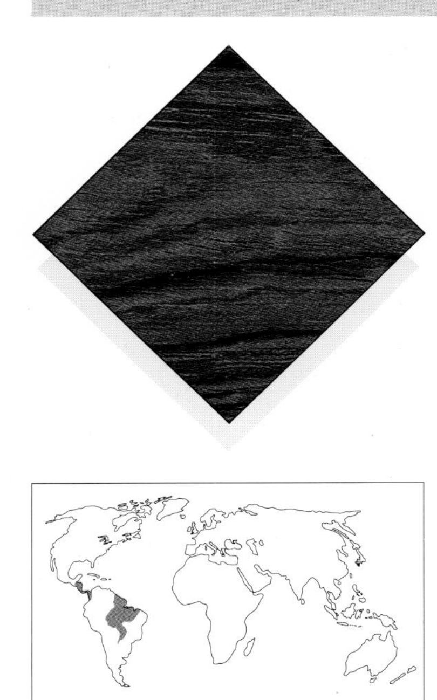

Where it grows
Brazilian rosewood has been one of the world's most treasured timbers for centuries. *Dalbergia nigra*, Fr., All. is sadly becoming rare. It is also known as Bahia rosewood or Rio rosewood in the UK and USA, and as jacaranda in Brazil; it grows to a height of about 38m (125ft) and the bole is irregular. When the sapwood is removed the heartwood is 0.5m (1½ft) in diameter. *Dalbergia stevensonii*, Standl. produces Honduras rosewood known as nogaed in the USA, and grows to 15-30m (50-100ft) and 1m (3ft) in diameter.

Appearance
The heartwood of Brazilian rosewood is a rich brown colour with variegated streaks of golden-to-chocolate brown and from violet to purple-black, sharply demarcated from the almost cream-coloured sapwood. The grain is mostly straight to wavy, the texture coarse, oily and gritty to the touch, and the timber has a mildly fragrant odour. Honduras rosewood has pinkish to purple-brown heartwood with irregular black markings; it is straight grained and has a medium-to-fine texture.

Properties
Brazilian rosewood weighs 850 kg/m³ (53 lb/ft³) when seasoned, and Honduras rosewood weighs around 930-1100 kg/m³ (58-68 lb/ft³). Both woods air dry fairly slowly, with a marked tendency to check, but are stable in service. They possess high strength in all categories but are low in stiffness, and have good steam-bending ratings. They are not unduly difficult to work, but have a severe blunting effect on cutting edges. Gluing is no problem if epoxy resin adhesives are used. A smooth, lustrous oil finish can be achieved. Both species are very durable, and the heartwood of both is extremely resistant to preservative treatment.

Uses
Rosewoods, both solid and veneer, have been highly prized for more than 200 years for very high-class furniture and superior cabinet making, for bank and shop fitting, boardroom panelling, tables and furniture. Also for piano cases and billiard tables. The wood is also in demand for wood carving and sculpture, and is excellent for turnery – fancy bowls, knife handles, brushbacks and so on. Specialized uses include musical instrument fingerboards, percussion bars for xylophones or marimba keys. Selected logs are sliced for highly decorative veneers.

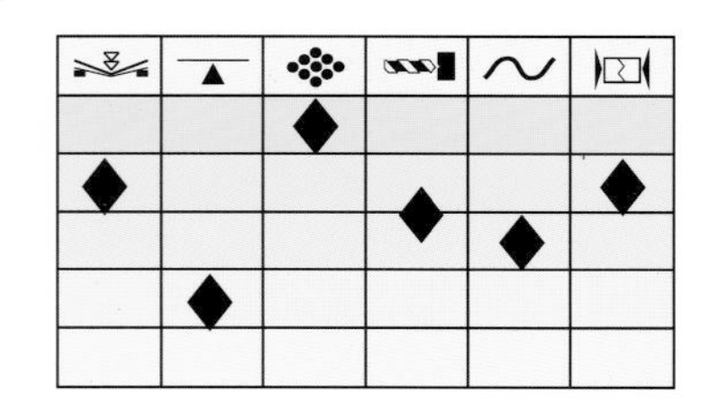

Dalbergia cearensis **Family:** *Leguminosae*
KINGWOOD

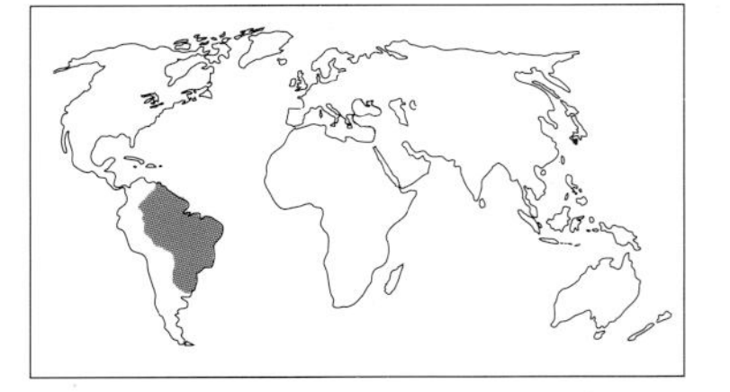

Where it grows
This member of the rosewood *Dalbergia* species occurs mainly in Brazil, a small to medium height tree. It is also known as violete (Brazil); violetta, violet wood (USA); and bois violet (France). It varies in height from 15-30m (50-100ft). It is exported in small logs from 1-3m (3-6ft) long and 0.1-0.25m (4-8in) diameter, with the sapwood removed.

Appearance
The heartwood is a rich violet-brown background colour shading almost to black, with variegated stripes or streaks of black, violet-black or violet-brown, and sometimes golden yellow. The almost white sapwood is clearly defined. The heartwood has a bright lustre, a fine uniform texture and a very smooth surface.

Properties
This very heavy wood weighs about 1200 kg/m³ (75 lb/ft³) when seasoned. Care is needed in air drying to avoid splits, but it kiln dries without degrade, and it is stable in use. It is very strong in all categories. Despite being hard and heavy, the wood works fairly well with both hand and machine tools, if cutters are kept very sharp. There is a moderate blunting effect on tools. The wood has good nail and screw-holding properties, but needs pre-boring. Care is needed in gluing, and the wood can be brought to a beautiful high polish because of its natural waxy properties. The timber is durable and extremely resistant to preservative treatment.

Uses
Kingwood is well named, for it certainly is a king among woods. This was a favourite of the Parisian *ébènistes* during the reigns of Louis XIV and XV of France and in the Georgian period of English furniture; it is in great demand today for restoration and reproduction of antique furniture, and for decorative work generally. Unfortunately its use is restricted because of the small sizes available. The wood is excellent for turnery for bowls and small fancy items, and is also treasured for sculpture. It is usually saw cut into veneers as the billets are too small for slicing. It is in demand in veneer form for inlay and marquetry work, and oyster veneering for antique repairs and restoration.

Dalbergia frutescens **Family:** *Leguminosae*
BRAZILIAN TULIPWOOD

Where it grows
This small, often misshapen tree grows in north east Brazil and around Bahia and Pernambuco. It is exported in small billets without sapwood 0.6-1.2m (2-4ft) long and 0.06 -0.2m (2-8in) diameter. It is also known as pau rosa, jacaranda rosa and pau de fuso (Brazil); pinkwood (USA); and bois de rose (France). It is not to be confused with the tulip tree, *Liriodendron tulipifera*.

Appearance
The very attractive pink-yellow heartwood has a pronounced striped variegated figure in shades of violet-red, salmon pink and rose red. In effect, the colouring of a red and yellow tulip blossom is reproduced by this beautiful wood. As it matures it tends to lose its original vividness, but it remains a strikingly beautiful timber. The grain is usually interlocked and irregular because of its twisted growth, with a medium to fine texture and a pleasantly mild fragrant scent.

Properties
The heartwood weighs about 960 kg/m³ (60 lb/ft³) when seasoned. It dries fairly easily and is stable in use. It is a very hard, dense and compact wood and liable to split after conversion. It is rather wasteful in conversion as it is extremely hard to work; it is fissile and tends to splinter. There is severe blunting of cutting edges. It also requires pre-boring for screwing or nailing but it glues well. The wood possesses a very high natural lustre and provides an excellent finish. It is non-durable and resists insect and fungal attacks, but is highly resistant to preservative.

Uses
This historical cabinet wood was known as bois de rose and used extensively in the furniture of the French Kings Louis XV and XVI, and classical English furniture of the 18th century. The billets are usually saw cut for decorative panel cross-bandings and marquetry inlay bandings in the restoration and repair of antiques. It is used in Brazil as a turnery wood for brushbacks, and for marimba keys, caskets, jewellery boxes and fancy goods.

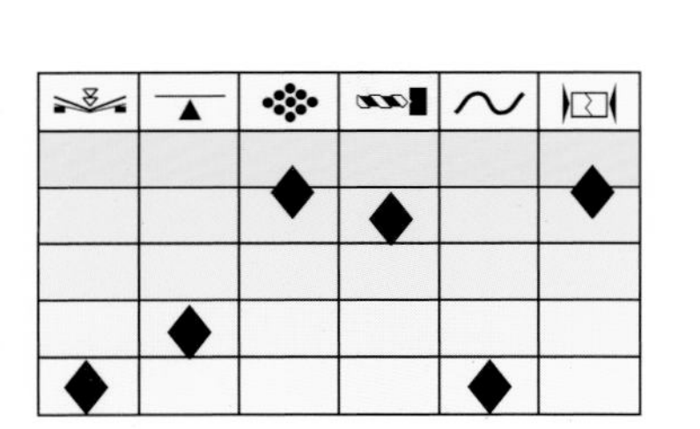

Dalbergia latifolia **Family:** *Leguminosae*
INDIAN ROSEWOOD

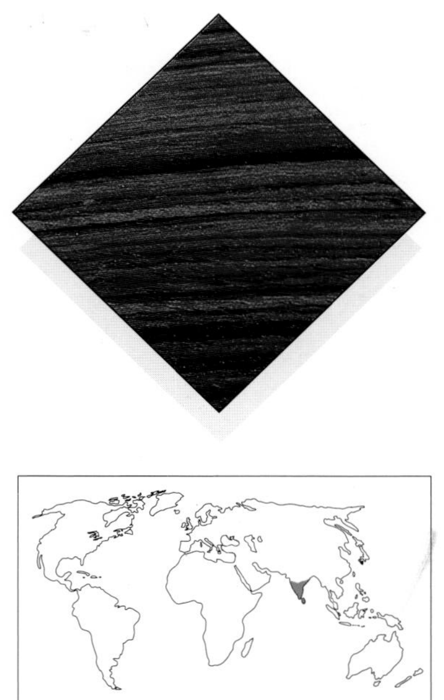

Where it grows
This is a large tree which varies considerably in size according to locality. It grows its best in southern India where it reaches a height of 25m (80ft) with diameters varying from 0.3-1.5m (1-5ft). It averages 0.8m (2½ft) with a cylindrical, fairly straight bole of 6-15m (20-50ft). It is also known as East Indian rosewood and Bombay rosewood in the UK, and as Bombay blackwood, shisham, sissoo, biti, ervadi and kalaruk in India.

Appearance
The narrow, pale yellow-cream coloured sapwood has a purplish tinge and is clearly defined from the heartwood, which is medium-to-dark purple-brown with darker streaks terminating the growth zones. Together with the narrowly banded interlocked grain, this produces an attractive ribbon-striped grain figure on quartered surfaces. The texture is uniform and moderately coarse, and the wood has a fragrant scent.

Properties
When seasoned, the weight averages about 850 kg/m³ (53 lb/ft³). This timber responds best by air drying in the round or as square baulks prior to conversion. The wood should be allowed to dry slowly, which improves the colour. It is stable in use. This very dense wood has high bending and crushing strengths with medium resistance to shock loads and low stiffness, and a good bending rating. It is fairly difficult to machine or work with hand tools, severely blunting the cutting edges with calcareous deposits in some of the vessels. It is unsuitable for nailing or screwing, but glues satisfactorily, and when filled, can be polished or waxed to provide an excellent finish. The sapwood is liable to attack by the powder post beetle but the heartwood is very durable and classified as resistant to preservative treatment.

Uses
This very decorative wood is used for top-class furniture and cabinet work, shop, office and bank fitting. It makes an excellent turnery wood for small fancy goods, handles and bowls. It is used for parts of musical instruments, particularly for guitar backs and sides, and is highly valued for sculpture and carving. In India it is also used for exterior joinery, decorative flooring, boat building, hammer heads and brake blocks. Selected logs are sliced for highly decorative veneers.

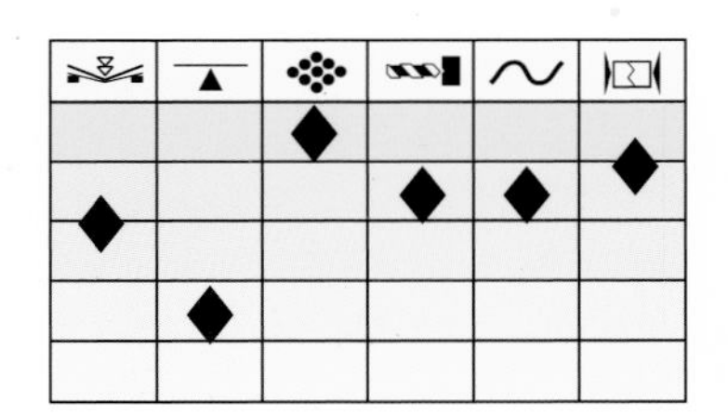

Dalbergia melanoxylon **Family:** *Leguminosae*
AFRICAN BLACKWOOD

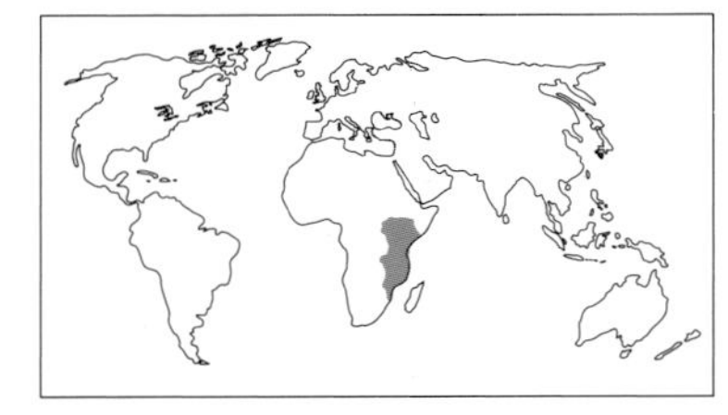

Where it grows
Often misshapen, this multi-stemmed small tree occurs in East Africa – chiefly Mozambique, where it is called Mozambique ebony, and Tanzania, where it is known as mpingo. It is not an ebony (*Diospyros*), and is closely related to the rosewoods as a member of the *Dalbergia* species. It grows to a height of 4.5-7.5m (15-25 ft), yielding only short, fluted logs about 1-1.5m (3-5 ft) long and 0.3m (10in) in diameter.

Appearance
The narrow white sapwood is clearly defined from the heartwood, which is dark purplish-brown with predominating black streaks which give the wood a black appearance. It has straight to irregular grain with an extremely fine texture oily to the touch.

Properties
This exceptionally hard and dense species weighs about 1200 kg/m^3 (75 lb/ft^3) when seasoned. It has to be partially dried in log or billet form before conversion, end-coated before stacking under cover, and allowed to dry extremely slowly. The timber commonly suffers from heart shakes and needs to be seasoned very carefully to avoid checking. It is extremely stable in service and holds its dimensions well. It is very durable and strong in all categories and selected straight-grained stock has good steam-bending properties. Stellite or tungsten carbide tipped saws are advised to minimize severe and rapid blunting. It requires pre-boring for nailing or screwing. The wood glues satisfactorily and polishes to a very smooth finish. The sapwood is liable to attack by the powder post beetle; the heartwood is very durable and extremely resistant to preservative treatment.

Uses
The natural oiliness of blackwood results in excellent resonant tonal properties, and resistance to climatic changes makes it preferable to ebony for many purposes. Its extreme stability and hardness enables it to be tapped for the screw-threads of the metal pillars of wind instrument keys, and it is used for clarinets, oboes, piccolos, flutes, recorders and the chanters of bagpipes. It is ideal for ornamental and plain turnery for brushbacks, knife handles and truncheons, and for bearings and pulley blocks. In Africa it is extensively used for carved animal and human figures, chessmen and walking-sticks. It is also used for inlay work.

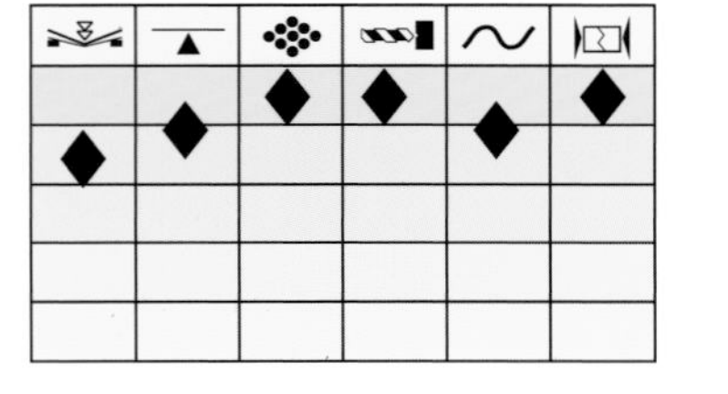

Dalbergia retusa **Family:** *Leguminosae*
COCOBOLO

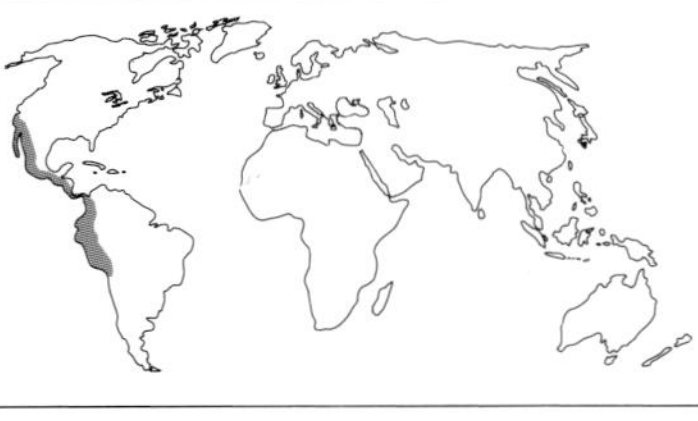

Where it grows
Cocobolo, also known as granadillo, occurs along the Pacific seaboard of Central America from Mexico to Panama. It is a medium sized tree with a fluted trunk which grows to 20-30m (65-98ft) with a diameter of about 1m (3ft). It is shipped in small, round billets, mainly from Costa Rica and Nicaragua.

Appearance
When freshly cut, the heartwood is an array of colours from lemon-orange to deep rich red with many variegated streaks and zones of yellow, orange and brick red. These mature to a deep mellow orange-red with darker stripes and mottling. The sapwood, clearly defined from the heartwood, is almost white in colour. The grain varies from straight to irregular and is sometimes wavy. It has a fine, medium, uniform texture.

Properties
The weight varies from about 990 to 1200 kg/m^3 (61-75 lb/ft^3) with a fair average weight of 1100 kg/m^3 (68 lb/ft^3) when seasoned. The wood dries out very slowly, with a tendency to check and split, but it is very stable in service. It is not unduly difficult to work with both hand and machine tools, but there is moderate blunting of cutting edges which should be kept very sharp. When machined, the wood gives off a mild fragrance from its natural oil. The fine dust can be an irritant and cause a form of dermatitis, staining the skin an orange colour. This very tough timber has high mechanical strength in all categories, but this is not important because of the end use. It is difficult to glue owing to its very high natural oiliness, but with care it can be brought to an excellent smooth finish which feels cold, like marble, to the touch. The heartwood is very durable and resistant to preservative treatment.

Uses
Cocobolo is ideal for turnery, and is traditionally used for cutlery handles, knife and tool handles, brushbacks, truncheons and bowling bowls. It is also highly valued for sculpture and carving, chessmen and small decorative items such as inlaid boxes and wooden jewellery. Selected logs are sliced for highly decorative veneers for inlay work, for the decoration of furniture and for panelling.

		◆	◆	◆	◆
◆	◆				

Diospyrus spp. **Family:** *Ebenaceae*
AFRICAN EBONY

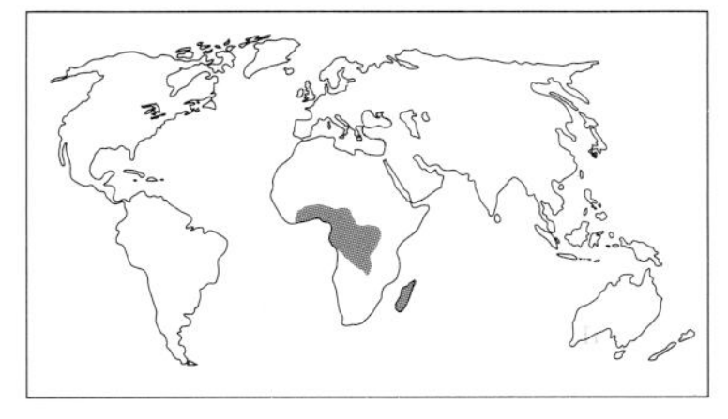

Where it grows
The name ebony covers all species of *Diospyros* with predominantly black heartwood. African ebony includes *D.crassiflora*, Hiern, and *D.piscatoria*, Gürke., which occurs mostly in southern Nigeria, Ghana, Cameroons and Zaire. It is named after its country or port of origin, e.g. Cameroon ebony, Kribi ebony, Gaboon ebony, Madagascar ebony or Nigerian ebony. 'White ebony' refers to the sapwood. The tree is small to medium in size, growing to a height of 15-18m (50-60ft) with an average diameter of about 0.6m (2ft). It is exported as short billets.

Appearance
Since ancient Eygptian times the black heartwood of ebony has been in great demand. *D.crassiflora* is considered to be the most jet black. The other species have a very attractive black and dark brown striped heartwood. It is usually straight grained to slightly interlocked or curly, and the texture is very fine and even.

Properties
This timber weighs about 1030 kg/m^3(64 lb/ft^3) when seasoned. The wood air dries fairly rapidly but is liable to surface checking, but kiln drying produces very little degrade. The wood is very stable in service. It is very dense, has very high strength properties in all categories, and a good steam-bending rating. Ebony is difficult to work with either hand or machine tools, as there is a severe blunting of edges. It is also inclined to be brittle, and needs a reduced cutting angle when planing the curly grain of quartered stock. It requires pre-boring for screwing or nailing. The heartwood requires care in gluing, but can be polished to a beautiful finish. The heartwood is very durable and extremely resistant to preservative treatment.

Uses
Ebony has always been used for sculpture and carving, and is an excellent turnery wood for the handles of tools, table cutlery and pocket knives. It appears as door knobs, brushbacks and the butt ends of billiard cues, also for the facings of tee squares. Other specialized uses include piano and organ keys, organ stops, violin finger boards and pegs, bagpipe chanters, castanets, and guitar backs and sides. It is used for luxury cabinet work, marquetry and inlay lines and bandings, and is saw-cut into veneers for antique repairs.

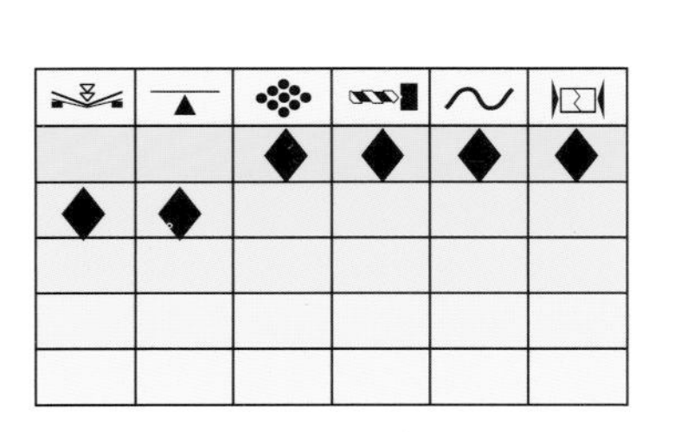

Diospyrus spp. **Family:** *Ebenaceae*
EAST INDIAN EBONY

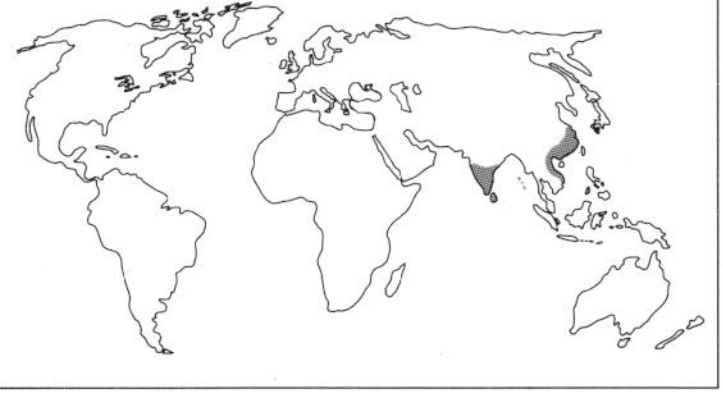

Where it grows
Ceylon ebony, *D.ebenum*, Koen., is known as true ebony as its heartwood has a uniform jet black colour. It occurs in Sri Lanka and southern India where it is known as tendu, tuki or ebans. Other ebonies are named after their country or port of origin. *D.melanoxylon*, Roxb. and *D.tomentosa*, Roxb. provide Indian ebony; *D.celebica*, Bakh., Macassar ebony from the Celebes Islands; *D.marmorata*, Park. provides Andaman marblewood from the Andaman Islands. The various species are smallish trees, with boles 4.5-6m (15-20ft) long and diameters of 0.3-0.6m (1-2ft).

Appearance
The sapwood is a light grey, but the heartwood of Ceylon ebony is jet black. Macassar ebony is medium brown in colour with beige and black stripes. Coromandel and calamander ebony refer to heart-wood with grey or brown mottled figure. The grain varies from straight to irregular or wavy, with a fine, even texture and a metallic lustre.

Properties
Ceylon ebony weighs 1190 kg/m^3 (73 lb/ft^3); other ebonies weigh between 1030 and 1090 kg/m^3 (64-68 lb/ft^3);Indian ebony is much lighter and weighs 880 kg/m^3 (55 lb/ft^3) when seasoned. Ebony is difficult to air dry and unless dried very slowly develops deep cracks and checks. These exceptionally heavy, dense timbers have very high strength in all categories, but are inclined to be brittle. The sapwood has a good steam-bending rating. The heartwood is extremely difficult to machine with a very severe blunting effect, and increased pressure should be applied to prevent the wood rising or chattering on the cutters. Care is required in gluing, but all species provide an excellent finish. The wood is subject to beetle attack but is very durable and extremely resistant to preservative treatment.

Uses
Ever since the ancient courts of Persia and India, ebony has been used for luxury furniture, wood sculpture and carving. It is used for turnery as tool, cutlery and knife handles, brushbacks, billiard cues, snuff boxes and combs and also for musical instruments parts – piano keys, finger boards, tail pieces and saddles for stringed instruments. Ebony also appears as inlay lines and bandings for antique repairs.

		◆	◆	◆	◆
◆	◆				

Diospyrus virginiana **Family:** *Ebenaceae*
PERSIMMON

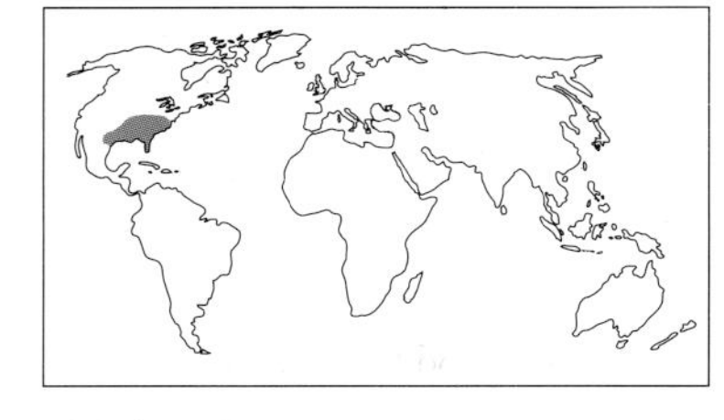

Where it grows
This member of the ebony family is known as 'white ebony', as the timber of commercial interest consists almost entirely of pale straw coloured sapwood. This small to medium sized tree occurs in the central and southern states of the USA where it is known as bara bara, boa wood, butter wood, possum wood and Virginia date palm. It grows to a height of about 30m (100ft), with a diameter of no more than 0.8m ($2\frac{1}{2}$ft).

Appearance
Persimmon has a very small heartwood core with variegated streaks of yellow-brown, orange-brown, dark brown or black. But the valuable sapwood is off-white with a grey tint, and straight grained with a fine even texture.

Properties
Persimmon weighs 830 kg/m^3 (52 lb/ft^3) seasoned; there is large movement in service with changes of humidity. The wood dries fairly rapidly, with some tendency to check. It is very dense, has high bending and crushing strengths, and medium stiffness and shock resistance. It is suitable for steam-bending to a moderate radius. This is a very tough timber which works readily with both hand and machine tools, but has a moderate blunting effect on cutting edges, which must be kept sharp. The wood requires pre-boring for screwing or nailing. It can be glued without problems and polished to an exceptionally smooth and high lustrous finish. The sapwood is liable to attack by powder post beetle; the heartwood is durable and classified as resistant to preservative treatment.

Uses
Persimmon sapwood is used for textile shuttles, as it can be machined to the intricate detail and very smooth finish required, and is very resistant to wear. It is claimed these shuttles can be used for more than 1000 hours without replacement. It is also ideal for golf-club heads, as it is highly resistant to impact, and for shoe lasts. It is used as a turnery wood for striking tool handles. Selected flitches are cut from logs which contain the variegated heartwood core, showing streaks from orange-brown to black with a wavy grain. These are sliced for ornamental veneers displaying variegated striped or roe figure, for cabinets and architectural panelling.

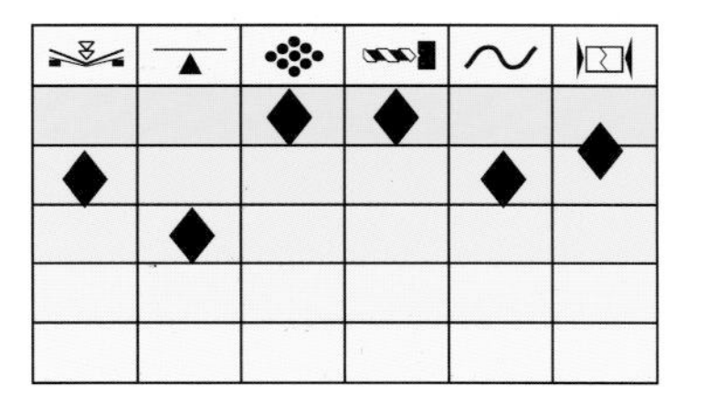

Dipterocarpus spp. **Family:** *Dipterocarpaceae*
KERUING/GURJUN/APITONG/YANG

Where it grows
These evergreen hardwood trees grow throughout south east Asia – in India, Burma, Thailand, Malaysia, Indonesia and the Philippines. There are more than 70 different species of *Dipterocarpus*, producing timber of the keruing type. They are marketed collectively under a group name, according to the country of origin. The principal species are Malaysian, Sarawak, Sabah and Indonesian keruing; and Malaysian, Indian or Andaman gurjun; Sri Lankan hora; Philippines apitong; Burma gurjun and kanyin and Thailand yang; South Vietnam, Khmer Republic dau; Burma eng or in. The trees grow to a height of 30-60m (100-200ft) and a diameter of 1-1.8m (3-6ft).

Appearance
The heartwood varies in colour from pink-brown to dark brown, and is rather plain in appearance. The grain is usually straight to shallowly interlocked, with a moderately coarse but even texture. Resin exudation a problem.

Properties
Weight is between 720-800 kg/m^3 (45-50 lb/ft^3) when seasoned. Eng is about 20 per cent heavier than keruing or gurjun. The wood is difficult to air dry without degrade. There is medium to large movement in service. These dense woods have high bending and crushing strength, high stiffness and medium shock resistance. Keruing from Sabah has moderate bending rating, but others exude resin, which causes difficulty in machining; tungsten carbide tipped saws are best. Despite moderate to severe blunting of tools, straight-grained timber can be machined with a fibrous finish. The resin content requires care in gluing and finishing, and also penetrates most paints and varnishes. The sapwood is liable to attack by the powder post beetle; the heartwood is moderately durable and varies from resistant to extremely resistant to preservative treatment.

Uses
These timbers are used for construction, frames, sides and flooring for vehicles, and flooring in domestic and public buildings if the material chosen is free from resin. They are not suitable for heavy-duty floors. When treated, they are used for exterior construction, wharf decking, bridges and boat building, but they are liable to exude resin when exposed to strong sunlight. Selected logs are rotary cut for plywood and sliced for veneers.

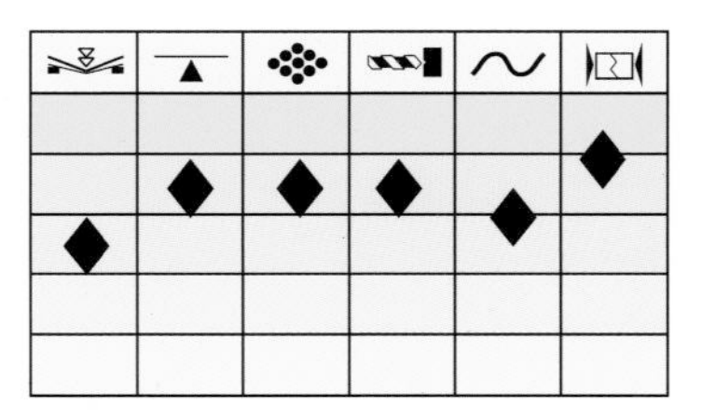

Distemonanthus benthamianus **Family:** *Leguminosae*
AYAN

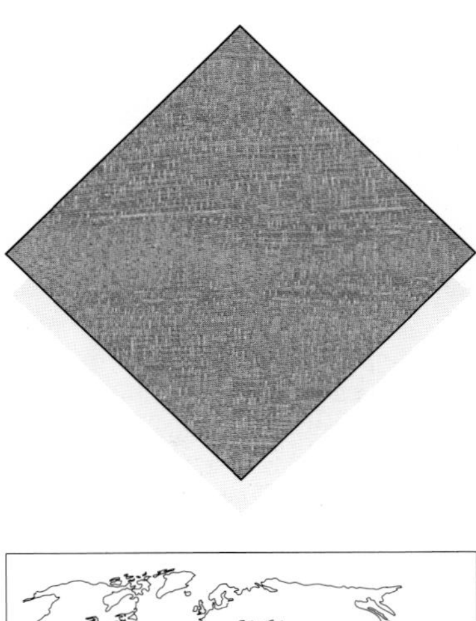

Where it grows
This tree occurs throughout tropical West Africa from the Ivory Coast to Gabon and Zaire. It grows to between 27-38m (90-125ft), with a diameter of 0.8-1.4m ($2\frac{1}{2}$-$4\frac{1}{2}$ft). The bole is straight and cylindrical. It is also known as movingue or distemonanthus (UK); Nigerian satinwood (USA and UK); barré (Ivory Coast); ayanran (Nigeria); bonsamdua (Ghana); eyen (Cameroon); oguéminia (Gaboon); and okpe (Togo).

Appearance
The sapwood is pale yellow in colour and not clearly demarcated from the heartwood, which varies in colour from lemon-yellow to golden-brown, sometimes with dark streaks. The grain is often irregular and interlocked, sometimes wavy. It may contain silica. The texture is fine and even, and the surface lustrous.

Properties
The weight varies from 600 to 770 kg/m^3 (37-48 lb/ft^3), averaging about 670 kg/m^3 (42 lb/ft^3) when seasoned. The heavier wood tends to be slightly darker in colour. It dries fairly rapidly and well, with little tendency to split or warp, and has very good dimensional stability. It is dense, has medium bending strength and high crushing strength, with low stiffness and shock resistance and good compression strength along the grain. It has a moderate bending classification. The material is fairly difficult to machine, with moderate to severe blunting of tools caused by silica in the wood; gum build-up on saws requires an increased set. Nailing requires pre-boring. The timber glues well and if the grain is first filled, a very good finish can be obtained. It is moderately durable, showing some resistance to termites in West Africa. The heartwood is resistant to preservative treatment.

Uses
Ayan is used for exterior joinery, doors, window frames and sills, and ships fittings; also for interior joinery, furniture and cabinet work, and for road and railway vehicle building. Its resilience makes it ideal for domestic and gymnasium floors. Logs are rotary cut for plywood manufacture and sliced for highly decorative veneers.

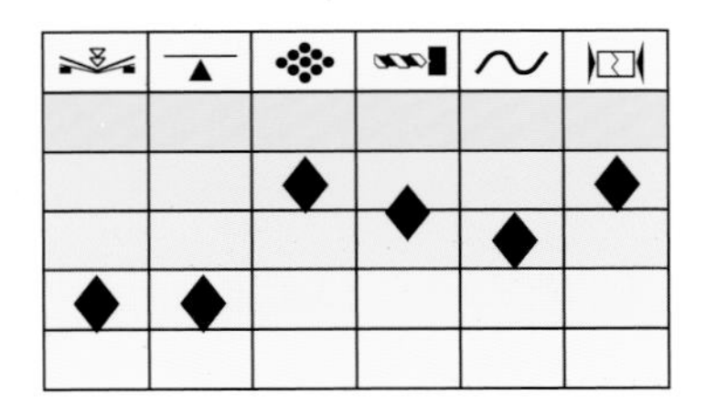

Dryobalanops spp. **Family:** *Dipterocarpaceae*
KAPUR

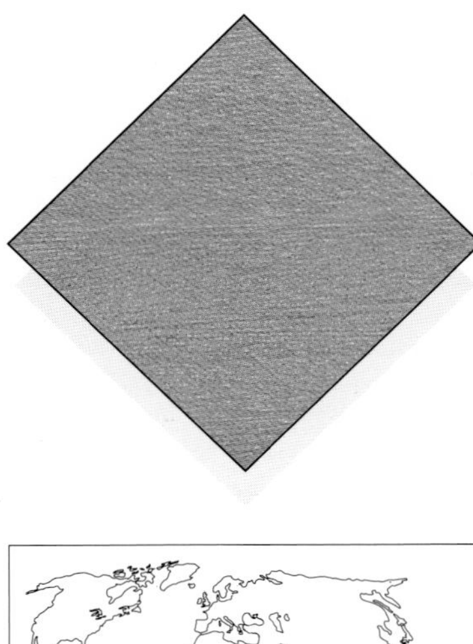

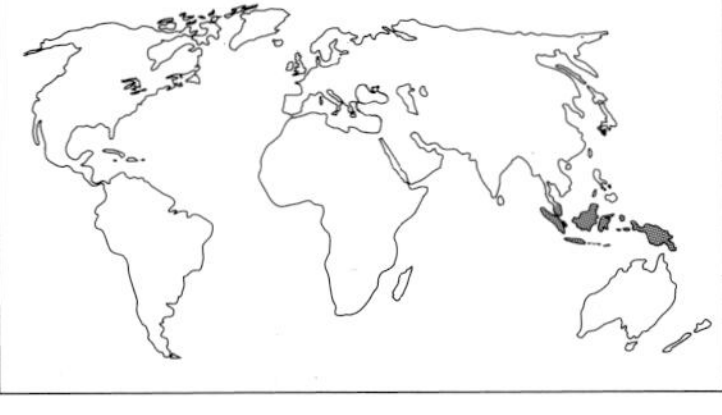

Where it grows
Several species of *Dryobalanops* are marketed together as kapur. The important commercial species are *Dryobalanops aromatica*, Gaertn.f. and *D.oblongifolia*, Dyer., which produce Malayan kapur for export, but the latter species is known locally as keladan. These two species also produce Indonesian kapur shipped from Sumatra. *D.aromatica* and *D.lanceolata*, Burck, produce Sarawak kapur and Kalimantan kapur. The latter species and *D.beccarii*, Dyer. produce Sabah kapur (kapor); *D.aromatica* and *D.fusca*, V.Sl. produce Indonesian kapur from Borneo. Other names include 'Borneo camphorwood', Kalimantan kapur, Sabah kapor and Indonesian kapoer. The various species reach a height of around 60m (200ft), and a diameter of usually 1-1.5m (3-5ft), with tapering boles above the buttresses.

Appearance
The sapwood is usually white to yellowish-brown in colour and sharply defined from the heartwood, which is light red-brown to deep red-brown, with straight to shallowly interlocked grain and fairly coarse but even texture. It contains resin but this does not exude on to the wood surface. The wood has a camphor-like odour.

Properties
Weight is 700-770 kg/m^3 (44-48 lb/ft^3) when seasoned. The wood dries fairly slowly but well with little degrade, and there is medium shrinkage in use. This dense wood has high bending and crushing strengths and medium stiffness and shock resistance, with a moderate steam-bending rating because of resin exudation. Machining can produce a fibrous finish, and there is a moderate to severe blunting effect of cutting edges. It glues well and provides a good finish when filled. Kaipur is liable to blue stain if it is in contact with iron compounds in damp conditions, and the acidic character may induce the corrosion of metals. The sapwood is durable but liable to attack by powder post beetle; the heartwood is very durable and extremely resistant to preservative treatment.

Uses
Kapur is a good constructional timber for estate and farm buildings, and for wharf decking, exterior joinery, windows, doors, frames, stairs, cladding, garden seats and vehicle floors. It is used for wharves, dock work and bridges above the water line. It is rotary cut for plywood manufacture and selected logs are sliced for decorative veneers.

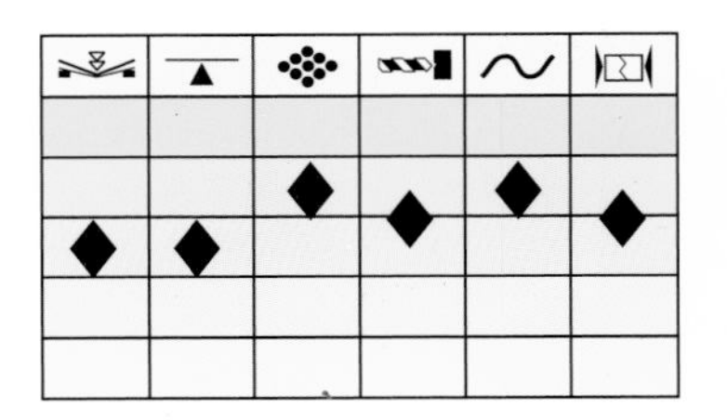

Dyera costulata **Family:** *Apocynaceae*
JELUTONG

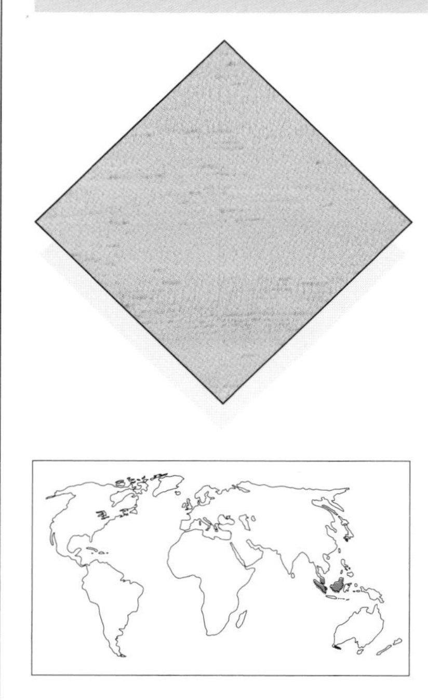

Where it grows
The tree is widely distributed throughout Malaysia and in the Indonesian islands of Sumatra and East Kalimantan. It grows to a substantial size, up to 60m (200ft) in height, with straight, cylindrical boles up to 27m (90ft) long with a diameter of 2.5m (8ft).

Appearance
Both the sapwood and heartwood are the same creamy-white colour when first cut, maturing to a pale straw colour on exposure, often stained by fungi after the tree has been tapped for latex. The wood is plain, almost straight grained, with a fine and even texture.

Properties
This figureless wood is very lustrous, but contains slit-like radial latex passages on tangential surfaces, in clusters or rows about 1m (3ft) apart. These passages or canals, which appear lens-shaped on flat-sawn surfaces, about 6mm (1/4in) wide and 12mm (1/2in) long, rule out the possibility of using jelutong where sizable pieces are required or where appearance is important. These defects are eliminated in conversion to relatively smaller dimensions. The weight is about 460 kg/m^3 (29 lb/ft^3) when seasoned. The wood dries fairly easily without degrade, but is difficult to dry in thick stock without staining. There is little shrinkage in service. This soft, weak and rather brittle timber is perishable with low strength in most categories and a very poor steam-bending classification. The wood works easily with both hand and machine tools with only a slight blunting effect, and provides a very smooth surface, taking screws and nails without difficulty. It can be glued easily and takes stain well and can be polished or varnished to a good finish. It is non-durable, liable to attack by powder post beetle, but permeable for preservative treatment.

Uses
Ease of working makes jelutong an excellent wood for sculpture and carving; for pattern making, architectural models, drawing boards, wooden clogs and handicraft work. It has specialized uses for battery separators and match splints, and also for lightweight partitions and some parts for interior joinery and fitments. The logs are also rotary cut for corestock for flush doors, plywood and laminated boards. The latex is extracted for the manufacture of chewing gum.

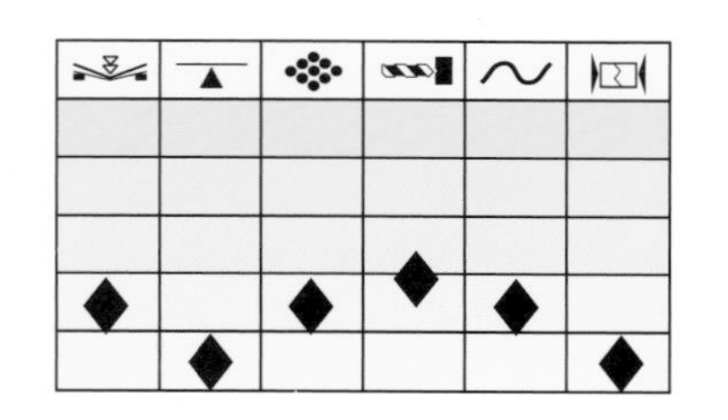

Endiandra palmerstonii **Family:** *Lauracae*
'QUEENSLAND WALNUT'

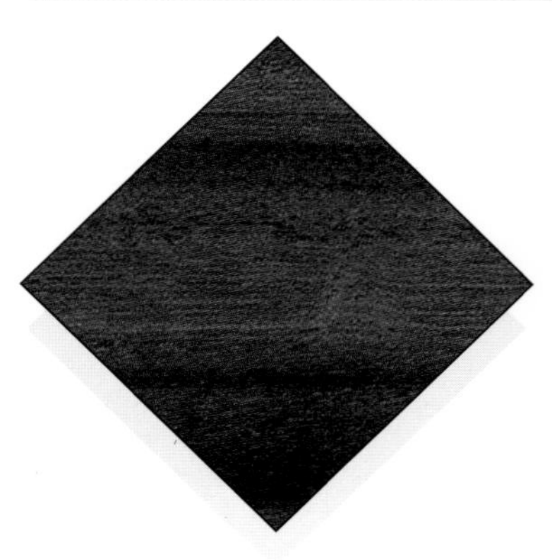

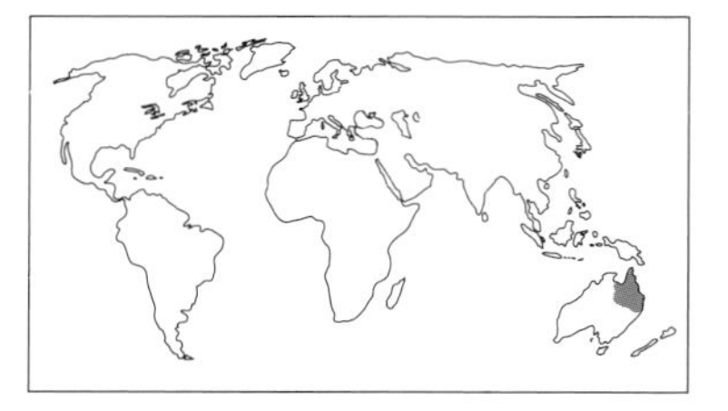

Where it grows
This large buttressed Australian tree, not a true walnut (*Juglans spp.*), grows abundantly in northern Queensland to a height of 37-43m (120-140ft) and a diameter of up to 1.8m (6ft), with a clean bole up to 25m (80ft). It is also known as 'Australian walnut', walnut bean or oriental wood.

Appearance
The wood bears a striking resemblance to European walnut, but has a more prominent striped figure. The pinkish sapwood is narrow, and the heartwood varies from grey to pale mid-brown, and often to dark brown, with pinkish, grey-green or purple-black streaks. The grain is interlocked and often wavy, giving a chequered or broken striped figure on quartered surfaces, with a medium texture.

Properties
Deposits of silica are sometimes present in the form of crystalline aggregates in ray cells. The wood weighs about 680 kg/m^3 (42 lb/ft^3) when seasoned. It tends to air dry fairly rapidly with some tendency to split at the ends, and kiln dries fairly rapidly in thinner sizes without checking, but tends to warp; slight collapse is possible. Thicker boards are liable to split unless quarter sawn. There is medium movement in service. This heavy timber has medium bending strength and shock resistance, low stiffness, and high crushing strength, with a moderate steam-bending classification. It is a rather difficult timber to work, requiring tipped saws and high-speed cutters to overcome the severe blunting of cutting edges caused by the silica. Gluing is satisfactory and holding properties are good. The wood takes stain and polish well and can be brought to an excellent standard of finish.

The heartwood is non-durable and resistant to preservative treatment but the sapwood is permeable.

Uses
The wood has high insulation properties and is used for shop, office and bank fitting, high-class cabinets and furniture, interior joinery and many forms of decorative work. As a flooring timber it is moderately resistant to wear. Logs are rotary cut for plywood faces, and selected logs are sliced for very attractive striped decorative veneers for high-quality cabinets and panelling.

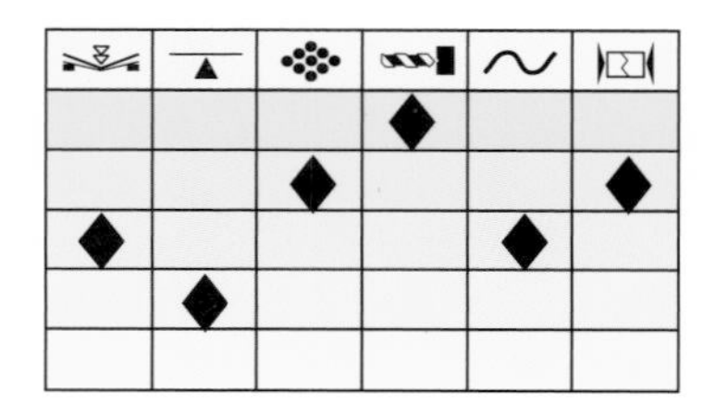

Entandrophragma angolense **Family:** *Meliaceae*

GEDU NOHOR

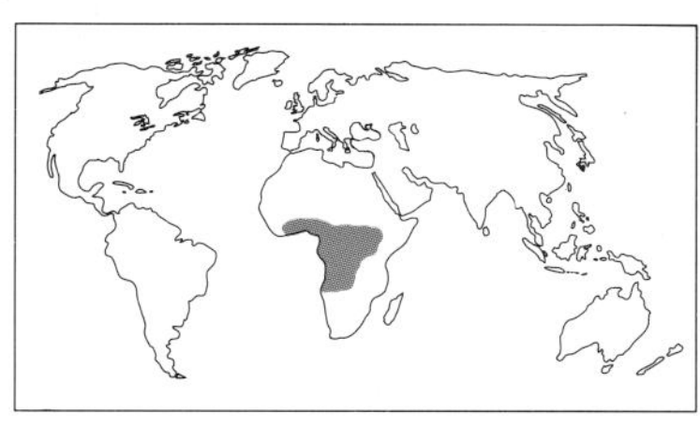

Where it grows

This large deciduous tree grows in the semi-evergreen forests of West, central and East Africa, from Uganda in the east across to Angola and the Ivory Coast in the west. Other names include edinam (UK and Ghana); tiama (Ivory Coast); kalungi (Congo and Zaire); gedu lohor and gedu noha (Nigeria); and edoussie (Cameroon). It grows to a height of 50m (160ft) with an average diameter of 1.2-1.5m (4-5ft). It often has strong winged buttresses extending for 6m (20ft) up the trunk of the tree.

Appearance

Typically the heartwood has a rather dull, plain appearance of light reddish-brown colour with a fairly interlocked grain, producing a weak strip on quartered surfaces and with a moderately coarse texture. Some logs are a much lighter, pale pinkish-grey colour similar to the pinkish-grey sapwood, which is sometimes about 100mm (4in) wide.

Properties

Weight averages 540 kg/m^3 (34 lb/ft^3) when seasoned, and the wood is stable in service. It dries fairly rapidly with quite a bit of distortion. It has medium density and crushing strength, low bending strength and shock resistance, and very low stiffness with a poor steam-bending classification. It works fairly easily with hand and machine tools but the interlocked grain has a moderate blunting effect on tools. Nailing is satisfactory; the wood glues and takes stain well and can be brought to a good finish. The sapwood is liable to attack by powder post beetle. The heartwood is moderately durable but extremely resistant to preservative treatment.

Uses

Gedu Nohor is extensively used as a substitute for mahogany in furniture and cabinet making for interior parts, partitions, edge lippings and facings; also for interior and exterior joinery and shop and office fitting. It makes a good domestic flooring timber. In boat building it is used for cabins, planking, furniture and fitments. It is also used for bus bodies and railway carriage construction. Wide boards are used for coffins. The logs are rotary cut for plywood and sliced for decorative veneers.

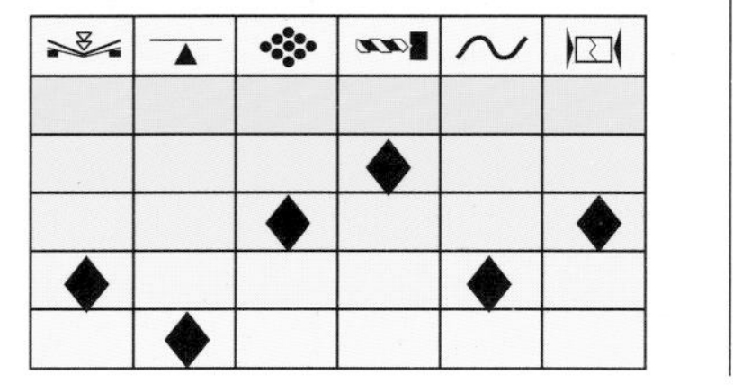

Entandrophragma cylindricum **Family:** *Meliaceae*

SAPELE

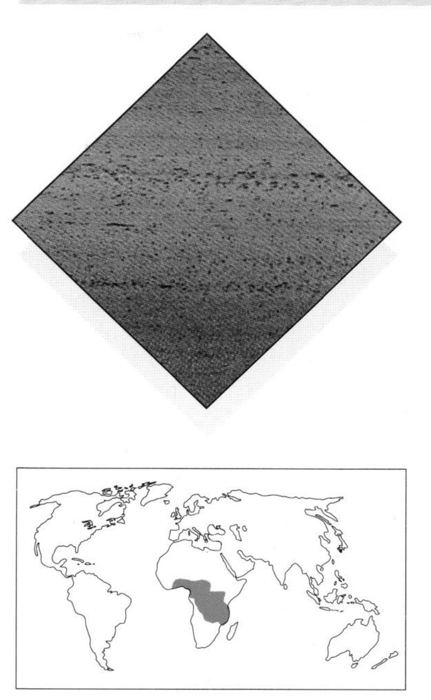

Where it grows

Sapele grows widely in the tropical rain forests of West, central and East Africa, from the Ivory Coast through Ghana and Nigeria to the Cameroons, and eastwards to Uganda, Zaire and Tanzania. It is known as aboudikro on the Ivory Coast. It grows to 45-60m (150-200ft) with a diameter of about 1m (3ft), and a clean bole for 30m (100ft).

Appearance

The narrow sapwood is pale yellow-white and the heartwood is salmon pink when freshly cut, maturing into reddish-brown. It has a closely interlocked grain, resulting in a pronounced and regular pencil striped or roe figure on quartered surfaces. Wavy grain yields a highly decorative fiddleback or mottled figure with a fine and even texture.

Properties

Weight is 560-690 kg/m^3 (35-43 lb/ft^3), averaging about 620 kg/m^3 (39 lb/ft^3) when seasoned. The wood dries fairly rapidly, with a marked tendency to distort. There is medium movement in service. Sapele has medium density, bending and shock resistance, high crushing strength and low stiffness, and a poor steam-bending rating. It works fairly well with both hand and machine tools, with moderate blunting of cutting edges caused by the interlocked grain. Nailing and gluing are satisfactory, and care is required when staining. When filled the surface can be brought to an excellent finish. The sapwood is liable to attack by powder post beetle and moderately resistant to impregnation. The heartwood is moderately durable but extremely resistant to preservative treatment.

Uses

Sapele enjoys a world wide reputation as a handsome wood for high-quality furniture and cabinet making, interior and exterior joinery, window frames, shop, office and bank fitting, counter tops and solid doors. It is widely used for boat and vehicle building, and for piano cases and sports goods. The wood is ideal for decorative flooring for domestic and public buildings. Logs are rotary cut for plywood and selected logs sliced for quartered ribbon striped decorative veneers, used in cabinets and panelling.

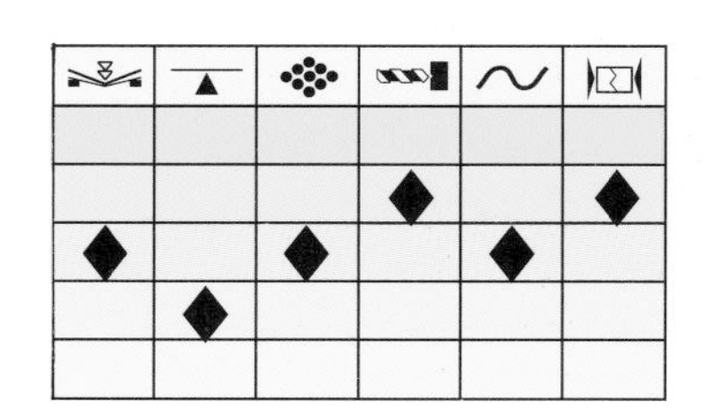

Entandrophragma utile **Family:** *Meliaceae*
UTILE

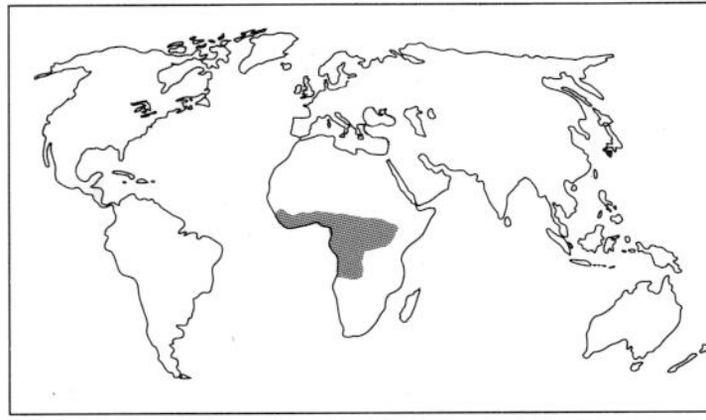

Where it grows
The tree grows in the moist, deciduous high forests of tropical West and East Africa; Sierra Leone, Cameroons, Liberia, Gabon, Uganda and Angola, with commercial supplies coming from the Ivory Coast and Ghana. It is also known as sipo and mebrou zuiri (Ivory Coast); assié (Cameroon);Tshimaje rosso (Zaire); kosi-kosi (Gabon); and afau-konkonti (Ghana). It is a very large tree, reaching a height of 45-60m (150-200ft), with a diameter above the small narrow buttresses of about 2.5m (8ft).

Appearance
The light brown sapwood is distinct from the heartwood, which is a uniform reddish-brown mahogany colour with an interlocked and rather irregular grain. The quartered surfaces do not possess such a fine ribbon striped figure as sapele, and the texture is more open.

Properties
Weight varies from about 550 to 750 kg/m^3 (34-47 lb/ft^3) with an average of 660 kg/m^3 (41 lb/ft^3) when seasoned. It dries at a moderate rate with some distortion but this is not severe. There is medium movement in service. Utile is a dense wood, with high crushing strength and medium bending strength, low stiffness and shock resistance, and a very poor steam-bending rating. Utile works well with both hand and machine tools, with only slight to moderate blunting of tools. The wood takes screws and nails satisfactorily, and is easy to glue. When the grain is filled, it takes stain and polish well and provides an excellent finish. The heartwood is durable but the sapwood is liable to attack by powder post beetle. The heartwood is extremely resistant both to decay and to preservatives.

Uses
Utile is extensively used for furniture and cabinets, counter tops, interior and exterior joinery for doors and window frames, and for interior construction work. It is also used for shop and office fitting, domestic flooring, vehicle and boat building and for musical instruments and sports goods. It is rotary cut for plywood and sliced for decorative furniture and panelling veneers.

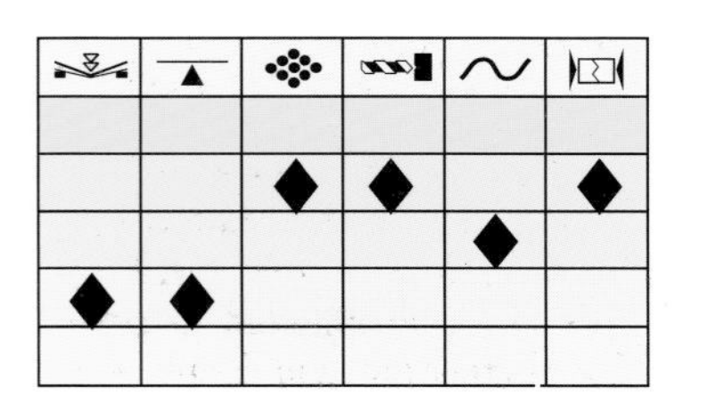

Eucalyptus delegatensis **Family:** *Myrtaceae*
'TASMANIAN OAK'

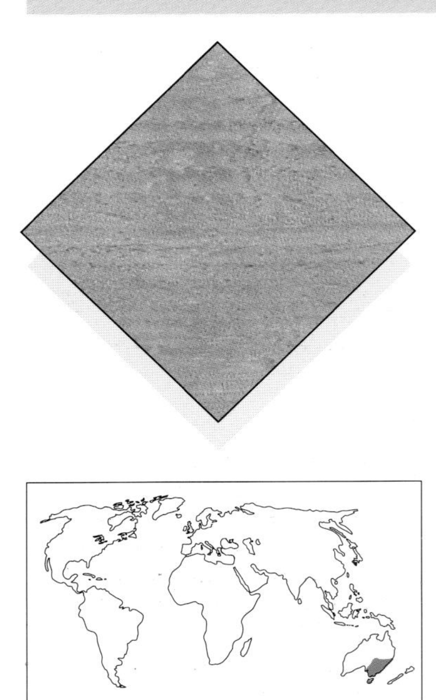

Where it grows
Three species of *Eucalyptus* occur in south eastern Australia from New South Wales to Tasmania; *E.delegatensis* is sold as alpine ash, white-top or gum-top stringybark, and woollybutt; *E.obliqua*, L'Hérit, produces messmate stringybark and brown-top stringybark; *E.regnans*, F. Muell. produces mountain ash, Victorian ash, stringy gum and swamp gum. All three species, known as 'Australian oak', are marketed together, but are not botanically related either to oak or ash. The trees reach a height of 60-90m (200-300ft) with a diameter of 1-2m (3-7ft) or more.

Appearance
The heartwood colour of these species varies from a pale biscuit to light brown with a pinkish tinge, and they have a narrow, indistinct, paler sapwood. The wood is usually straight grained, but sometimes interlocked or wavy and with a coarse texture. Hard gum or kino veins are often present, especially in *E.obliqua*.

Properties
E.delegatensis averages 640 kg/m^3 (40 lb/ft^3) in weight; *E.obliqua* 780 kg/m^3 (49 lb/ft^3); and *E.regnans* 620 kg/m^3 (39 lb/ft^3) when seasoned. The wood dries readily and fairly quickly but is liable to develop surface checks and to distort. There is medium shrinkage in use. It has medium bending strength, shock resistance and stiffness with high crushing strength. Only *E.obliqua* has a moderate steam-bending rating. The timbers work satisfactorily with both hand and machine tools, and hold nails and screws well. They can be glued easily, take stain and polish well, and can be brought to an excellent finish. The sapwood is liable to attack by powder post beetle; the heartwood is moderately durable and resistant to preservative treatment, but the sapwood is permeable.

Uses
This timber is extensively used for interior and exterior joinery and building construction, and for cladding and weatherboards. It also appears in furniture, cooperage, agricultural implements, handles, coach and truck building, sports goods and domestic flooring. It is rotary cut in Australia for the manufacture of plywood and sliced for export in decorative veneer form.

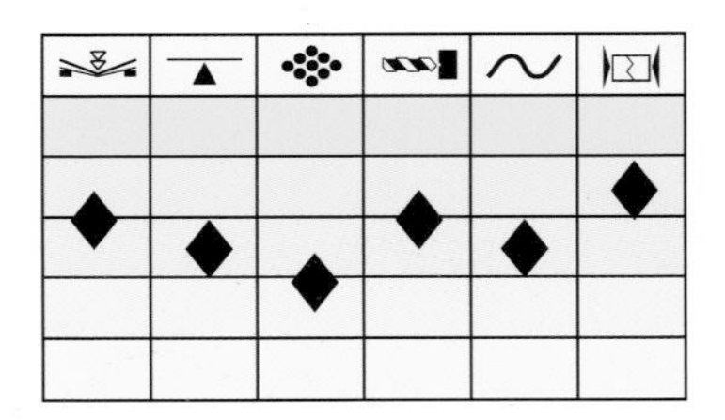

Eucalyptus diversicolor **Family:** *Myrtaceae*
KARRI

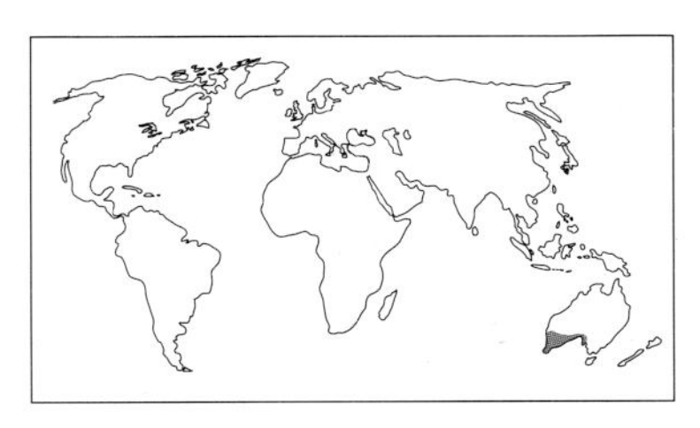

Where it grows
This important tree of south western Australia grows to an immense 85m (279ft), with a diameter of 1.8-3m (6-10 ft), clear of branches for about 25-30m (80-100ft).

Appearance
The heartwood of karri is a uniform reddish-brown in colour with an interlocked grain, producing a striped figure on quartered surfaces. The texture is moderately coarse but even. Jarrah and karri can be distinguished by a 'splinter-test'; a small burnt splinter of karri forms a thick white ash, while jarrah burns to a black, ashless charcoal.

Properties
The average weight is 880 kg/m³ (55 lb/ft³) when seasoned. Karri requires great care in drying and is prone to deep checking in thick stock and distortion in thin stock. There is also a large movement in service. This very heavy timber is high in all strength categories and has a moderate steam-bending classification, but cannot be bent if small knots are present. It is a difficult wood to work with hand tools and fairly difficult to machine, as the wavy and interlocked grain has a moderate to severe blunting effect on tools. It requires pre-boring for nailing, but glues well and when filled can be brought to an excellent finish. The heartwood is durable and extremely resistant to preservative treatment, but the sapwood is classified as permeable.

Uses
Karri is stronger than jarrah but inferior for underground use, if exposed to fungal, marine borer or termite attack, or in contact with water for dock and harbour work. It is therefore used above water for wharf and bridge construction, for superstructures and ship building. It is ideal for building construction as joists, rafters and heavy beams. It is also used for agricultural implements. Selected pieces are used for furniture, cabinet fittings and domestic flooring. When treated it is used for railway sleepers, poles and piles. It is rotary cut for plywood manufacture and sliced for decorative cabinet and high-quality panelling veneers.

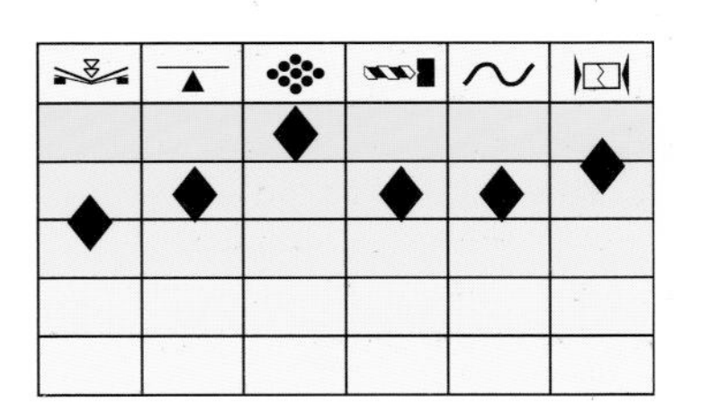

Eucalyptus marginata **Family:** *Myrtaceae*
JARRAH

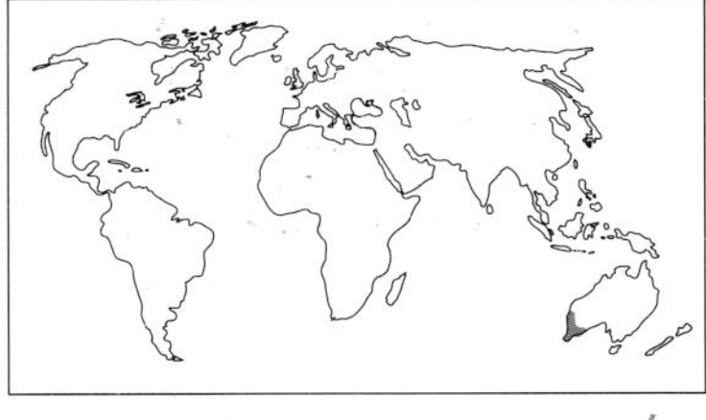

Where it grows
More jarrah is cut than any other Australian commercial timber. It is found in the coastal region south of Perth in Western Australia, and huge quantities are exported. Trees reach an unremarkable (for eucalyptus) height of 30-45m (100-150ft), with a diameter of 1-1.5m (3-5ft).

Appearance
The heartwood is a rich dark reddish-brown, often with occasional gum veins and pockets that detract from the appearance, and boat-shaped flecks caused by fungus that enhance it. The grain is usually fairly straight, sometimes interlocked or wavy. The texture is moderately coarse but even.

Properties
Weight varies between 690 and 1040 kg/m³ (43-65 lb/ft³), averaging around 820 kg/m³ (51 lb/ft³) when seasoned. It will distort unless air dried before kilning, particularly in large sizes which are often air dried only. There is medium shrinkage in service. It has medium bending strength and stiffness, high crushing strength and a moderate steam-bending rating. It is fairly hard to machine and not really suitable for working with hand tools because of its hardness and density. There is high resistance in cutting with a moderate blunting effect on tools. Pre-boring is necessary for nailing and screwing. It glues and finishes well. Jarrah has a very durable heartwood and is highly resistant to insect attack and preservative treatment, but the sapwood is classified as permeable.

Uses
Jarrah is an ideal constructional timber, used in Australia for marine structures like dock pilings and harbour work, wharf, bridge building and sea defences; other marine uses include ship building for rails and decking. It is used throughout the world for railway sleepers. Jarrah makes a good flooring timber and is used for shingles and weatherboards, rafters and joists, also for interior joinery and furniture, chemical vats and filter presses. As a turnery wood it is highly valued for striking tool handles. Selected logs are sliced for highly decorative veneers for architectural panelling.

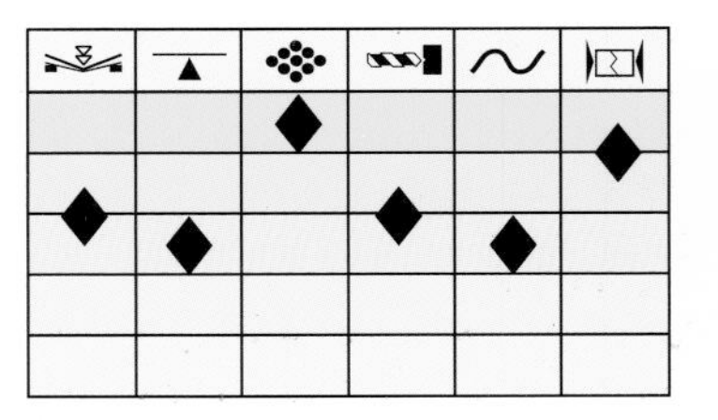

Fagara flava **Family:** *Rutaceae*

WEST INDIAN SATINWOOD

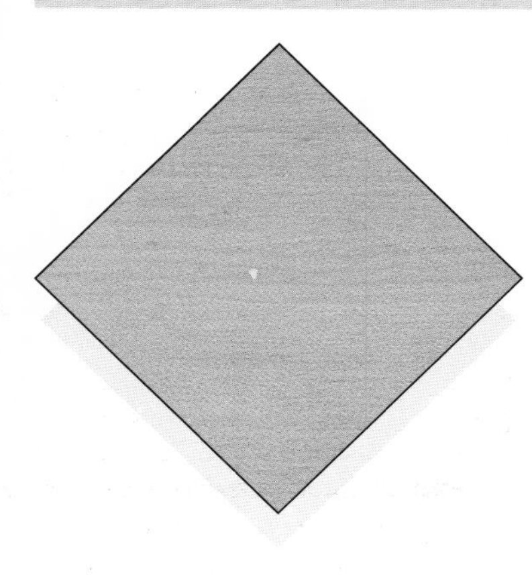

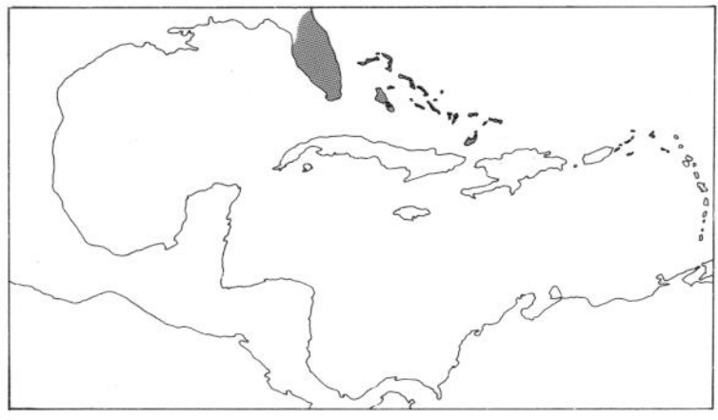

Where it grows

Often confused with Sri Lanka satinwood – Ceylon satinwood (*Chloroxylon swietenia*) – this tree grows in Bermuda, the Bahamas and southern Florida. It reaches its best development in Jamaica, where it grows to a height of about 12m (40ft) with a diameter of about 0.5m ($1^1/_2$ft). It is also known as Jamaica satinwood and San Domingan satinwood (USA); yellow sanders (West Indies); and aceitillo (Cuba).

Appearance

The heartwood is rich cream to golden yellow, with a straight to interlocked, wavy or irregular grain, producing bee's wing cross-mottled, roe or broken striped figure on quartered surfaces. The texture is fine and even with a bright, satin lustre. It smells like coconut oil when freshly cut.

Properties

This very heavy and dense timber weighs about 900 kg/m^3 (56 lb/ft^3) when seasoned. It requires care in drying to avoid distortion. It works well with both hand or machine tools but there is high resistance in cutting. Cutting edges have to be kept very sharp where irregular grain is present and a reduced cutting angle is recommended. The fine dust produced in machining operations is liable to cause dermatitis. The timber requires pre-boring for screwing or nailing, but glues well and takes an excellent polish. The timber is non-durable and extremely resistant to impregnation.

Uses

West Indian satinwood was used in England for more than a century before mahogany became most popular for Georgian furniture, and during the 18th century 'Golden Age of Satinwood' this beautiful timber was used extensively for the high-class cabinet work of Adam, Sheraton and Hepplewhite. Today it is exported in only small quantities and used for furniture, high-class cabinet making and reproduction and restoration of period furniture. It is an excellent turnery wood for brushbacks, hand-mirrors, bobbins and fancy goods. Selected logs are sliced in a wide range of attractively figured decorative veneers for cabinets and panelling and for inlay and marquetry work.

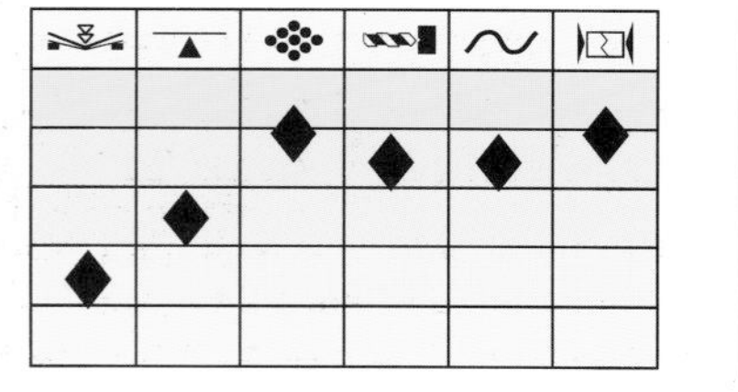

Fagus spp. **Family:** *Fagaceae*

BEECH

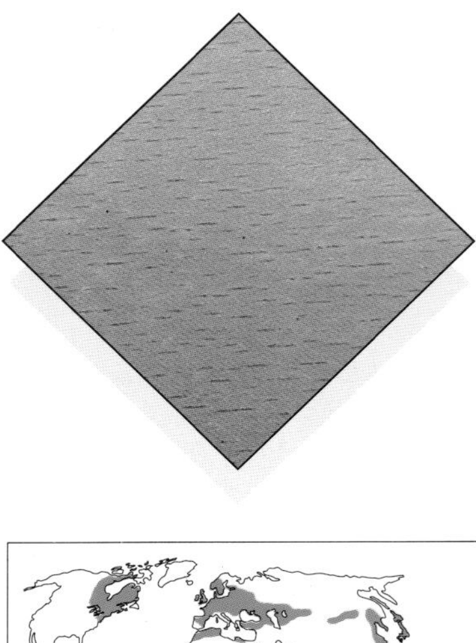

Where it grows

Beech grows in the northern temperate regions of Europe, Canada and America, western Asia, Japan and northern Africa. It is used in larger quantities than any other hardwood in the United Kingdom. It is known as 'the Mother of the Forest' because other hardwoods in mixed broadleaved forests would have a struggle to survive without it; its leaf drip kills weeds and leaf fall provides rich humus for the soil. *F.sylvatica*, L. produces European beech; *F.grandifolia*, Ehrh., American beech; *F.orientalis*, Lipsky, Turkish beech; *F.crenata*, Bl., Japanese beech. Each is named according to its country of origin. They grow to an average height of 45m (150ft), with a diameter of about 1.2m (4ft).

Appearance

The timber is very pale cream to pinkish-brown, and is often 'weathered' to a deep reddish-bronze-brown after steaming. It is typically straight grained with broad rays, and has a fine, even texture.

Properties

Japanese beech, the lightest, weighs 620 kg/m^3 (39 lb/ft^3). Slavonian beech weighs about 670 kg/m^3 (42 lb/ft^3); European beech 720 kg/m^3 (45 lb/ft^3); and American beech 740 kg/m^3 (46 lb/ft^3) when seasoned. Special care is needed in drying as it dries fairly rapidly and well but is moderately refractory and shrinks considerably in service. The wood has medium strength in bending, stiffness and shock resistance, and high crushing strength, with an exceptionally good steam-bending rating. It works readily with hand or machine tools and has good holding properties; it glues very easily and can be brought to an excellent finish. The heartwood is perishable and liable to attack by the common furniture beetle and death watch beetle. The sapwood is affected by the longhorn beetle. Beech is classified as permeable for preservative treatment.

Uses

Beech is perhaps the most popular general purpose timber for furniture, chairs, school desks and the like, and interior joinery and, when treated, for exterior joinery. It is a turnery wood for tool handles, brushbacks, bobbins, domestic woodware, sports goods, parts of musical instruments and domestic flooring. It is also used for bent work and cooperage. It is rotary cut for utility plywood and corestock and sliced for unremarkable decorative veneers.

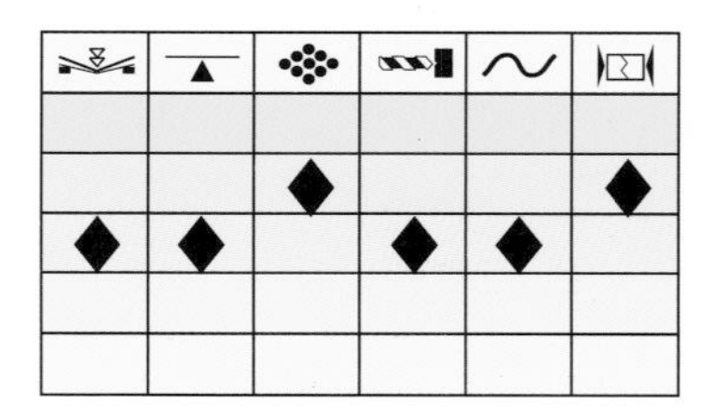

Flindersia spp. **Family:** *Rutaceae*
'QUEENSLAND MAPLE'

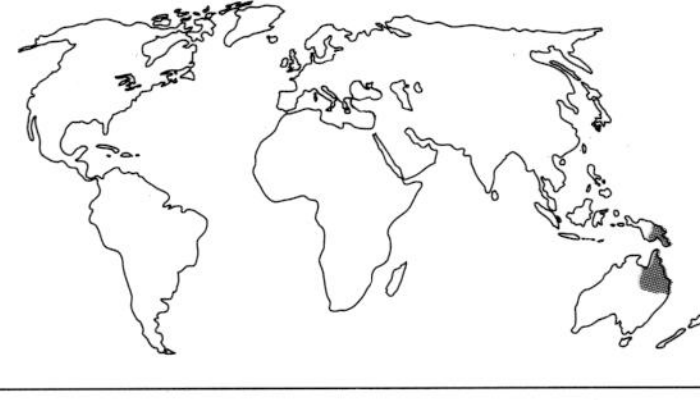

Where it grows
This decorative timber, not a true maple, is highly valued in northern Queensland, Australia. It also occurs in Papua New Guinea. There are a few closely related species; *Flindersia brayleyana*, F.Muell. produces 'Australian maple', or maple silkwood; *F.pimenteliana*, F.Muell. and *F.laevicarpa*, var. *heterophylla* are both known as scented maple. The trees grow to about 30m (100ft) with a diameter of 0.9-1.2 m (3-4ft).

Appearance
The heartwood is pale brown to pink with a silken lustre, which matures to a pale brown. The grain is often interlocked, sometimes wavy or curly, producing a range of attractive figures, with a medium texture.

Properties
The average weight of *F.brayleyana* and *F.pimenteliana* is 550 kg/m^3 (34 lb/ft^3) and *F.laevicarpa*, Papua New Guinea, weighs 690 kg/m^3 (43 lb/ft^3) when seasoned. They air and kiln dry satisfactorily but with some tendency to collapse; distortion may need kiln reconditioning. There is medium shrinkage in service. The wood has medium bending and crushing strengths, low stiffness and resistance to shock loads, and a poor steam-bending rating. It works readily with both hand or machine tools but has a moderate blunting effect on cutting edges. Interlocked grain tends to pick up when working on quartered stock, and a reduced angle is necessary. The timber holds screws well, and glues satisfactorily. When filled, stained and polished it can be brought to an excellent finish. The heartwood is moderately durable and resistant to preservative treatment, and is not subject to insect attack.

Uses
Queensland maple is used for high-class cabinets and furniture, for interior fittings and mouldings, and interior joinery. It has many specialist uses such as vehicle body work, ornamental rifle and gunstocks and the bases of printing blocks. In boat building it has a wide use for parts, oars and superstructures. It is also a very good turnery wood. It is often rotary cut for plywood and sliced to produce 'watered-silk' moiré, block mottle, fiddleback, striped and bird's eye figured veneers for cabinets and panelling.

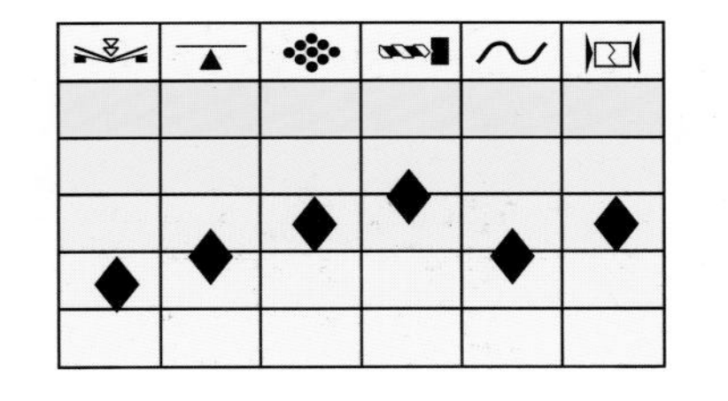

Flindersia schottiana **Family:** *Rutaceae*
'SOUTHERN SILVER ASH'

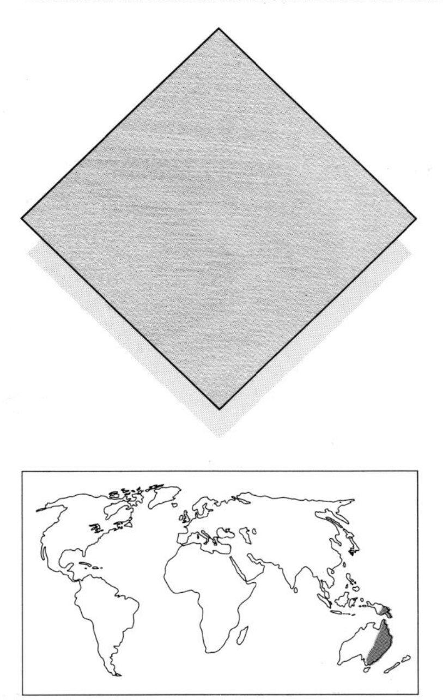

Where it grows
There are many *Flindersia* species trees growing in Australia producing timbers with names such as 'ash' and 'maple', but which are botanically unrelated to these timbers. *F.schottiana* produces 'Southern silver ash', also known as 'bumpy ash' or cudgerie, and is found in New South Wales, Queensland and Papua New Guinea; *F.bourjotiana* produces Queensland silver ash; *F.pubescens*, northern silver ash also in New South Wales and Queensland. They grow to about 30m (100ft), with a diameter of about 0.75m (2$\frac{1}{2}$ft).

Appearance
The pale biscuit-coloured heartwood is mostly straight grained, but sometimes shallowly interlocked or wavy, with a medium texture.

Properties
Weight varies, but an average is 560 kg/m^3 (35 lb/ft^3) when seasoned. The wood dries rather slowly and shows a slight tendency to warp, but it can be air or kiln dried satisfactorily, up to about 50mm (2in) thick, with little degrade. It is stable in service. It has medium bending and crushing strength, low stiffness and resistance to shock loads. This tough, resilient and elastic wood has a very good steam-bending classification. It works easily with both hand and machine tools, with only a moderate blunting effect. Quartered surfaces tend to pick up interlocked grain when planing or moulding. The material has good nail- and screw-holding properties, and it can be glued without difficulty. If care is taken in grain filling and staining, it can be polished to an excellent finish. The heartwood is very durable above ground and resistant to preservative treatment, but the sapwood is classified as permeable.

Uses
Silver ash is used in Australia for high-class cabinets and furniture, interior trim and fitments, and interior and exterior joinery. It is used for food containers as it is odourless and will not taint. It is extensively used as a structural timber, for ship and boat building, and vehicle bodies. Its specialist uses are for sporting goods and musical instruments, and it is popular for carving and turnery for domestic woodware. It is also rotary cut for plywood and sliced into very attractive decorative veneers for cabinets and panelling.

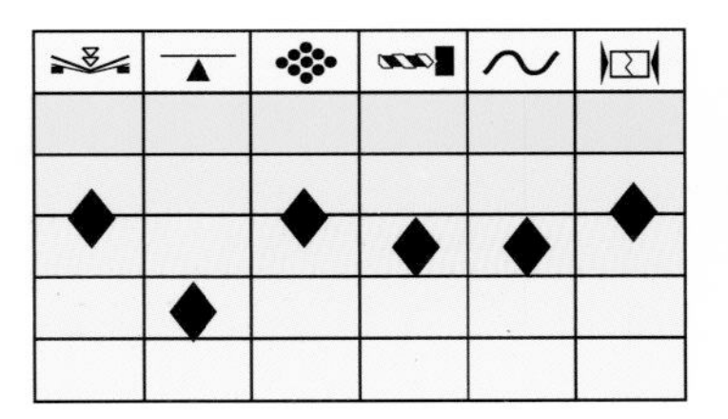

Fraxinus spp. **Family:** *Oleaceae*
ASH

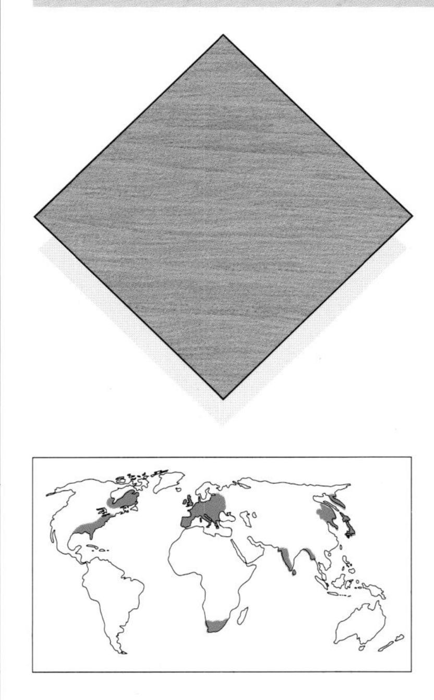

Where it grows
Ash thrives throughout Europe, North America and Japan. *F.americana*, L. is known as white ash (Canada); *F.pennsylvania*, Marsh., American ash, is known as green ash (USA) or red ash (Canada); *F.exclesior*, L., European ash, is named after the country of origin as English, French, Polish, Slavonian and so on in the UK. The tree grows to 25-36m (80-120ft), with a diameter of 0.6-1.5m (2-5ft).

Appearance
American ash is grey-brown in colour with a reddish tinge. European ash is cream white to light brown, sometimes with a sound dark brown to black heart which is marketed separately as 'olive ash'. It is straight grained and coarse textured, and the growth rings produce a very decorative figure on plain-sawn surfaces.

Properties
Weight varies as follows: *F.americana*, 660 kg/m³ (41 lb/ft³); *F.pennsylvania*, 690 kg/m³ (43 lb/ft³) and *F.exclesior*, 580 kg/m³ (36 lb/ft³) when seasoned. The timber dries fairly rapidly, with little degrade and medium shrinkage in use. It has medium bending and crushing strength and shock resistance, low stiffness, and an excellent steam-bending classification. It works satisfactorily with both hand and machine tools. Pre-boring is advised for nailing. It glues with ease and takes stains and polishes well to provide a good finish. Ash is non-durable and perishable. It is liable to attack by the powder post beetle and the furniture beetle. The heartwood is moderately resistant to preservative treatment, but the black heartwood is resistant.

Uses
Ash, one of the very best woods for bending, is used extensively for chair making and in cabinet making, furniture and interior joinery. Specialist uses include bent handles for umbrellas, shop fitting, vehicle building, wheelwrighting and agricultural implements. It is used in boat building for bent parts for frames for canoes and canvas boats; also for sports goods, tennis racquets, hockey sticks, baseball bats, billiard cues and gymnasium equipment. It is an excellent turnery wood for tool handles, shovels and pick axes. It is sliced for decorative furniture and panelling veneers.

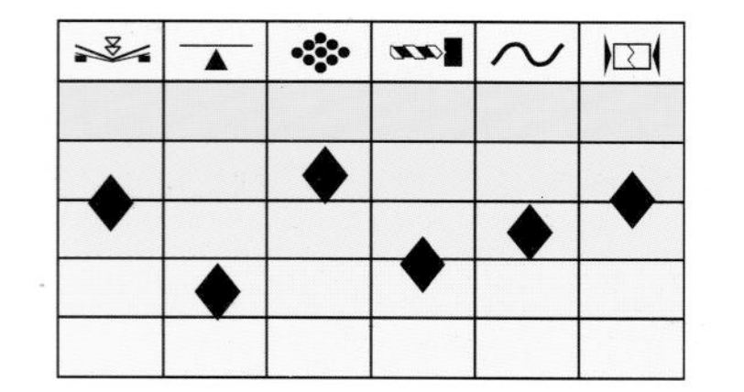

Gonystylus macrophyllum **Family:** *Gonystylaceae*
RAMIN

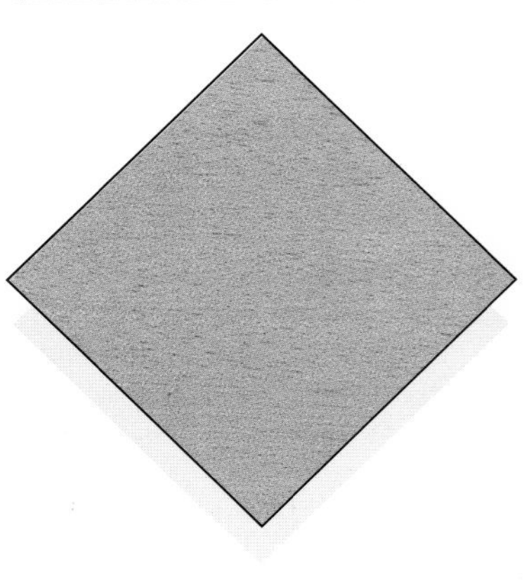

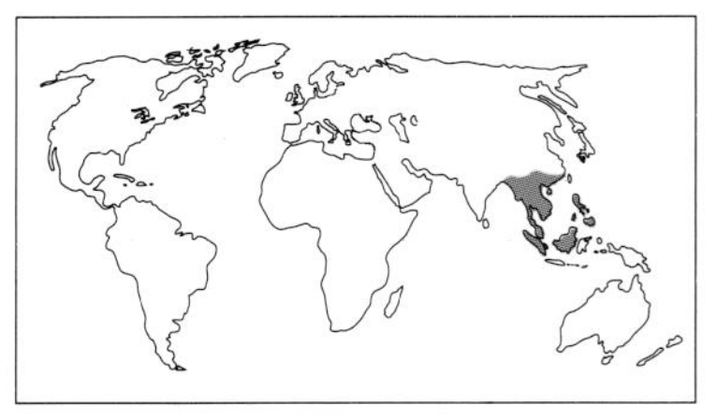

Where it grows
The tree grows in the freshwater swamps on the west coasts of Sarawak and throughout Malaysia and south east Asia. It is known as ramin telur in Sarawak; *G.macrophyllum* and *G.bancanus* (Miq.) Kurz. produces melawis in west Malaysia. The bulk of melawis comprises *G.macrophyllum*, but also includes *G.affinis* and *G.confusus*. These medium-sized trees attain a height of about 24m (78ft) with a diameter of 0.6-1.0m (2-3ft), and a clear cylindrical bole for 15-18m (50-60ft).

Appearance
The sapwood is up to 50mm (2in) wide, but is not clearly defined from the heartwood which is a uniform cream-white to pale straw colour, with a featureless straight to shallowly interlocked grain and a moderately fine and even texture.

Properties
Ramin weighs 640-720kg/m³ (40-45 lb/ft³), averaging about 660 kg/m³ (41 lb/ft³) when seasoned. Melawis containing *G.affinis* and *G.confusus* weigh about 705 kg/m³ (44lb/ft³). Air dry stock is easily kilned without degrade but is prone to staining and has to be dipped immediately after conversion. There is large movement in service. This dense wood has high bending and crushing strengths, low shock resistance and medium stiffness, with a very poor steam-bending rating. It works fairly easily by hand and machine, with a moderate blunting effect on tools, but the grain tends to tear on quartered material. It requires pre-boring for nailing, glues well and, when properly filled, can be stained and brought to a good finish. The heartwood is perishable; the sapwood is liable to attack by powder post beetle, but is permeable for preservation treatment.

Uses
Ramin is extensively used as a substitute for beech for furniture, fittings and picture-frame and other mouldings. It is used in carving and turnery, for handles of non-striking tools, for dowels; the wood is also used for interior joinery, shop and office fittings and wooden toys. It appears in light building work for skirtings, mouldings and domestic flooring. Selected logs are rotary cut for plywood and for corestock in laminated boards, and sliced for decorative veneers.

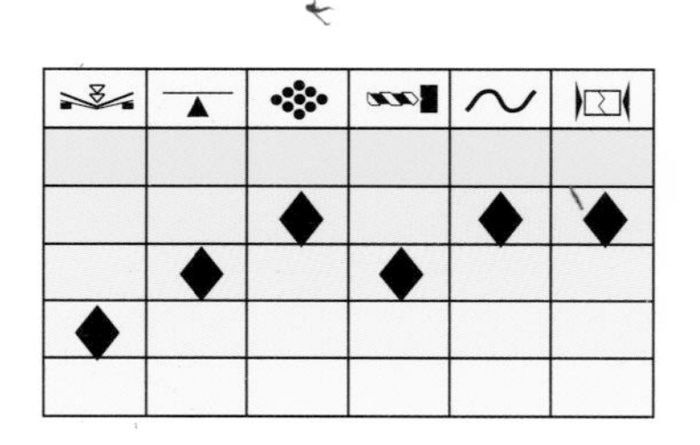

Gossweilerodendron balsamiferum **Family:** *Leguminosae*
AGBA

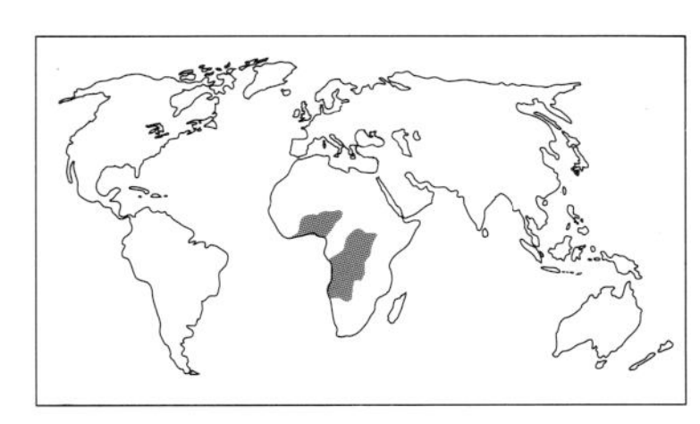

Where it grows
This is one of the largest trees in tropical West Africa, occurring mainly in western Nigeria and also in Angola and Zaire. It is known as tola, ntola and mutsekamambole (Zaire); moboron, tola branca and white tola (Angola); 'Nigerian cedar' in the UK; mutsekamambole (Nigeria); and nitola in the Congo. Trees reach a height of 60m (200ft) and a diameter of over 2m (7ft).

Appearance
The heartwood has a uniform pale straw to tan brown colour, with a straight to shallow interlocked or wavy grain which produces a broad striped figure on quartered surfaces. It has a fine texture.

Properties
The wood weighs about 510 kg/m³ (32 lb/ft³) seasoned. Drying is fairly rapid with little degrade, but exudation of dark oleo-resin can occur. This may also affect machining, which is otherwise satisfactory with only slight blunting of tools. The wood is stable with small movement in service. It has a moderately good steam-bending classification, with very low stiffness and low bending strength and shock resistance, and a medium crushing strength. Nailing is satisfactory, and it glues and finishes well if the grain is filled. This durable wood is very resistant to decay but the sapwood is liable to attack by the common furniture beetle. The heartwood is resistant to preservative treatment but the sapwood is permeable.

Uses
Abga is excellent for interior joinery for table tops, shop fitting, panelling, and chair seats. It is used as a substitute for oak for school desks, church furniture, including coffins, and for mouldings. It is also a good turnery wood for handles and dowels, domestic hardware and toymaking, but a slight resinous odour makes it unsuitable for items which may come into contact with food. As domestic flooring it is ideal over underfloor heating. It is also used for exterior joinery for cladding and planking, the laminated frames of boats, and for motor body coachwork, truck and trailer flooring. It is rotary cut for constructional veneers for marine plywood and boat building, and for decorative veneers.

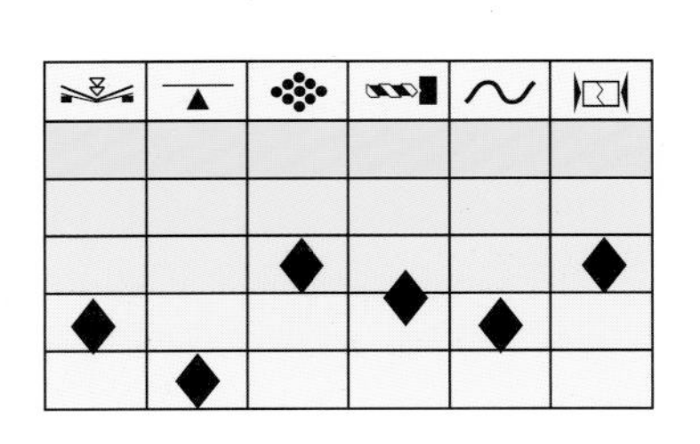

Gossypiospermum praecox **Family:** *Flacourtiaceae*
'MARACAIBO BOXWOOD'

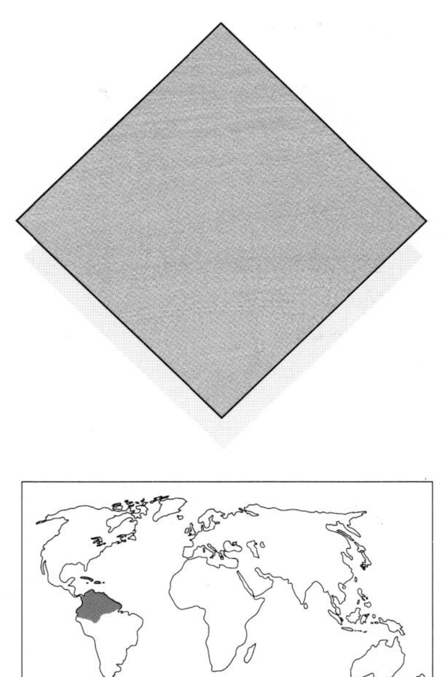

Where it grows
The trade name 'boxwood' originally applied to *Buxus sempervirens* from Europe and eastern Asia, but today covers a wide range of unrelated species. Although 'Maracaibo boxwood' occurs in Cuba and the Dominican Republic, commercial supplies come from Colombia and Venezuela. It is a small tree, producing logs 2.5-3.5m (8-12ft) long and 0.15-0.3m (6-12in) diameter. It is also known as 'Venezuelan boxwood', 'West Indian boxwood', 'Colombian boxwood' (UK); zapatero (Venezuela); palo blanco (Dominican Republic); and pau branco, castelo and zapateiro in Brazil.

Appearance
The timber is cream-white to lemon-yellow, with little difference between heartwood and sapwood, and a high lustre. Blue stain is common in timber stored in humid conditions. It is normally straight grained, featureless, with a very fine and uniform texture.

Properties
Maracaibo weighs 800-900 kg/m³ (50-56 lb/ft³) when seasoned. The timber dries very solowly, with a tendency to split surface check.
It is extremely stable in use. This dense wood is not used for its strength, but it has good steam-bending properties where discolouration is unimportant. It works satisfactorily with both hand and machine tools but there is high resistance in cutting. The wood glues well and can be brought to a very smooth, excellent finish. The sapwood is liable to fungal attack and from the common furniture beetle, but the hardwood is durable.. The sapwood is permeable for preservative treatment

Uses
Maracaibo boxwood is outstanding for its very fine, smooth texture and excellent turning properties. It is used for tool handles, skittles, croquet-mallet heads, textile rollers, silk industry shuttles, spindles, pulley blocks and fancy turnery such as chessmen. The wood is ideal for precision rules, drawing and measuring instruments and engravers' blocks because of its extreme stability. It is also used for some parts of musical instruments such as pianos. It is sliced for decorative cabinet veneers and dyed black for inlay lines and bandings.

		◆			
			◆		◆
◆	◆			◆	

Guaiacum spp. **Family:** *Zygophyllaceae*
LIGNUM VITAE

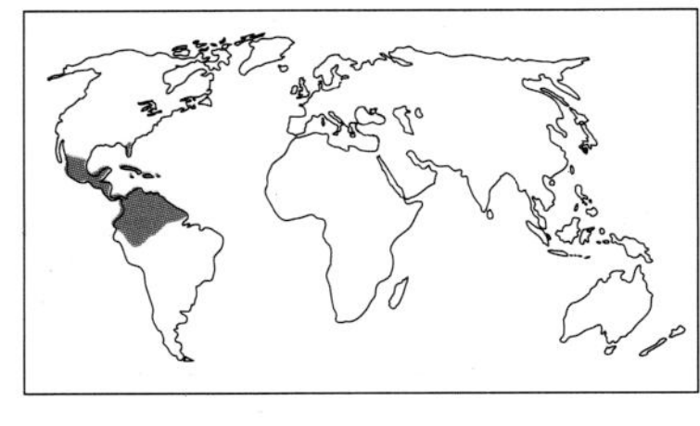

Where it grows
This wood, known as 'The Wood of Life' in the 16th century because its resin was believed to cure diseases, is one of the hardest and heaviest timbers in commerce. It is produced from three species: G.*officinale*, L., known as guayacan (Spain), bois de gaiac (France), guayacan negro and palo santo (Cuba) and ironwood (USA); G.*sanctum*, L., known as guayacan blanco, gaiac femelle or guayacancillo; and G.*guatemalense*, Planch, which occurs in Nicaragua. All three types are exported as lignum vitae. They occur from southern Florida and the Bahamas through Jamaica, Cuba and the West Indies, and from Mexico down through Central America to Colombia and Venezuela. It is a small, slow-growing tree about 9m (30ft) in height, with a diameter of 0.25-0.45m (10-18in).

Appearance
The timber is dark greenish-brown or nearly black, with a closely interlocked grain and a fine, even texture.

Properties
The wood weighs on average about 1230 kg/m^3 (77 lb/ft^3) when seasoned. It dries very slowly, is refractory and liable to check. There is medium movement in use. The wood has outstanding strength properties in all categories, particularly hardness, and has a very high crushing strengh. Unsuitable for bending, it is a very difficult wood to machine, with its very high resistance to cutting. Gluing can be somewhat difficult but the wood polishes well. It is extremely durable and resistant to preservation treatment.

Uses
The self-lubricating properties of lignum vitae, from its high oil content, make it ideal for marine equipment such as bushing blocks and bearings for ships' propeller shafts, pulley sheaves and dead-eyes, and as a replacement for metal thrust bearings in steel and tube works. It is used anywhere where lubrication is impractical or unreliable such as in wheels, guides, rollers and blocks; in the textile industry for cotton gins, polishing sticks and rollers. It is also used in die cutting. Lignum vitae has long been a favourite for wood sculpture and carving as well as an excellent turnery wood for mallet heads, and for bowling 'woods'.

◆	◆	◆	◆	◆	◆

Guarea cedrata **Family:** *Meliaceae*
GUAREA

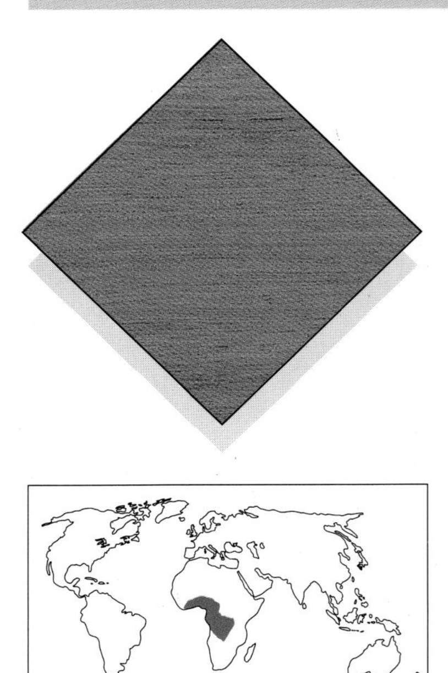

Where it grows
Guarea occurs in tropical West Africa, principally the Ivory Coast and Nigeria. G.*cedrata* produces white guarea, obobonufua (Nigeria); and scented guarea (UK). G.*thompsonii*, Sprauge and Hutch., produces black guarea, obobonekwi, or obobo (Nigeria); bossé (Ivory Coast); diambi (Zaire); ebanghemwa (Cameroon); and divuitii (Gabon). Although the two species are distinguished by the colour of their bark, they are sold under a single name. Both species grow to 50m (160ft), with a diameter of 0.9-1.2m (3-4 ft).

Appearance
The wide sapwood is slightly paler than the heartwood, which is a pale pinkish-brown mahogany colour with straight to interlocked grain. G.*cedrata* produces a mottled or curly figure, and G.*thompsonii* is straighter-grained and plainer in appearance, though both have a fine texture.

Properties
G.*cedrata* weighs 580 kg/m^3 (36 lb/ft^3) and often contains silica and resin, while silica-free G.*thompsonii* weighs about 620 kg/m^3 (39lb/ft^3) when seasoned. It has a lower resin content. They dry fairly rapidly with little tendency to warp, but G.*cedrata* exudes resin, which can mar the appearance, and G.*thompsonii*, tending to split, requires greater care in drying. They are stable in use. Guarea has medium bending strength, low stiffness and resistance to shock loads. G.*cedrata* has medium crushing strength and a good bending rating, while G.*thompsonii* has high crushing strength and a moderate bending rating. Both work fairly easily with hand and machine tools, but are inclined to be woolly; both woods glue well, but care is needed in finishing as resin may exude. Machining dust is highly irritant, and efficient dust extraction is important. It is a very durable timber, extremely resistant to preservative treatment, but the sapwood is permeable.

Uses
Guarea, a mahogany-type timber, is used in furniture for chairs, drawer sides and rails, interior fittings, edge lippings and facings, for high-class joinery, shop fitting, boat building, vehicle construction, floor boards and planking in caravans. It also occurs in sports goods, rifle butts, marine piling, exterior plywood and decorative veneers.

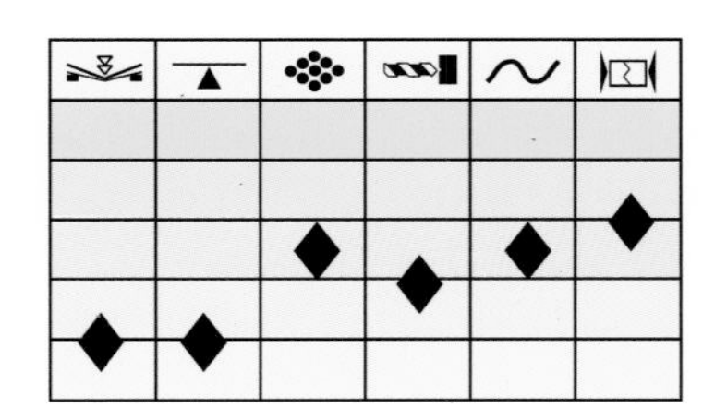

Guibourtia ehie **Family:** *Leguminosae*

OVANGKOL

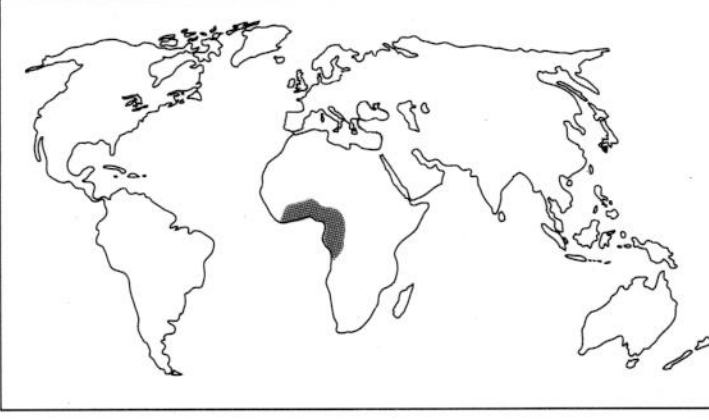

Where it grows
This tall tree occurs in tropical West Africa, mainly the Ivory Coast, where it is also known as amazoué or amazakoué, and Ghana, where it is known as ehie, anokye or hyeduanini. Supplies of Ovangkol also come from southern Nigeria and Gabon. The tree reaches a height of about 30m (100ft), with a diameter of about 0.8m (2½ft). Older stems have a growth of slightly raised horizontal rings, which affect the figure in the log.

Appearance
The heartwood is yellow-brown to deep chocolate-brown with greyish-black stripes. The grain is interlocked and the texture is moderately coarse.

Properties
Average weight is about 800 kg/m³ (50 lb/ft³) when seasoned. The wood dries rapidly and fairly well, but care is needed in kilning thick stock to avoid collapse. There is medium movement in service. The wood has medium bending and crushing strengths and shock resistance; despite low stiffness it has a poor steam-bending rating, and only shallow bends are possible. The wood presents a moderate resistance and blunting to cutting edges because of the silica content, and tools should be kept very sharp. The wood saws slowly but well and the cutting angle should be reduced, especially when planing or moulding quartered stock, to avoid picking up or tearing the interlocked grain. It has good holding properties, glues without difficulty, takes stain and polish well and can be brought to an excellent finish. The heartwood is moderately durable and resistant to preservative treatment but the sapwood is permeable.

Uses
Ovangkol is a very attractive wood suitable for high-class furniture and cabinet making, interior joinery and decorative work. It makes hard-wearing domestic flooring and is used for shop, office and bank fitting. It is an excellent turnery wood for handles and fancy items. Ovangkol logs are rotary cut for plywood faces, and selected logs are sliced for decorative veneers for cabinets, flush doors and panelling.

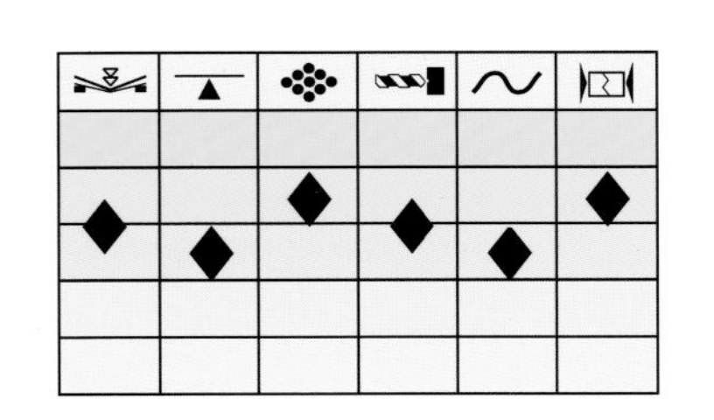

Heritiera spp. **Family:** *Sterculiaceae*

MENGKULANG

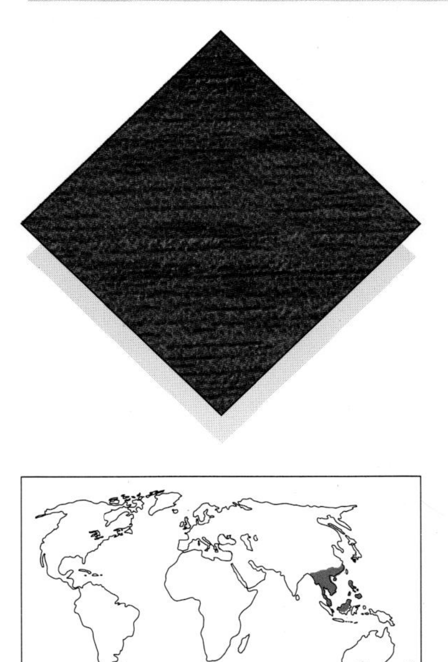

Where it grows
A wide number of species of *Heritiera* occur in the tropics, including Niangon from West Africa and chumprak in Thailand. *H.javanica* is the principal species found in the Philippines. Mengkulang is a Malayan name for half a dozen or more species which occur in south east Asia. Commercial timber is produced from *H.simplicifolia* (Mast.) Kosterm; *H.javanica* (Bl.) Kosterm and *H.borneensis*. They are also known as kembang (Sabah); chumprak or chumprag (Thailand); lumbayao (Philippines); and gisang (Malaya). Trees reach a height of 30-45m (100-150ft), with a diameter of 0.6-1.0m (2-3ft).

Appearance
The sapwood is pale orange, blending into the heartwood, which varies from mid-pinkish-brown to dark red-brown, often with dark streaks on longitudinal surfaces. The grain is interlocked and sometimes irregular, producing a broad striped figure on quartered surfaces, often with conspicuous reddish flecks. The texture is moderately coarse and fairly even.

Properties
The weight varies according to species but averages 640-720 kg/m³ (40-45 lb/ft³) when seasoned, although kembang is slightly lighter. The wood dries rapidly and well, with a tendency to warping or to surface checking in some species. This dense wood is stable in service, possesses medium bending strength, stiffness and resistance to shock loads, high crushing strength, and a very poor steam-bending rating. The timber works moderately well but there is severe blunting of cutting edges, especially saw teeth. It requires pre-boring for nailing, glues well, and with the grain filled can be brought to a good finish. It is perishable and the sapwood is liable to attack by powder post beetle. The heartwood is resistant to preservative treatment, the sapwood moderately resistant.

Uses
Mengkulang is used locally for cabinet and furniture fitments, wheelwrighting, vehicle framing, panelling, sills, sleepers, boat ribs and planking; also for domestic flooring. It also appears in general construction, carpentry and interior joinery. It is rotary cut for plywood for local structural use and sliced for decorative veneers.

Hymenaea courbaril **Family:** *Leguminosae*

COURBARIL

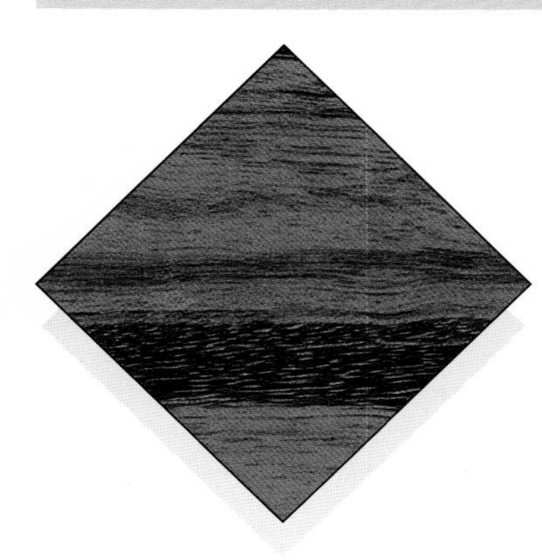

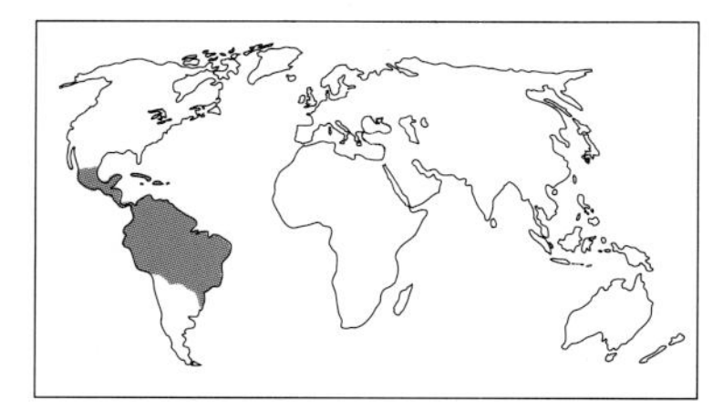

Where it grows

This large, almost evergreen tree grows from southern Mexico through central America and the West Indies, down as far as Brazil, Bolivia and Peru. It is also known as West Indian locust (UK); rode locust (Surinam); jutaby, jatoba or jatai amerelo (Brazil); algarrobo (Puerto Rico); guapinal (Mexico); copal (Equador); and marbre (Guadaloupe). *H.davisii* is locust in Guyana. Trees reach 31-46m (100-150ft), with a diameter from 1.0-1.5m (3-5ft).

Appearance

The heartwood matures to an orange-red to reddish-brown, marked with russet and dark brown streaks. It often has a golden lustre. The grain is usually interlocked with a medium to coarse texture.

Properties

This dense wood weighs about 910 kg/m³ (56 lb/ft³) when seasoned. It is a rather difficult timber to dry, with surface checking, warping and occasionally case-hardening. It is stable in service and very strong in all categories, but has low stiffness which gives it a moderate steam-bending classification. The timber is moderately difficult to work, with a severe blunting effect on tools. It requires pre-boring for nailing, but the wood has good screw-holding properties. It glues well, and can be stained and polished for a very good finish. Courbaril is moderately durable, but non-durable when a high proportion of sapwood is present. It is very resistant to termites and extremely resistant to preservative treatment.

Uses

Courbaril is used locally for general building construction, wheelwrighting, looms, carpentry and joinery. Its high shock resistance makes it suitable for sports goods and striking tool handles. It is also used for high-class cabinets and furniture, interior joinery and especially for flooring and stair treads, because of its great resistance to wear. It is ideal for ships' planking, gear cogs and so on. In boat building it is used for steam-bent parts. Externally it is used for lock gates in areas free from marine borers. Selected logs are sliced for highly decorative veneers suitable for cabinets and architectural panelling.

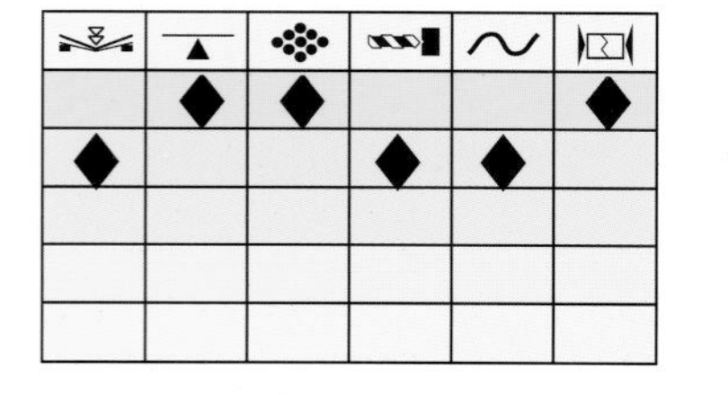

Ilex spp. **Family:** *Aquifoliaceae*

EUROPEAN HOLLY

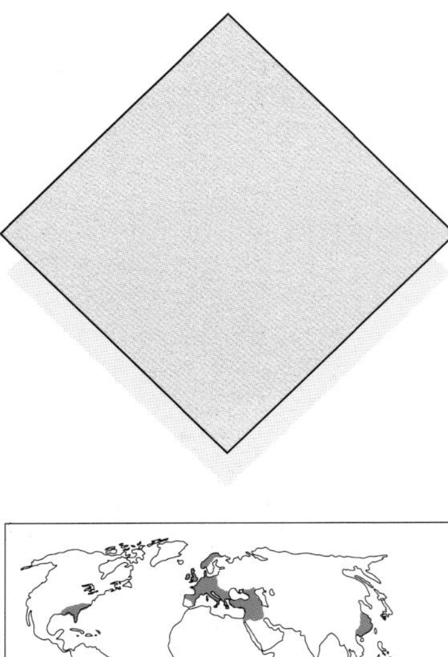

Where it grows

There are about 175 different species of holly, perhaps the whitest known wood. In the UK, *I.aquifolum* has become a hedgerow small tree, about 9m (30ft) tall, with a bole of 3m (10ft) and a diameter of 0.5m (18in). The tree occurs in Europe from Norway, Denmark and Germany down to the Mediterranean and west Asia. It reaches a height of 25m (80ft), with a diameter of about 0.6m (2ft). *I.opaca* produces holly in the USA.

Appearance

The heartwood is white to grey-white, sometimes with a slight greenish-grey cast, with little or no figure. The sapwood is not distinct from the heartwood. The grain tends to be irregular, with a very fine, even texture.

Properties

The weight averages about 780 kg/m³ (49 lb/ft³) when seasoned. Holly is fairly difficult to dry, and it is best to cut the stock into small dimensions then slowly air dry under a weighted-down pile. It is stable in use when dry. This heavy, dense wood is tough in all strength categories but not suitable for steam bending. It has a high resistance to cutting and sawing and a moderate blunting effect on tools, which should be kept very sharp, especially when working with the irregular grain. The wood turns well, requires pre-boring for screwing or nailing, glues easily and can be brought to an excellent, smooth finish. In the round, logs are liable to attack by forest longhorn or Buprestid beetles. The heartwood is perishable and the sapwood is liable to attack by the common furniture beetle, but is permeable for preservation treatment.

Uses

Holly is available in limited quantities, in small dimensions and narrow veneers. It is mainly used as a substitute for boxwood; when dyed black, it is a substitute for ebony for marquetry inlay motifs, lines, bandings and stringings in antique repair and restoration and reproduction furniture. It is excellent for fancy turnery and engraving blocks. It is used for musical instrument parts, piano and organ keys, parts of harpsichords and clavichords, and billiard cue butts.

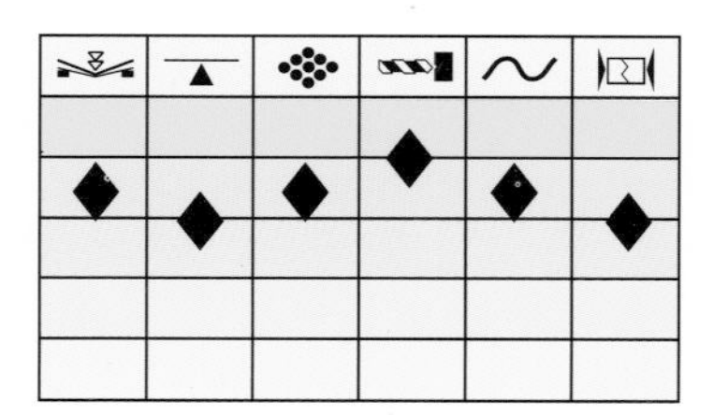

Intsia spp. **Family:** *Leguminosae*
MERBAU

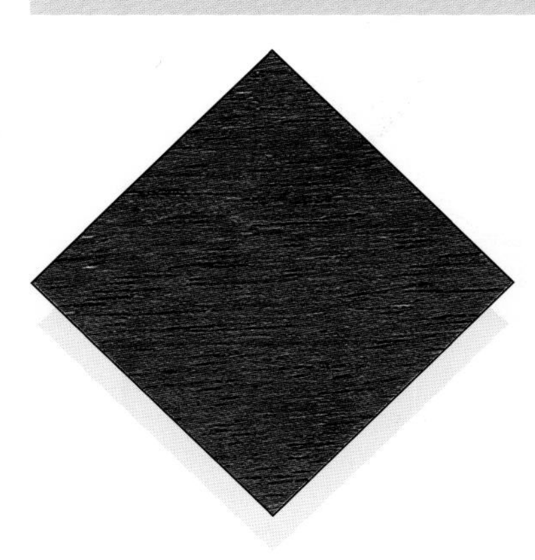

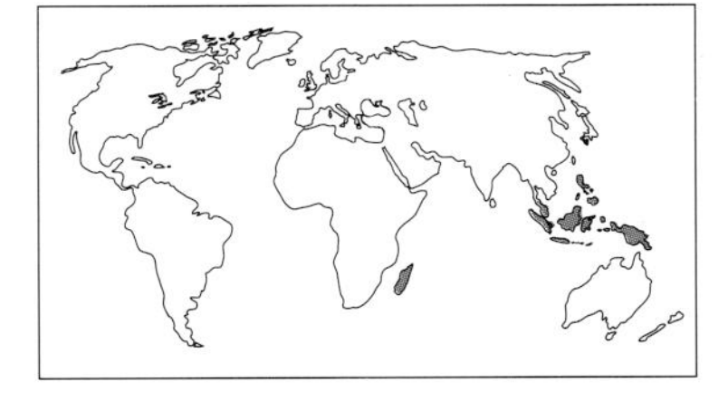

Where it grows
Intsia bijuga (Colebr.) O.Ktze. produces hintzy (Madagascar); ipil (Philippine Islands); and kwila (Papua New Guinea and New Britain): *I.palembanica*, Miq. produces merbau (Malaysia, Indonesia and Papua New Guinea) and mirabow (Sabah). These large trees with short, thick, often fluted boles, reach about 40m (130ft) and a diameter of about 1.8m (6ft).

Appearance
The sapwood is usually about 75mm (3in) wide, and clearly defined from the white-to-pale yellow heartwood, which matures into a medium-to-dark red-brown on exposure. The grain is interlocked and sometimes wavy, which produces a ribbon figure on quartered surfaces. Lighter parenchyma markings provide an attractive figure on tangential surfaces. The texture is rather coarse but even. The timber is liable to blue stain if in contact with iron compounds in damp conditions, and to corrode ferrous metals.

Properties
Average weight is 800 kg/m^3 (50 lb/ft^3) when seasoned. The wood dries fairly rapidly with little degrade, and is stable in service. It has medium bending and crushing strengths, stiffness and resistance to shock loads. It has a moderate steam-bending rating due to gum exudation, and a moderate blunting effect on tools, though gum tends to build up on saws. Pre-boring is necessary for nailing. Care is required in gluing, and the pores require considerable filling to get a good finish. The heartwood is durable and resistant to preservative treatment.

Uses
Merbau is used locally for heavy constructional work, interior joinery, panelling, furniture making, agricultural implements and axe and striking tool handles. The wood has a moderately high resistance to wear and is used for good-quality flooring. When treated it is used for railway sleepers, poles and so on. Selected logs are sliced for decorative veneers for doors and panelling. The pores may contain a yellow dye which can leach out and stain stonework or fabrics unless they are protected.

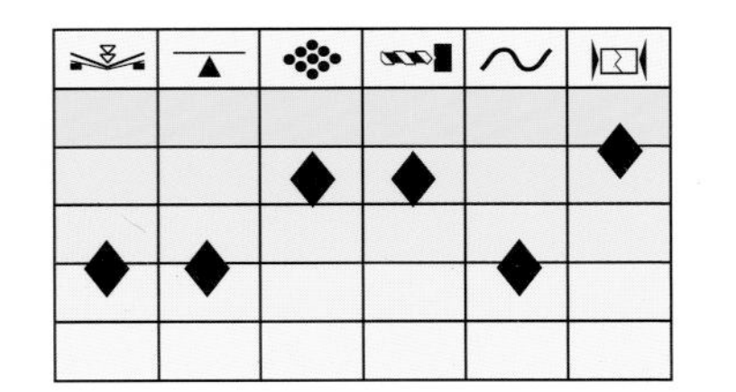

Juglans nigra **Family:** *Juglandaceae*
AMERICAN WALNUT

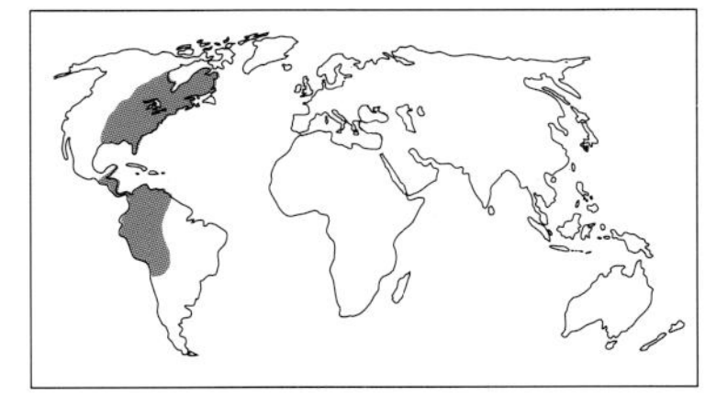

Where it grows
This is one of the true walnut trees, widely distributed throughout North America from southern Ontario in Canada down to Texas, and in the east from Maine to Florida. It is also known as black American walnut, Virginia walnut (UK); walnut, canaletto, black hickory nut or walnut tree (USA); and Canadian walnut (Canada and USA). In favourable conditions the trees reach a height of 30-45m (100-150ft), with a diameter of about 1.2-1.8m (4-6ft).

Appearance
The attractive heartwood matures into a rich dark brown to purplish-black colour. It is usually straight grained, but sometimes wavy or curly. Texture is rather coarse, but uniform.

Properties
The weight is about 640 kg/m^3 (40 lb/ft^3) when seasoned. The wood requires care in drying to avoid checking and degradation. There is small movement in service. Walnut is of medium density, bending and crushing strength, with low stiffness and shock resistance; it has a good steam-bending rating. The timber works well with hand and machine tools, with a moderate blunting effect on cutting edges. It holds nails and screws well, and can be glued satisfactorily. It is a delight to work, takes stain and polish with ease and can be brought to an excellent finish. Walnut is very durable. The sapwood is liable to attack by powder post beetle, the heartwood is resistant to preservative treatment and biodegradation, but the sapwood is permeable.

Uses
All species are extensively used for rifle butts and gunstocks, high-class cabinets and furniture, interior joinery, boat building, musical instruments, clockcases, turnery, carving and wood sculpture. It is a major timber for plywood manufacture, and selected logs are sliced for decorative veneers of all kinds for cabinets and panelling. Related species: *J.neotropica*, *J.columbiensis* and *J.australis* produce South American walnut in Peru, Colombia, Ecuador, Venezuela, Argentina and Mexico (called Peruvian walnut in the UK and USA). *J.sieboldiana* produces Japanese walnut, known as Japanese claro walnut in the UK.

Juglans regia **Family:** *Juglandaceae*
EUROPEAN WALNUT

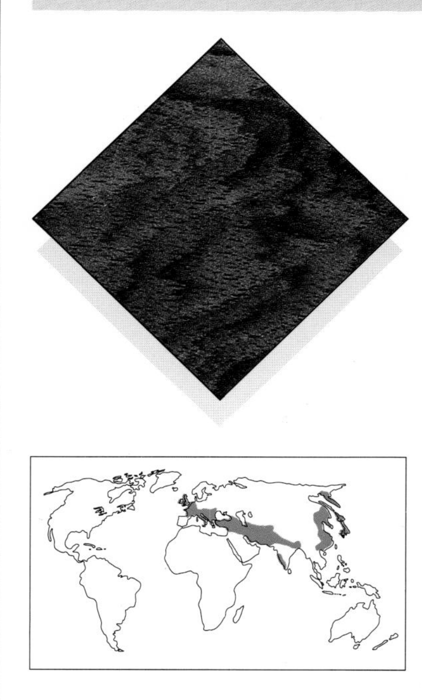

Where it grows
The walnut tree originated in the Himalayas, Iran, Lebanon and Asia Minor, but today commercial supplies come from France, Italy, Turkey, Yugoslavia and south west Asia. It is known as Ancona walnut, Black Sea, Circassian, English, French, Italian or Persian walnut, according to the country of origin. In favourable conditions it attains an average height of 25-30m (80-100ft), a diameter of 0.6-1.5m (2-5ft), with a rugged bole about 6m (20ft) long.

Appearance
The heartwood is usually grey-brown, with irregular dark brown streaks accentuated by a natural wavy grain. This highly figured wood often forms a central core, sharply defined from the remaining plain heartwood; this is more pronounced in Italian than in English walnut, and French is even paler and greyer than English. It is straight to wavy grained, with a rather coarse texture.

Properties
The weight averages about 640 kg/m³ (40 lb/ft³) seasoned. Blue-black stains occur if the wood is in contact with iron compounds in damp conditions. Walnut dries well but with a tendency for checks to occur in thicker material. There is medium shrinkage in service. It has medium bending strength and resistance to shock loads, with a high crushing strength and low stiffness. It has a very good steam-bending rating. The wood works easily and well with hand and machine tools, glues satisfactorily, and can be brought to an excellent finish. The heartwood is moderately durable. The sapwood is liable to attack by powder post beetle and the common furniture beetle. The heartwood is resistant to preservative treatment but the sapwood is permeable.

Uses
Ever since the 'Age of Walnut' this beautiful timber has been used for high-class cabinets and furniture; interior joinery, bank, office and shop fittings. It is also popular for attractive rifle butts and gunstocks, and for all sorts of sports goods. It is a favourite for carving, wood sculpture, turnery and fascias and cappings of expensive cars. Highly decorative figured veneers of stumpwood, crotches and burrs (burls) are used for plywood, doors and panelling.

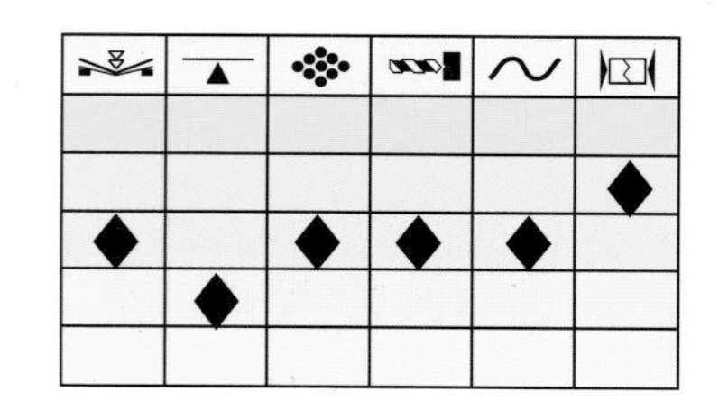

Juniperus spp. **Family:** *Cupressaceae*
'PENCIL CEDAR' (S)

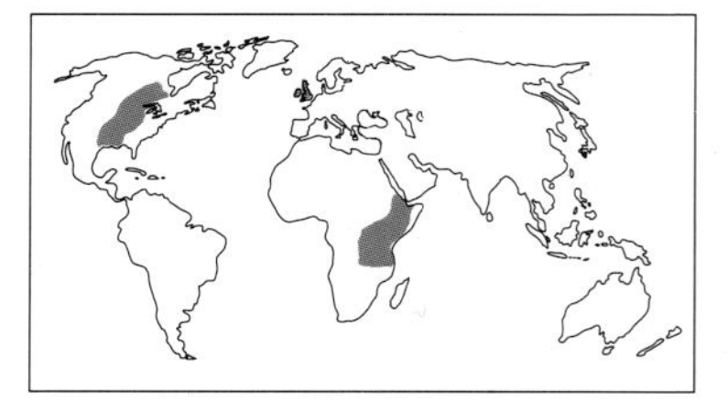

Where it grows
'East African pencil cedar' (*Juniperus procera*) grows in Kenya, Uganda, Tanzania and Abyssinia in the higher forests, reaching a height of 24.4-30.5m (80-100ft) with a diameter of 1.2-3m (4-10ft). 'Virginian pencil cedar' (*J.virginiana*), found in Canada in southern Ontario and in the United States down to eastern Texas, is known as 'Eastern red cedar' (USA) and 'pencil cedar' in Great Britain. It varies in height from 15.3-30.5m (50-100ft) with a diameter of 0.3-1.2m (1-4ft). These species are not true cedars.

Appearance
Both species mature to a uniform reddish-brown. *J.procera* is marked with wide bands of dense tissue; *J.virginiana* with a thin dark line of latewood marking the boundary of each growth ring. Timber from old trees shows uniformly narrow rings, but small trees are knotty. Both species are soft and straight grained, with a fine, even texture, and have the characteristic scent.

Properties
J.procera weighs an average of 580 kg/m³ (36 lb/ft³), while *J.virginiana* weighs 530 kg/m³ (33 lb/ft³) when seasoned. Pencil cedar needs to dry slowly to avoid checking and end splitting. It is stable in service, and possesses medium bending and crushing strengths, very low stiffness and shock resistance, and a poor steam-bending rating. Straight-grained stock works readily with both hand and machine tools, but knotty stock needs care in planing to avoid tearing. It requires pre-boring for nailing, but glues well and can be brought to an excellent finish. The heartwood is durable and extremely resistant to preservative treatment, and the sapwood moderately resistant.

Uses
'Pencil cedar' is the standard timber for making slats for lead pencils. Locally it is used for joinery, furniture and carpentry. It has a very strong but fragrant aroma and is used for linen and blanket chests and interior furniture linings. The leaves and shavings are recovered and distilled for essential oils. Pencil cedar is used in cigar boxes, ship building and coffins.

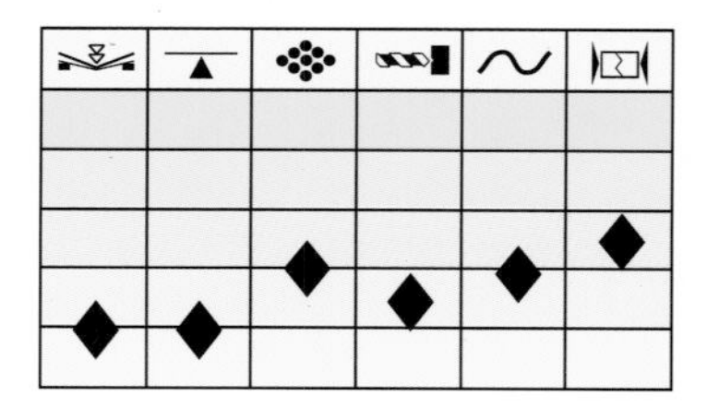

Khaya spp. **Family:** *Meliaceae*
'AFRICAN MAHOGANY'

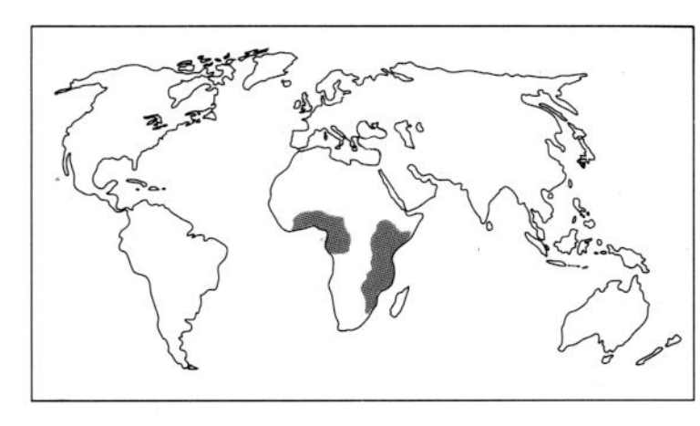

Where it grows
The name 'African mahogany' covers all trees of the Khaya species. The bulk of commercial timber is produced by: *K.ivorensis*, A.Chév., which occurs in the coastal rain forests of West Africa from the Ivory Coast to the Cameroons and Gabon, known as Benin, Degema, Lagos or Nigerian mahogany; *K.anthotheca* (Welw.) C.DC., which is not found in the coastal belt of West Africa and grows in areas of lower rainfall. It occurs in Uganda and Tanzania and is known as krala (Ivory Coast), mangona (Cameroon), munyama (Uganda), mbaua (Mozambique), mbawa (Malawi) and mkangazi (Tanzania). *K.nyasica*, Stapf. ex Bakerf., occurs in Uganda and Tanzania. African mahogany reaches a height of 55-60m (180-200ft) and a diameter of 1.2-1.8m (4-6ft).

Appearance
The tree has a typically reddish-brown heartwood. The grain can be straight but is usually interlocked, producing a striped or roe figure on quartered surfaces.

Properties
The weight of *K.ivorensis* averages about 530 kg/m^3 (33 lb/ft^3); *K.anthotheca*, 540 kg/m^3 (34 lb/ft^3); *K.nyasica*, 590 kg/m^3 (37 lb/ft^3) when seasoned. The wood dries fairly rapidly with little degrade and is stable in use. The timber is of medium density and crushing strength, has a low bending strength, very low stiffness and resistance to shock loads, and a very poor steam-bending rating. Mahogany works easily with both hand and machine tools. Nailing is satisfactory, the wood glues well, and can be stained and polished to an excellent finish. The heartwood is moderately durable and the sapwood, liable to attack by powder post beetle, is resistant to impregnation.

Uses
This is a very important timber for furniture, office desks, cabinet making, shop and bank fitting, and for high-quality joinery for staircases, banisters, handrails and panelling; also for domestic flooring, boat building and vehicle bodies. Logs are rotary cut for plywood and selected logs sliced for decorative veneers for cabinets and panelling.

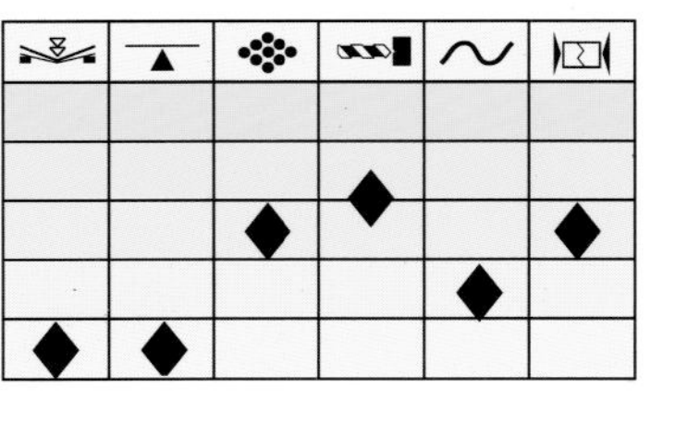

Knightia excelsa **Family:** *Proteaceae*
REWAREWA

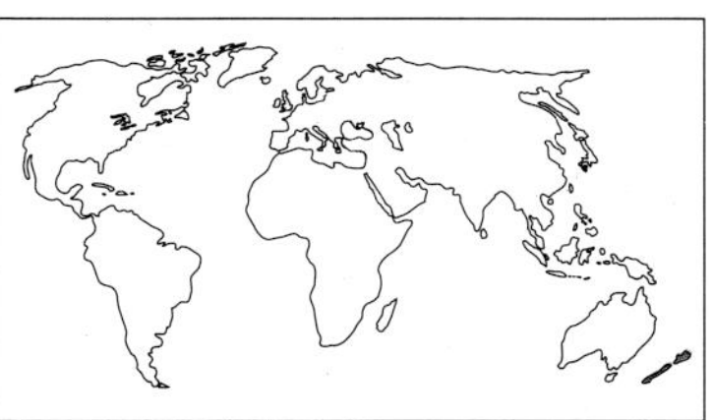

Where it grows
This crooked tree, also known as New Zealand honeysuckle, provides only clean, straight, cylindrical boles of about 6m (20ft), with a diameter of about 0.4m (16in).

Appearance
The heartwood is a deep red colour, strongly marked with dark red-brown ray figure. It is strikingly mottled on the quartered surfaces, with a more subdued growth ring figure on the flat-sawn ones. The grain is irregular and the texture is fine and lustrous.

Properties
The weight is 736 kg/m^3 (46 lb/ft^3) when seasoned. Reaction wood is often present, making it a rather difficult wood to dry. It requires very accurate conversion to avoid serious distortion. There is large movement in service. The wood has medium density and bending strength, but high stiffness, shock resistance and crushing strength. It is not suitable for steam bending. It works readily with both hand and machine tools, with moderately severe blunting of cutting edges. Nailing requires pre-boring but it has good holding properties for screws. It glues satisfactorily. Oil finishes and varnish should be avoided, as they are absorbed to the detriment of the finely marked grain. With care, the wood can be brought to an excellent finish. The heartwood is non-durable and resistant to preservative treatment, but the sapwood is permeable.

Uses
This is one of the most attractive hardwoods in New Zealand, in demand for ornamental cabinet and furniture making. It is also widely used for interior joinery, flooring and staircases, banisters and handrails. It is often used for carved fireplaces because it is incombustible. It is a firm favourite for ornamental turnery for fancy items, handles and bowls. Logs given preservative treatment are used for house blocks, decking, gates, rails, piling and railway sleepers. Selected logs are rotary cut for plywood faces and for corestock of laminated boards, but rewarewa is best when sliced into the most beautiful decorative veneers for cabinets, panelling and marquetry.

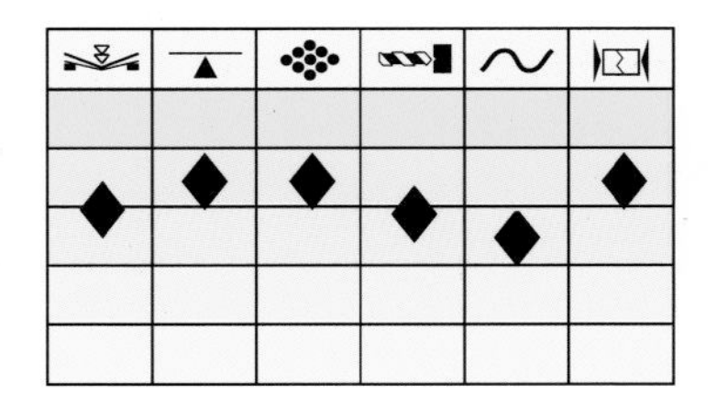

Larix spp. **Family:** *Pinaceae*
LARCH (S)

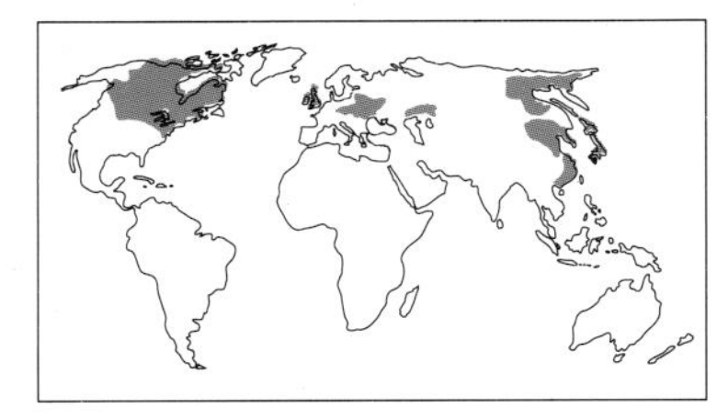

Where it grows
Larch is unusual among the softwoods because it casts its leaves in winter. It occurs in mountainous districts from the Swiss Alps to the Carpathians of Russia. Other varieties occur in Canada, USA and Japan. The chief species are: *L.decidua*, Mill. (*L.europaea DC*), European larch; *L.kaempferi* (Lamb.), Carr., Japanese larch or red larch (UK); *L.eurolepis*, A. Henry Dunkeld larch; *L.laricina* (Du Roi), K.Koch., Tamarak larch (Canada and USA); eastern larch (USA); *L.russica* (Endl.), Sabine,Siberian larch (north eastern Russia); *L.occidentalis*, Nutt., western larch, (British Columbia and western USA). Height averages 30.5-45.7m (100-150ft), with a diameter of 0.9-1.2m (3-4ft).

Appearance
The heartwood is pale reddish-brown to brick red in colour, with clearly marked annual rings; larch has straight grain and a fine uniform texture.

Properties
Weight varies from 480 kg/m^3 (30 lb/ft^3) for the lightest to about 610 kg/m^3 (38 lb/ft^3) for the heaviest species when seasoned. This resinous wood dries fairly rapidly with some distortion but is stable in use. It has medium bending strength, low stiffness, medium crushing strength and resistance to shock loads, and a moderate steam-bending rating. It works fairly readily with both hand and machine tools. It requires pre-boring for nailing and takes stain, paint or varnish satisfactorily for a good finish. The wood is moderately durable and liable to insect attack by pinhole borer beetle, longhorn beetle, sometimes *Sirex*, and the common furniture beetle. The heartwood is resistant to preservative treatment and the sapwood is moderately resistant.

Uses
European larch is chiefly used externally when treated for pit props, stakes, transmission poles and piles. The heartwood is suitable for work where durability is of prime importance, such as boat planking, bridge construction, railway sleepers and exterior joinery in contact with the ground. Selected logs are also rotary cut for plywood faces and sliced to provide very ornamental veneers for flush doors and architectural panelling.

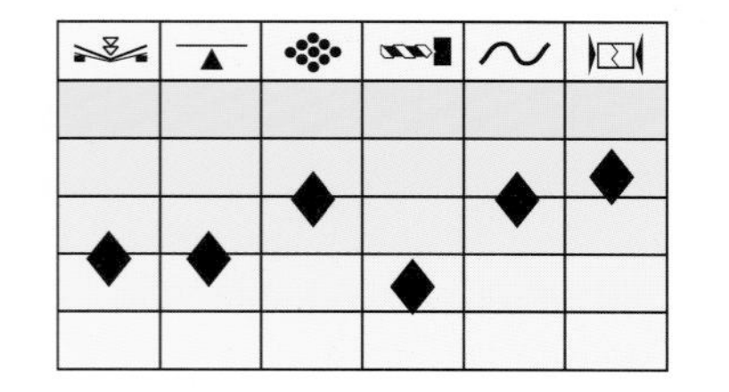

Laurelia sempervirens **Family:** *Monimiaceae*
'CHILEAN LAUREL'

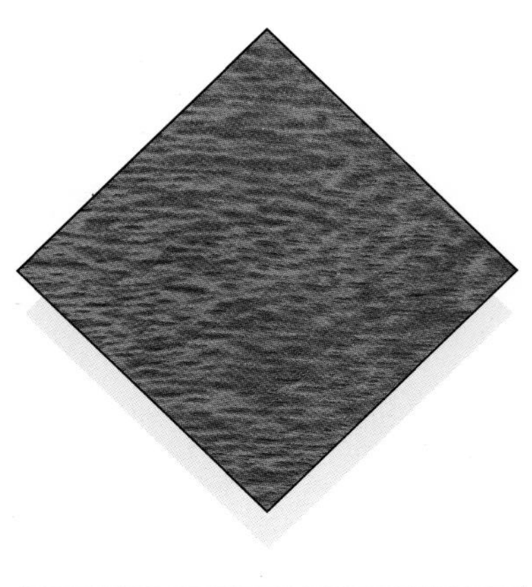

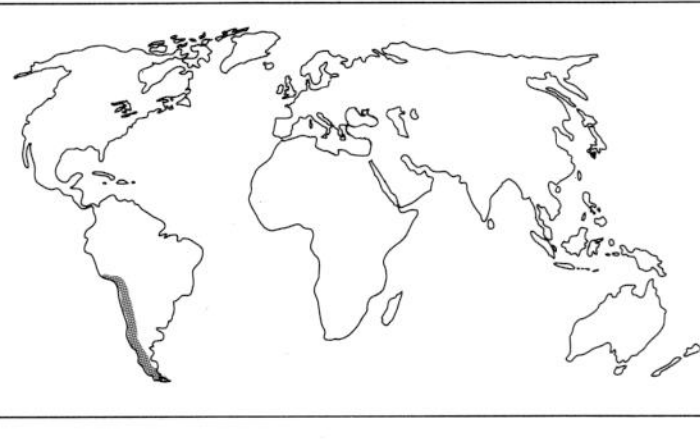

Where it grows
This relatively small tree occurs in small stands in the mixed forests of Chile and is also known as huahuan. It grows to a height of 14-15m (45-50ft) with a diameter of 0.6-0.9m (2-3ft).

Appearance
The sapwood is a uniform greyish-brown, and the heartwood a yellowish-brown with darker streaks of green brown, grey or purple; the timber has straight grain and a moderately fine texture.

Properties
The weight averages about 510 kg/m^3 (32 lb/ft^3) when seasoned. The wood dries fairly rapidly, but with a strong tendency to collapse; it responds well to reconditioning in the kiln. There is large movement in service. The wood has medium bending and crushing strengths with very low stiffness and resistance to shock loads. It has a moderate steam-bending classification; it is suitable for bends of a medium radius, but at the risk of distortion and fracture. The timber works readily with both hand and machine tools, but cutters should be kept very sharp for a clean finish. It can be nailed and glued without problems and can be stained and polished satisfactorily to a good finish. The heartwood is non-durable and moderately resistant to preservation treatment, although the sapwood is permeable.

Uses
This is a very good general-purpose timber suitable for the interior parts of furniture, drawer sides, partitions, shelves, and for edge lipping and facings of panels. It is also widely used for interior joinery, doors, light domestic flooring, light construction, and corestock for plywood and laminated board manufacture. Selected logs are sliced for very attractive decorative veneers for panelling.

Related species: *L.philippiana* Looser. (*L.serrata* Ph.) produces tepa (or laurela) from Chile, used locally for making beehives, boxes and crates, turnery and plywood. It is a major source of wood pulp.

Liquidambar styraciflua **Family:** *Hamamelidaceae*

AMERICAN RED GUM

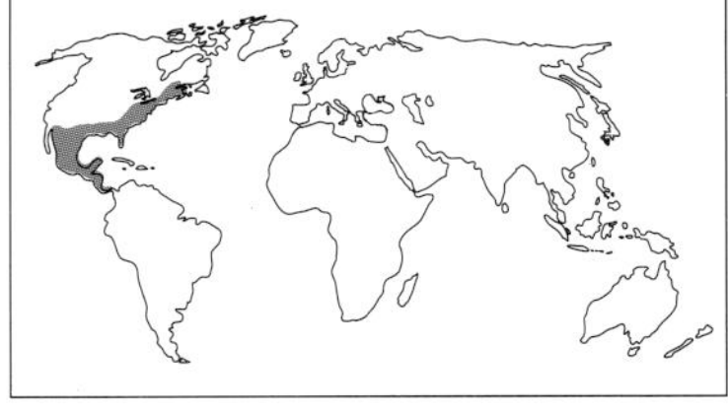

Where it grows
This important timber occurs from New England to Mexico and into Central America. It is also known in the US as gum, sweet gum, bilsted, red gum (heartwood), or sap gum (sapwood). In the UK the names 'hazel pine' (sapwood) and 'satin walnut' (heartwood) can be misleading. The average height is 31-46m (100-150ft), and the diameter 1.0m (3ft).

Appearance
The creamy-white sapwood is sold separately as sap gum, while the heartwood, sold as red gum, varies from pink-brown to a deeper reddish-brown, often with darker streaks with a marbled appearance. The grain is usually irregular; it has a fine, uniform texture and a satin-sheen lustre. It has a very attractive mottled figure on quartered surfaces.

Properties
The weight is about 560 kg/m^3 (35 lb/ft^3) when seasoned. The wood dries fairly rapidly, with a pronounced tendency to warp and twist, and if not dried correctly can suffer acute shrinkage, swelling or splitting. Quarter-sawn stock is less prone to this problem. The material has medium strength in all categories and a very poor steam-bending rating. Red gum works easily with both hand and machine tools, with only slight blunting of cutting edges. The material takes screws and nails without difficulty, is easy to glue, takes stain and polish well and can be brought to an excellent finish. It is non-durable and liable to insect attack. The heartwood is moderately resistant to preservative treatment but the sapwood is permeable.

Uses
American red gum is used either as red (sweet) gum or in sapwood form as sap gum. It is extensively used for furniture, interior trim and fittings, doors and panelling and for interior joinery; also for dry cooperage, packing cases, crates and pallets. Logs are rotary cut for plywood and sliced for decorative veneers. It also produces a vanilla-scented resin formed in the bark by wound simulation, which is a source of storax or styrax, a balsam used for medicine and perfumery.

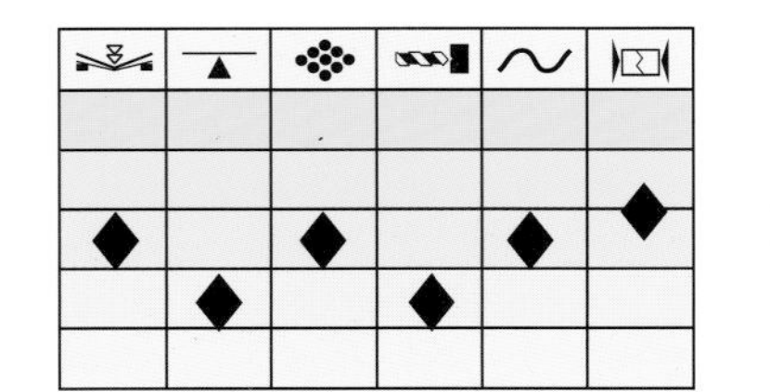

Liriodendron tulipifera **Family:** *Magnoliaceae*

AMERICAN WHITEWOOD

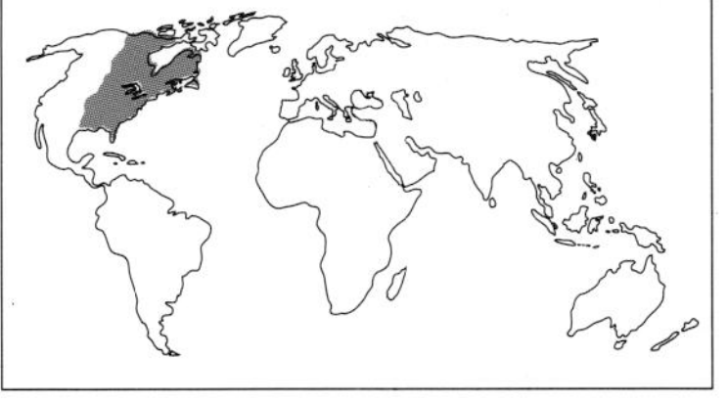

Where it grows
This large 'yellow poplar' tree, not a true poplar (*Populus*), occurs in eastern USA and Canada where it is known as canary wood. In the UK it is canary whitewood; tulip tree in the UK and USA; poplar and yellow poplar in the USA. The magnificent burr (burl) from this tree is marketed as 'green cypress burr' in the UK. It grows to a height of 30-50m (100-150ft) with a long, clear, cylindrical bole of 1.8-2.5m (6-8ft) in diameter.

Appearance
The wide sapwood is creamy-white, the heartwood varies from yellow-brown to pale olive-brown, streaked with olive-green, dark grey, black, pinkish-brown, red, and sometimes mineral stains of steel blue. A wide band of parenchyma shows as pale veins on flat-sawn surfaces. It is straight grained, with a fine, even texture.

Properties
The weight is 450-510 kg/m^3 (28-32 lb/ft^3) when seasoned. The wood kiln dries easily and well and air dries easily with little degrade. There is little movement in service. The timber has medium crushing strength, low bending strength, stiffness and resistance to shock loads, and a medium steam-bending rating. It works easily by hand and machine, has good nailing properties, glued joints hold well, and it can be stained, painted or polished to a good finish. The heartwood is non-durable, and the sapwood is liable to attack by the common furniture beetle. It is moderately resistant to preservative treatment, but the sapwood is permeable.

Uses
American whitewood is a favourite wood for pattern making, sculpture and wood carving. It is used for interior parts of furniture, joinery and doors, for dry cooperage, packaging and pallets, and interior trim for boats. It is extensively used for plywood manufacture and corestock for laminated boards. Sliced for decorative veneers, it appears in cabinets, marquetry and panelling. When treated, it is used for external joinery not in contact with the ground; it is also used for wood pulp and wood flour.

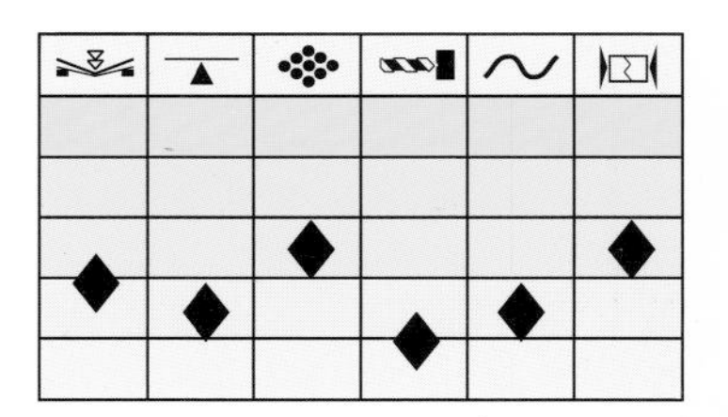

Litsea calicaris **Family:** *Lauraceae*
MANGEAO

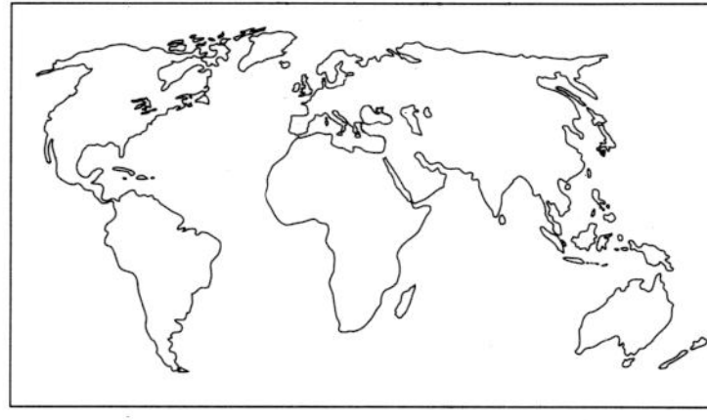

Where it grows
Mangeao – also known as tangeao and one of New Zealand's most valuable trees – is confined to the extreme northern part of the North Island, growing mainly on the coasts of Auckland Province. The average height is 31-43m (100-140ft), and the diameter 1m (3ft).

Appearance
The heartwood is cream to pale brown, with darker vessel lines visible on flat-sawn surfaces. Mangeao is straight grained, with a fine, even texture and a dull lustre.

Properties
Weight averages about 640 kg/m^3 (40 lb/ft^3) when seasoned. The wood dries well without degrade, and there is small movement in service. It is of medium density, bending and crushing strengths, low stiffness and has high resistance to shock loads. It has a very good steam-bending rating. Mangeao works easily with both hand and machine tools, with only a slight blunting effect on tools. It takes nails and screws without difficulty, glues well, stains and polishes satisfactorily and can be brought to an excellent finish. The wood is durable and immune from insect or fungal attack.

Uses
Mangeao is suitable for a variety of purposes which require strength, toughness and elasticity combined with light weight. It is used locally in agricultural machinery, as frameworks requiring transverse strength and durability, and for working parts. Ships' blocks, wheelwright's work, bullock yokes, cooperage hawles for striking tools and axes, picks all use mangeao; also gun and rifle stocks and sporting goods. Other specialized uses include vehicle body building and panelling for railway carriages. The timber is excellent for turnery for ornamental bowls and dowels, and it also makes first-class heavy-duty flooring for public transport, dance halls, gymnasia and factories. It is used for exterior joinery, boat building, bridge decking, stockyard gates, poles, scaffolding, mining timber and pit props and railway sleepers. Logs are rotary cut for plywood in New Zealand and selected logs are sliced and exported in the form of highly decorative veneers.

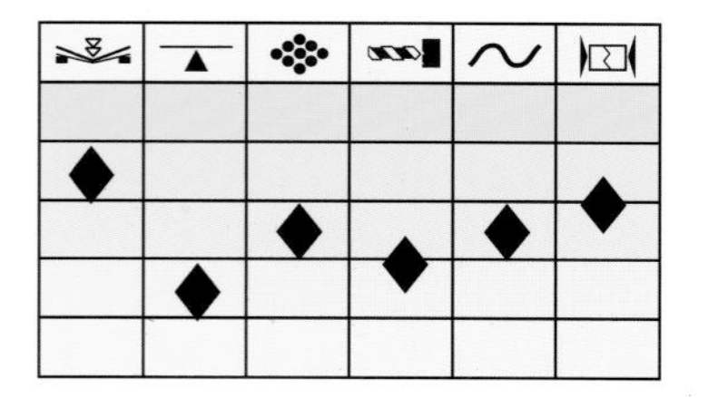

Lophira alata **Family:** *Ochnaceae*
EKKI

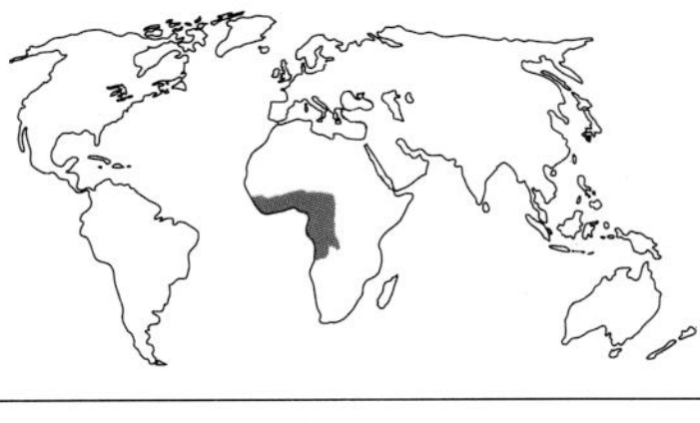

Where it grows
This timber grows from Sierra Leone to Nigeria and the Cameroons in the heavy rain forests and swamps. It is also known as kaku (Ghana), eba (Nigeria), azobé (Ivory Coast), bongossi (Cameroons), akoura (Gabon), hendui (Sierra Leone), and red ironwood (USA). Trees grow to a height of about 55m (180ft), with a diameter of 1.5-1.8m (5-6ft).

Appearance
The pale pink sapwood is clearly demarcated from the valuable heartwood, which varies from dark purple-brown to chocolate-brown with conspicuous white flecked deposits in the pores. The grain is usually interlocked, sometimes irregular, and with a coarse, uneven texture.

Properties
The weight is 950-1100 kg/m^3 (59-69 lb/ft^3) when seasoned. Ekki is extremely refractory and difficult to dry, and shakes badly. Serious degrade is possible in the form of surface checking and end splitting, and the timber needs to be piled in stick with great care. There is large shrinkage in service. Ekki is exceptionally heavy, with very high bending strength and stiffness, high crushing strength and very high resistance to shock loads. It is unsuitable for steam bending and very difficult to work with hand tools, but can be tackled with machine tools. It has a great resistance to cutting and a severe blunting effect on tools. Pre-boring is required for screwing or nailing; gluing has variable results and the grain needs filling for staining and polishing. The timber is very resistant to decay, insect and fungal attack and is one of the most durable woods grown in Africa. It is extremely resistant to preservative treatment.

Uses
Ekki is good for heavy construction, wharves, bridge building and decking, dock and river piling, marine work, sea defences, groynes, jetties, mine shaft guides and so on. Ekki has a high resistance to wear and makes an excellent heavy-duty flooring where surface smoothness is not essential. Acid resistance makes it suitable for filter press plates and frames. It is used as the running track for the rubber-wheeled trains of the Paris Métro.

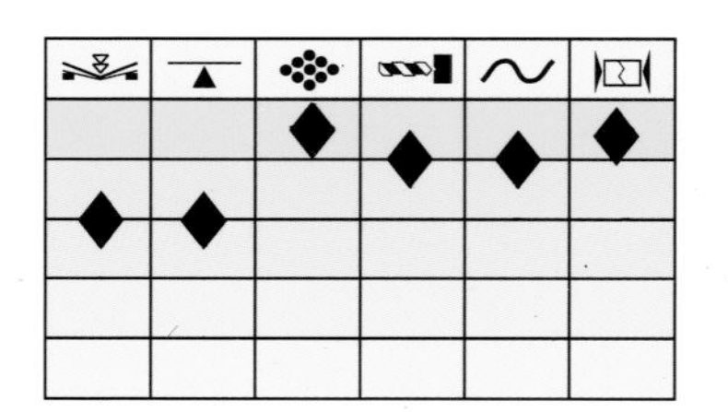

Lovoa trichilioides **Family:** *Meliaceae*
'AFRICAN WALNUT'

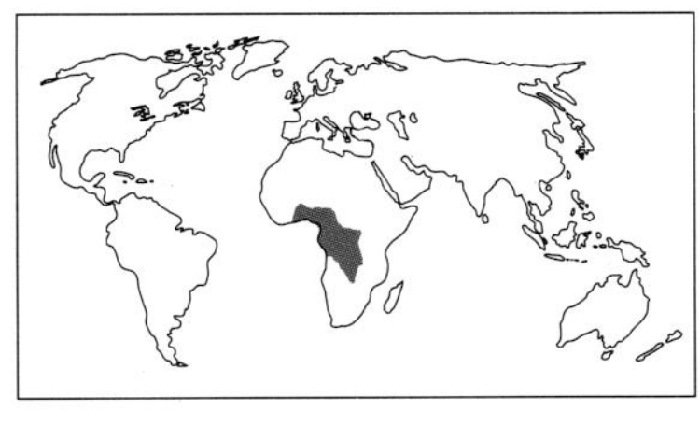

Where it grows
This tall tree, not a true walnut (*Juglans spp.*), occurs in Nigeria, Ghana, Cameroons, Zaire and Gabon. It is also known as 'Benin walnut', 'Nigerian golden walnut', 'Nigerian walnut' and 'Ghana walnut' (UK); apopo and sida (Nigeria); bibolo (Cameroons); dibétou, noyer d'Afrique and noyer de Gabon (Ivory Coast); eyan (Gabon); nvero and embero (Spanish Guinea), alona wood, congowood, lovoa wood and tigerwood (USA); bombolu (Zaire); and dilolo (France). It grows to a height of 45m (150ft) and 1.2m (4ft) in diameter, with a clear cylindrical bole of 18m (60ft).

Appearance
The sapwood is pale brown to buff, clearly demarcated from the golden brown-bronze heartwood, which is marked with black streaks caused by 'gum lines'. The grain is interlocked, sometimes spiral, producing a beautiful ribbon striped figure on quartered surfaces. The texture is moderately fine and lustrous.

Properties
The weight averages 560 kg/m^3 ($35lb/ft^3$) when seasoned. The wood dries fairly well; existing shakes may extend slightly and some distortion may occur, but the degrade is not serious. There is small movement in service. The timber has low bending strength and resistance to shock loads, very low stiffness, medium crushing strength, and a moderate steam-bending classification. It works well with hand or machine tools. The wood tends to split when nailed and needs pre-boring. By sanding and scraping before filling, the surface can be brought to an excellent finish. African walnut is fairly durable, and the sapwood is liable to attack by powder post beetle and dry wood termites in Africa. It is extremely resistant to preservative treatment, the sapwood moderately resistant.

Uses
African walnut is a decorative timber used extensively for furniture and cabinet making, edge lippings and facings, billiard tables and chairs. It also appears in flush doors, decorative interior joinery, panelling and domestic flooring. It is an ideal turnery wood for bowls and lamp-standards; it is used for gun and rifle stocks, car window and door cappings, and sliced for decorative veneers.

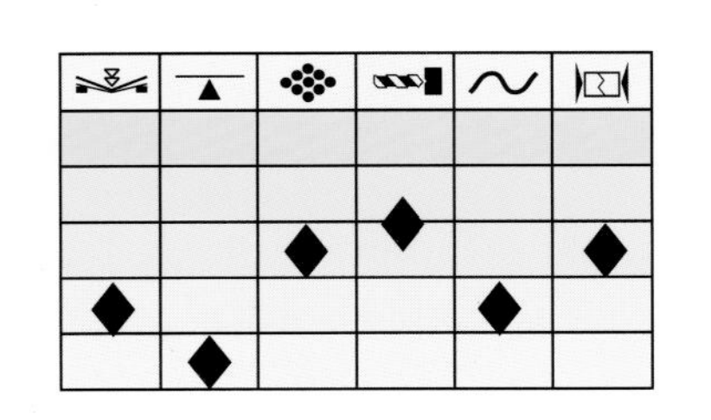

Mansonia altissima **Family:** *Triplochitonaceae*
MANSONIA

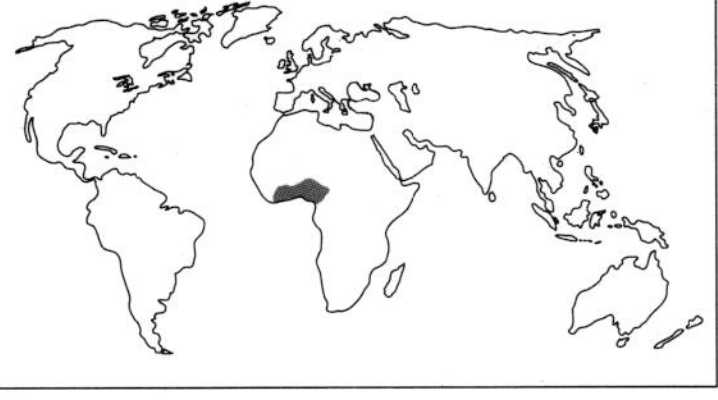

Where it grows
This tree ocurs in tropical West Africa, mainly in southern Nigeria, the Ivory Coast and Ghana. It is also known as bété (Ivory Coast and Cameroon), aprono (Ghana), and ofun (Nigeria). It grows to a height of about 30m (100ft) with a diameter of 0.6-0.8m (2-2½ft).

Appearance
The sapwood is creamy-white and the heartwood varies from dark grey-brown to mauve-brown, with a strong purple tint. It often has lighter or darker bands, but fades with maturity to a dull pale purplish-brown. It is usually straight grained, sometimes interlocked, and the texture is fine, smooth and even.

Properties
The wood weighs on average 590 kg/m^3 (37 lb/ft^3) when seasoned and dries fairly rapidly and well with some distortion in the length. There is medium shrinkage in service. It has high bending strength, low stiffness, medium shock resistance and high crushing strength. It has a good steam-bending rating only if there are no knots. It is best to bend this timber green. It works easily with both hand and machine tools, with only a slight blunting effect on cutters. Mansonia produces a fine machine dust highly irritant to the skin, nose, eyes and throat. Good dust extraction, face masks and barrier creams are necessary. It screws and nails without difficulty, glues well, takes stain and polish and can be brought to an excellent finish. The sapwood is rarely liable to attack by powder post beetle; the heartwood is very durable and extremely resistant to preservation treatment, and the sapwood is permeable.

Uses
Mansonia is an excellent timber for high-quality cabinets and furniture making, chairs, radio and television cabinets, interior joinery, shop and office fitting, pianos and musical instruments. It is a good turnery wood for fancy bowls and the like. It is also used in the car industry for window cappings, dashboards and fascias. Selected logs are sliced for decorative veneers for cabinets, marquetry and panelling.

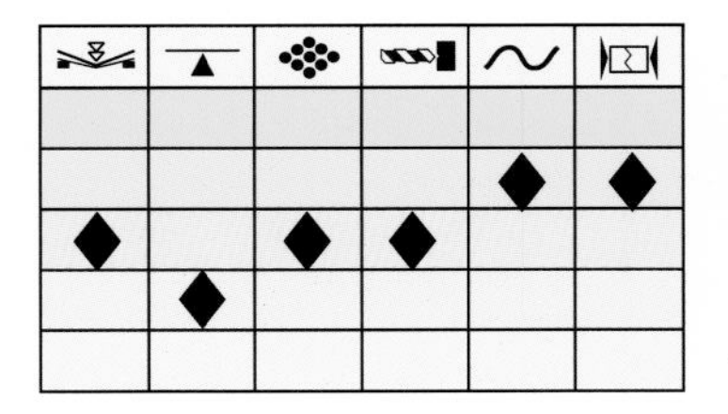

Microberlinia brazzavillensis **Family:** *Leguminosae*
ZEBRANO

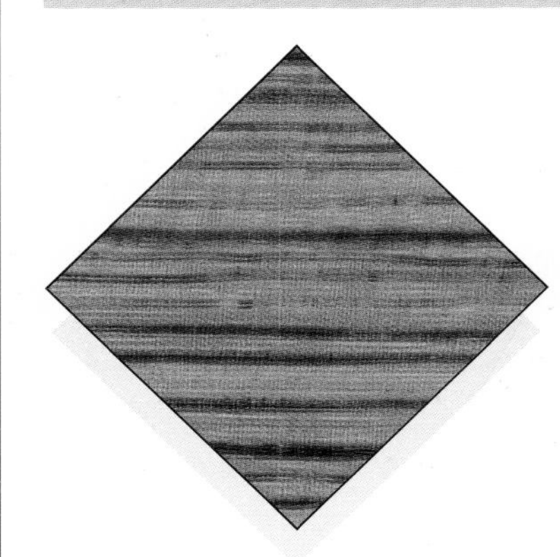

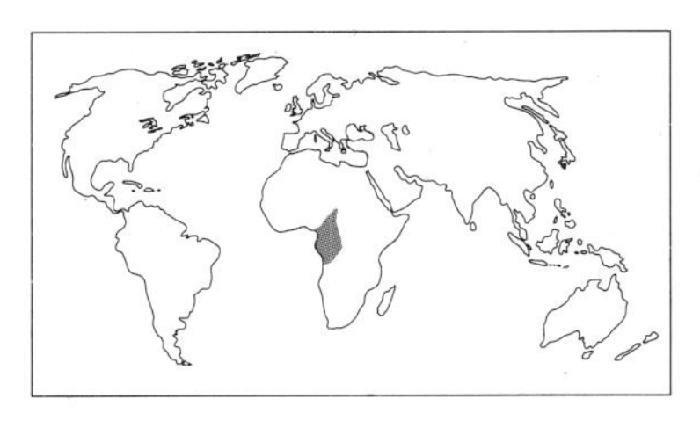

Where it grows
This distinctive, decorative wood comes from two species, chiefly in the Cameroons and Gabon. It is also known as zingana (Cameroons and Gabon); allene, ele and amouk (Cameroon) and zebrawood (UK, USA).

Appearance
On quartered surfaces the heartwood has a light golden-yellow background, with narrow regular and parallel veining of dark brown to almost black producing a zebra-striped appearance. On flat sawn or rotary surfaces this gives the wood a wild streaked pattern. The grain is interlocked or wavy and the texture is coarse with a lustrous surface.

Properties
The weight averages 740 kg/m^3 (46 lb/ft^3) when seasoned. The wood is difficult to dry because of the alternate hard and soft grain and is liable to distortion. It is stable in service. This hard, heavy, dense wood is high in all strength properties and is noted for its very high stiffness. It is not suitable for steam bending. Zebrano can be worked readily with both hand and machine tools, but it is difficult to obtain a smooth finish because of the nature of the grain. Care is required in gluing and a clear filler should be used in polishing when the surface can be brought to an excellent finish. The wood is non-durable and resistant to preservative treatment, but the sapwood is permeable.

Uses
Zebrano is commonly supplied as sliced decorative veneers, usually quarter-cut to avoid buckling because of the alternating hard and soft grain. It is used for inlays and marquetry on cabinets and furniture in the form of cross bandings. Veneers kept in stock tend to buckle, and it is advisable to keep them weighted down. Zebrano in the solid is used for turnery for brushbacks, and small turned items such as fancy handles and tool handles. It is also used in sculpture and wood carving.

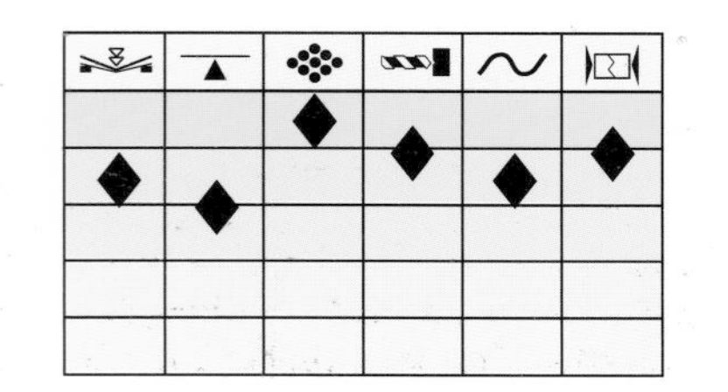

Millettia laurentii **Family:** *Leguminosae*
WENGE and PANGA-PANGA

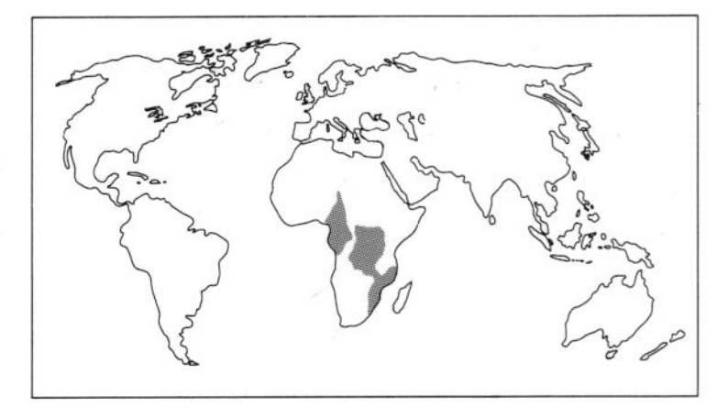

Where it grows
Wengé occurs mainly in Zaire, Cameroon and Gabon. A very closely related species, *Millettia stuhlmannii* Taub., producing panga-panga occurs in Mozambique. They are also known as palissandre du Congo (Congo); dikela, mibotu, bokonge and tshikalakala (Zaire); awong (Cameroon); nson-so (Gabon); and panga-panga (Mozambique). Trees reach about 18m (60ft) with a diameter of 0.6m (2ft).

Appearance
The sapwood is whitish, clearly defined from the heartwood; this is dark brown, with close black veining and alternate closely spaced whitish bands of light and dark parenchyma tissue, which produces a very decorative figure. The wood is straight grained and with an irregular coarse texture.

Properties
Wengé weighs 830-1000 kg/m^3 (52-62 lb/ft^3), and Panga-panga weighs 800 kg/m^3 (50 lb/ft^3) when seasoned. Both dry very slowly and require care to avoid surface checking, but generally the degrade is minimal. The wood is stable in service. Wengé has high bending strength and resistance to shock loads, and is especially noted for its shock resistance and medium stiffness. It has a poor steam-bending rating, but high resistance to abrasion. This durable wood works readily with both hand and machine tools, but cutters should be kept very sharp. It requires pre-boring for nailing. Resin cells in the wood structure can interfere with gluing and polishing, but filling is the answer for a very good finish. The wood is durable and extremely resistant to preservative treatment.

Uses
Wengé has a high natural resistance to abrasion, and is thus excellent for flooring for public buildings, or where there is heavy pedestrian traffic. The dark chocolate-brown makes for a dark floor, but this is not a disadvantage for certain types of hotel, showroom and boardroom. It is also used for all forms of interior and exterior joinery and general construction work. It makes a very good turnery wood and is ideal for wood sculpture. Selected logs are sliced for decorative veneers for cabinets and architectural panelling.

Mitragyna ciliata **Family:** *Rubiaceae*
ABURA

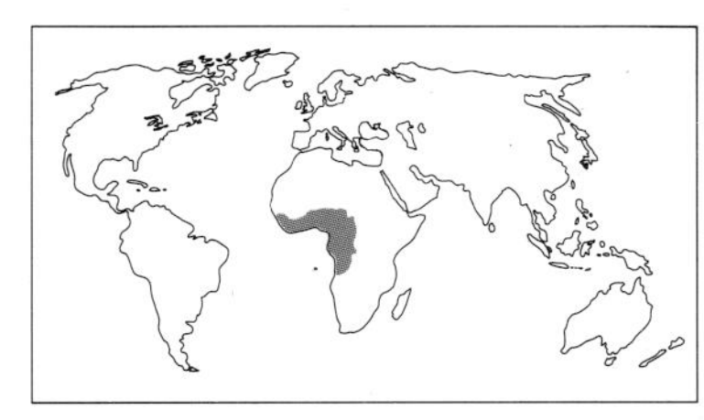

Where it grows
The tree occurs in the wet coastal swamp forests of tropical West Africa from Sierra Leone and Liberia to the Cameroons and Gabon, where it grows to a height of over 30m (100ft) with a diameter of 1.0-1.5m (3-5ft). Related species; M.*stipulosa* and M.*rubrostipulara* (Uganda) are sold as nzingu.

Appearance
The rather wide, plain sapwood is not clearly defined from the heartwood, which tones from a pale yellow to pinkish-brown to an orange-red to light brown. The irregular grey-brown spongy heart may have occasional streaks of dark red-brown in large logs and can give the appearance of stained sapwood. Grain is mostly straight, sometimes interlocked and occasionally spiral. The texture is moderately fine and very even.

Properties
Weight varies from 460-690 kg/m^3 (29-43 lb/ft^3) but averages 560 kg/m^3 (35 lb/ft^3) when seasoned. Abura air and kiln dries rapidly and very well. The wood is very stable with small movement in service when dry. It has very low stiffness, medium crushing strength and low shock resistance, with a very poor steam-bending classification. It works well and cleanly with hand and machine tools, but is sometimes siliceous and the blunting effect on cutting edges can be moderate or severe. Machining is satisfactory but cutting edges must be kept sharp to prevent a woolly surface. The timber glues well, stains easily and can be brought to a very good finish. Pre-boring is advised for nailing. It is perishable and liable to insect attack, but the sapwood is permeable.

Uses
Abura is one of the best West African timbers for mouldings. It is extensively used for interior joinery and furniture framings, edge lippings and drawer sides; its abrasive qualities make it ideal for decorative, hard-wearing flooring. Specialist uses include pattern making and vehicle bodywork; it is used for oil vats and laboratory fittings, and battery and accumulator boxes because it is acid-resistant. It is rotary cut for plywood and sliced for decorative veneers suitable for cabinets and panelling.

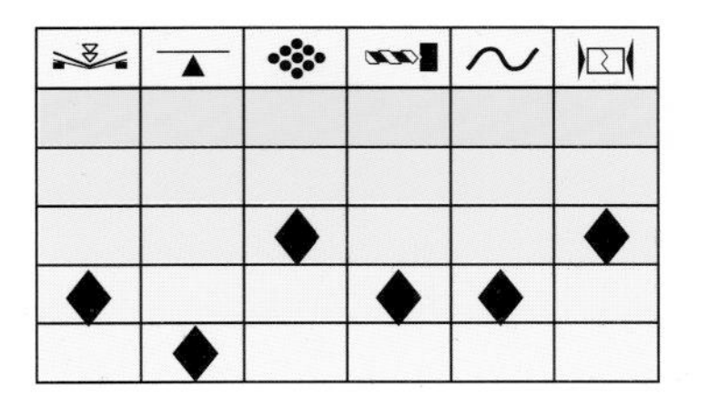

Nauclea diderrichii **Family:** *Rubiaceae*
OPEPE

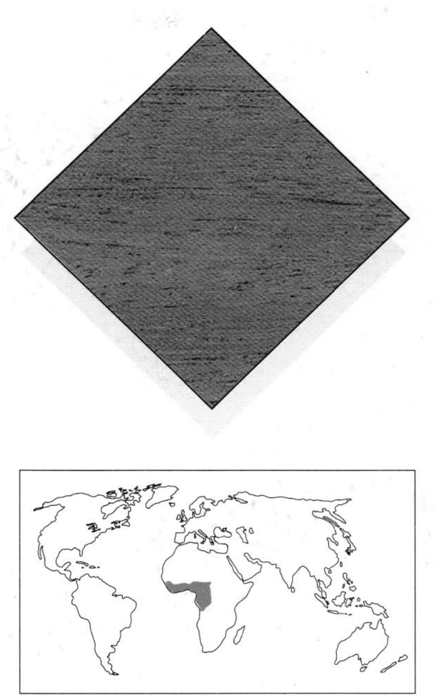

Where it grows
This tree occurs in the equatorial forests of West Africa – Guinea, Liberia, Ivory Coast, Ghana, Nigeria and the Cameroons. It is also known as kusia, kusiaba (Ghana); badi (Ivory Coast); and bilinga (Gabon). It reaches a height of about 50m (160ft), generally without buttresses, with a diameter of 1.5m (5ft).

Appearance
The creamy-pink sapwood, clearly demarcated from the heartwood, measures about 50mm (2in) in width. The heartwood is a distinctive golden-yellow maturing to a uniform orange-brown colour. The timber is sometimes straight grained, but usually interlocked or irregular, with a striped or roll figure on quartered surfaces. The texture is coarse.

Properties
The weight averages about 740 kg/m^3 (46 lb/ft^3) when seasoned. Quarter-sawn material dries fairly rapidly, with very little checking or distortion. Flat-sawn timber is much more refractory, and considerable checking and splitting may occur with serious distortion. There is small movement in service. Opepe has medium bending strength and stiffness, high crushing strength and low shock resistance, with a poor steam-bending classification. The wood works moderately easily with both hand and machine tools; quartered surfaces require a reduced cutting angle for planing. It also requires pre-boring for nailing, but it glues well, and when the grain is filled it can be brought to an excellent finish. The heartwood has a high resistance to marine borers and fungi. The sapwood is liable to insect attack but is permeable for preservation treatment

Uses
Opepe is particularly suitable for use in large sizes for exterior construction, piling and decking in wharves and jetties, docks and marine work. It is good for boat building – except for bent parts – exterior joinery, railway sleepers, wagon bottoms, sills and thresholds. It is also used for interior joinery, furniture and cabinets and shop fitting, and makes excellent domestic flooring. It is a good turnery wood. Selected logs are sliced for decorative veneers.

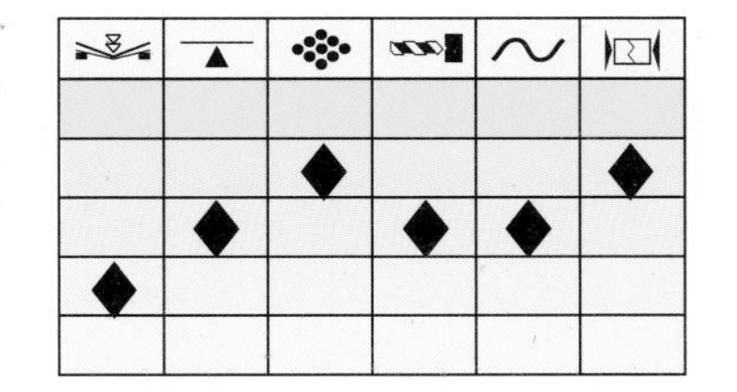

Nesogordonia papaverifera **Family:** *Tiliaceae*
DANTA

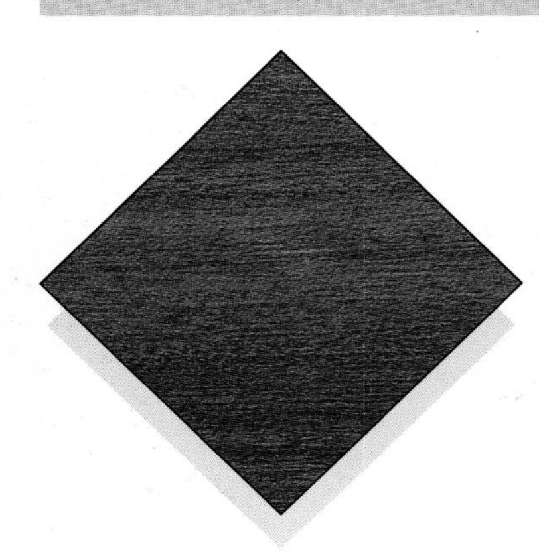

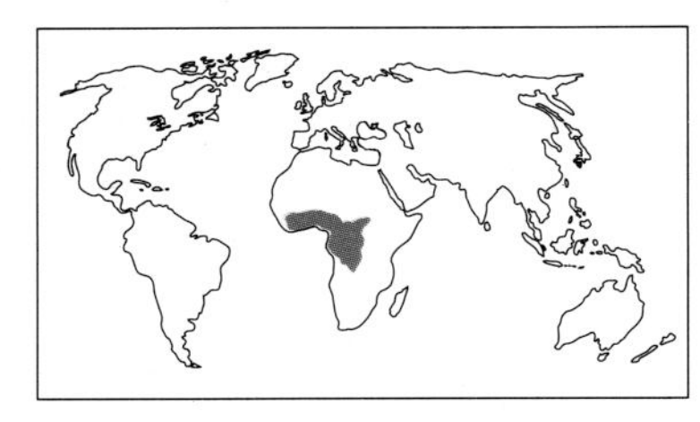

Where it grows
Danta occurs in the mixed deciduous forests of southern Nigeria, the Ivory Coast and Ghana. It grows to a height of 27-30m (90-100ft) above a sharp buttress, and with a diameter of 0.6-0.8m (2-2½ft). It is also known as kotibé (Ivory Coast); otutu (Nigeria); olborbora (Gabon); ovoué (Cameroons); eprou (Ghana); tsanya (Zaire).

Appearance
The sapwood is very pale brown with a pinkish tinge, clearly defined from the heartwood, which is a dark reddish-brown mahogany colour. It has a narrowly interlocked grain, which produces a striped figure when cut on the quarter. Sometimes small sound pin knots, or dark streaks of scar tissue can spoil the appearance and it has a greasy feel. The texture is fine.

Properties
The weight averages about 740 kg/m³ (46 lb/ft³), when seasoned. The wood dries well but rather slowly, with little degrade. There is medium movement in service. It has high bending and crushing strengths, low stiffness and medium resistance to shock loads, with a moderate steam-bending rating. The wood works easily with both hand and machine tools, with a tendency to pick up on quartered surfaces. It requires pre-boring for screwing or nailing but glues well and can be brought to an excellent finish. It is liable to attack by powder post beetle, and the heartwood is durable to marine borers.

Uses
Danta is a very strong and elastic timber. It is widely used for furniture, cabinet making, interior and exterior joinery, shop fitting and bench tops. It has excellent resistance to abrasion and is ideal for decorative flooring. Specialist uses include etching timber for graphic art; a turnery wood for tool handles; stylish rifle and gun stocks; strong framing for truck, coach and wagon carriage vehicle body work; and boat building bent work. When treated with preservative treatment, it is used for telegraph cross arms and railway sleepers. Selected logs are sliced for decorative veneers for cabinets and panels.

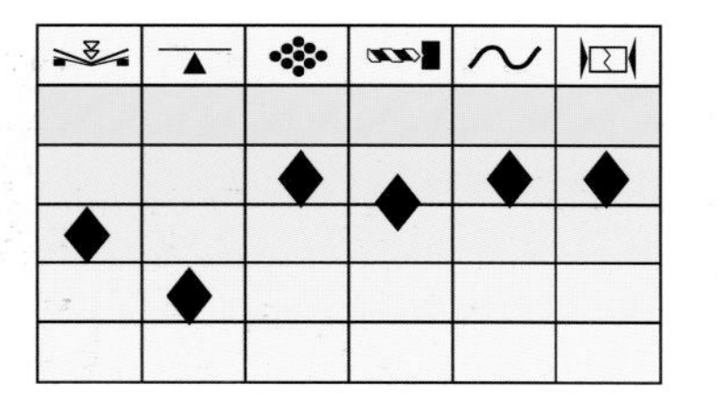

Nothofagus cunninghamii **Family:** *Fagaceae*
'TASMANIAN MYRTLE'

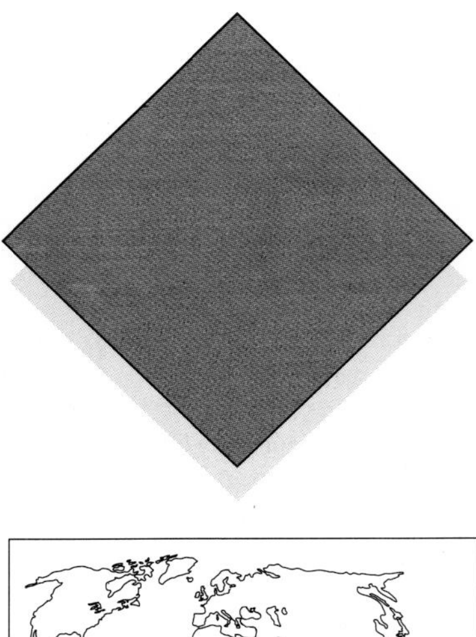

Where it grows
This Australian timber, closely related to northern hemisphere beech, (*Fagus spp*), occurs in Tasmania and Victoria. It reaches a height of 30-40m (100-130ft), occasionally up to 60m (200ft), with a diameter of 0.6-1.0m (2-3ft). It is known as 'myrtle beech', 'myrtle' and 'Tasmanian beech', but it is not a true myrtle or beech.

Appearance
The pink to reddish-brown inner heartwood is separated from the narrow white sapwood by a zone of intermediate colour. The grain is straight to slightly interlocked, ofter wavy, with a very fine, uniform even texture.

Properties
The weight averages 720 kg/m³ (45 lb/ft³) when seasoned. The outer zone lighter wood dries readily and well, but the darker inner zone heartwood is liable to honeycombing, severe internal checking and collapse. This can be restored by reconditioning. There is small movement in service. The wood has medium bending strength and stiffness, high crushing strength and low resistance to shock loads, with a good steam-bending classification. It works readily with both hand and machine tools, but has a moderate blunting effect on cutting edges. It has good holding properties for nailing or screwing, glues satisfactorily, and can be stained and polished to a good finish. The heartwood is non-durable and the sapwood is liable to attack by powder post beetle but is permeable for preservation treatment.

Uses
This very versatile timber is used locally for similar purposes to European beech. It appears extensively in cabinets and furniture, interior joinery and mouldings; it is very popular for domestic flooring and parquetry blocks. As a turnery wood it is used for brushbacks, bobbins and handles; also for shoe heels and food containers, butter boxes and so on. It is also used for motor vehicle body work and, when treated, for exterior heavy construction and joinery, bridge and wharf decking. It is rotary cut for plywood and laminated work, and sliced for decorative veneers for cabinets and doors.

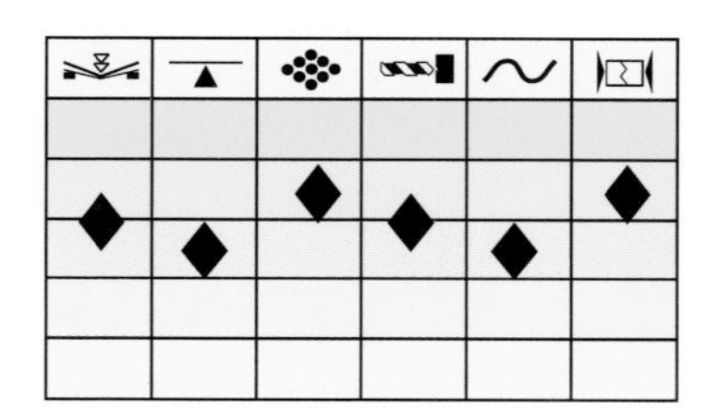

Nothofagus menziesii **Family:** *Fagaceae*
'SILVER BEECH'

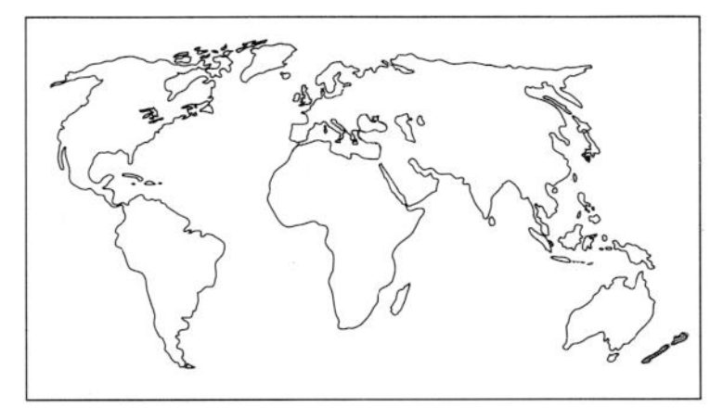

Where it grows
These New Zealand trees grow to about 30m (100ft) with a diameter of 0.6-1.5m (2-5ft). They are also known as 'Southland beech' (*N.menziesii*), 'red beech' (*N.fusca*), and 'hard beech' or clinker beech (*N.truncata*). They are not true beeches.

Appearance
The inner heartwood is a uniform pink-brown; there is a narrow sapwood and an intermediate zone of salmon pink false heartwood (commercially regarded as sapwood) separating them. It is mostly straight grained, sometimes curly, with a fine, even texture.

Properties
'Silver beech' weighs 530-740 kg/m³ (33-46 lb/ft³); 'Red beech' weighs 710 kg/m³ (44 lb/ft³) and 'hard beech' weighs 770 kg/m³ (48 lb/ft³) when seasoned. The wood dries fairly easily; there is a tendency to end splitting, but as a general rule distortion is comparatively slight. There is small movement in service. These woods have medium bending and crushing strengths, low stiffness and resistance to shock loads, and a good steam-bending classification. 'Silver beech' and 'red beech' work easily with both hand and machine tools, except where irregular grain is present on quartered stock, where a reduction in the cutting angle is recommended. Silica in the ray cells of 'hard beech' causes severe dulling of tools. They stain and glue well, and can be brought to a good finish. 'Silver beech' is non-durable, but 'red' and 'hard' beech are both durable. They are liable to attack by the common furniture beetle and powder post beetle. They are all extremely resistant to preservative treatment.

Uses
These timbers are used in New Zealand for cabinets and furniture, interior and exterior joinery and mouldings. Their abrasive qualities make them suitable for domestic flooring. They are good turnery woods used for tool handles, bobbins, shoe heels and brushware, and have many specialist uses such as boat building, building construction, food containers, butter boxes, laundry boxes and vehicle body work. Logs are rotary cut for plywood and sliced for decorative veneers for furniture and panelling.

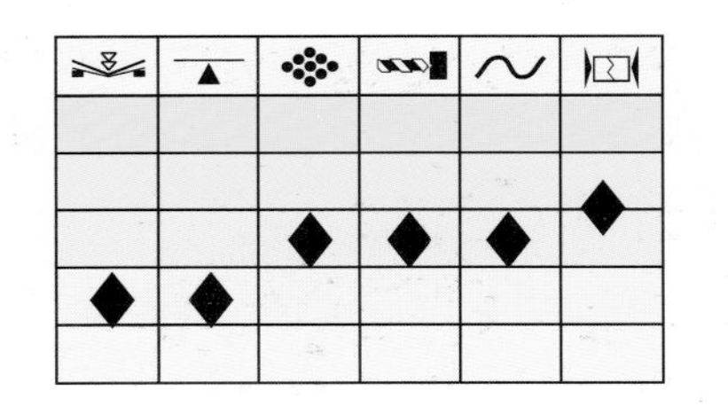

Nothofagus procera **Family:** *Fagaceae*
RAULI

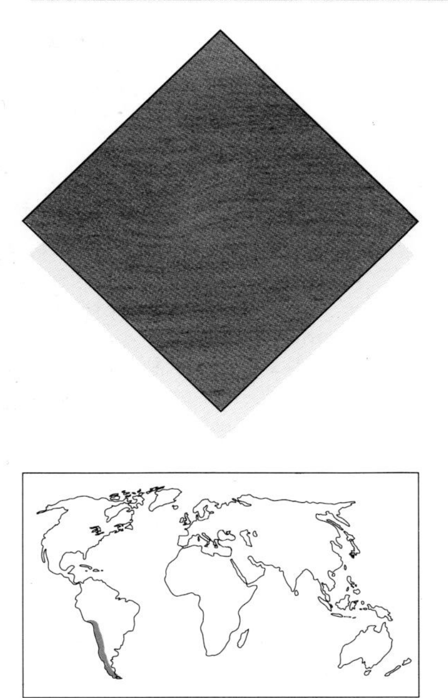

Where it grows
This once-abundant species, found in Chile, has been depleted by heavy felling in recent years. It is closely related to 'Tasmanian myrtle' and New Zealand 'silver', 'red' and 'hard' beech, although it is not a true beech. It is known as 'Chilean beech' or 'South American beech'. The trees grow to a height of 40m (130ft), with a diameter of about 0.8m (2½ft).

Appearance
The heartwood is a uniform reddish-brown to bright cherry red, and does not possess the prominent fleck figure, produced by rays on radial surfaces, of the true beeches. It sometimes shows a pore ring on the tangential face. It is straight grained, with a fine and uniform texture.

Properties
The average weight is about 540 kg/m³ (34 lb/ft³) when seasoned. Rauli dries rather slowly but with little degrade, and there is small movement in service. The timber is of medium density, bending and crushing strengths, with very low stiffness and low resistance to shock loads. It has a moderate steam-bending classification, but cannot be bent if there are pin knots present. It is not suitable for most types of bends required for furniture. Rauli is a pleasure to work with both hand and machine tools, and finishes cleanly like a very mild European beech. It has good nail- and screw-holding properties, can be glued easily and well, takes stain and polish satisfactorily and can be brought to an excellent finish. The heartwood is durable and the sapwood is liable to attack by powder post beetle. The timber is moderately resistant to preservation treatment.

Uses
Rauli is extensively used in Chile for cabinet work and furniture, interior trim, joinery, doors, window frames and domestic flooring. It is used in the UK as a substitute for beech (which is a third heavier), where strength and hardness are unimportant, for jigs, pattern making and framing for vehicle bodies. It is rotary cut for plywood in Chile but not exported. Selected are logs sliced for decorative veneers for cabinets, doors and panelling for export.

Ochroma pyramidale **Family:** *Bombacaceae*
BALSA

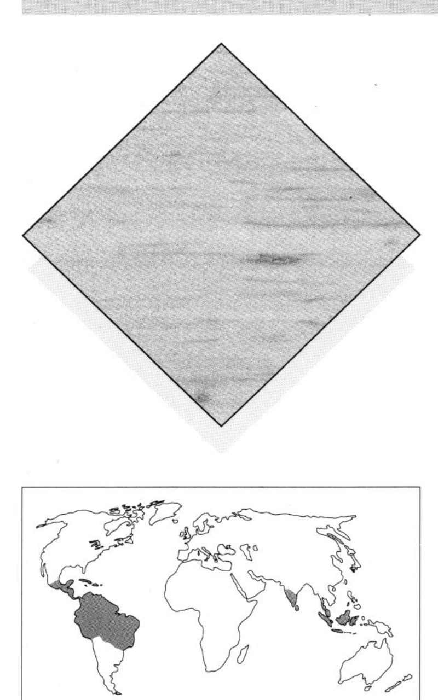

Where it grows
The word 'balsa' (Spanish for 'raft') was given to the wood by early colonists when they found the Indians using it to make rafts. It occurs from Cuba to Trinidad and from southern Mexico through Central America to Brazil. The bulk of the world's lightest wood comes from Ecuador. Supplies also come from India and Indonesia where it has been planted. The trees grow rapidly to a height of 20m (70ft) with a diameter of about 0.6m (2ft). They reach maturity in 12-15 years, then deteriorate rapidly.

Appearance
The bulk of commercial timber is the sapwood, which is white to oatmeal in colour, often with a pinkish or yellow tinge. The central core of heartwood is pale brown. It is straight grained, the texture fine, even and lustrous.

Properties
The weight ranges between 80 kg/m^3 (5 lb/ft^3) to 250 kg/m^3 (16 lb/ft^3), averaging 160 kg/m^3 (10 lb/ft^3) when seasoned. It is a very difficult wood to dry and should be converted immediately after it is felled. Kiln drying of converted stock is much preferred to air drying to mimimize splitting and warping. It is stable in use. Balsa is the weakest of all commercial timbers in all categories, and has a very poor steam-bending rating. It is easy to work with both hand and machine thin-edged tools if they are kept sharp. Gluing is the best way to fasten balsa; it can be stained and polished but is very absorbent. The timber is perishable and liable to insect attack, but is permeable for preservation treatment.

Uses
Balsa is highly valued for heat insulation in refrigerated ships, cold stores and so on; for buoyancy in life-belts, rafts, floats, buoys and water sports equipment. It is valued for its resilience in protective packaging; its lightness in model making and theatre props; and for its sound and vibration insulation. It is used for corestock in lightweight metal-faced sandwich construction sheets for laminated aircraft floors and partitions.

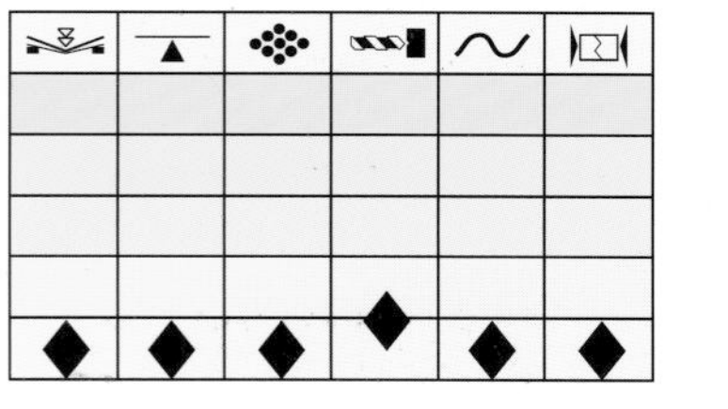

Ocotea bullata **Family:** *Lauraceae*
STINKWOOD

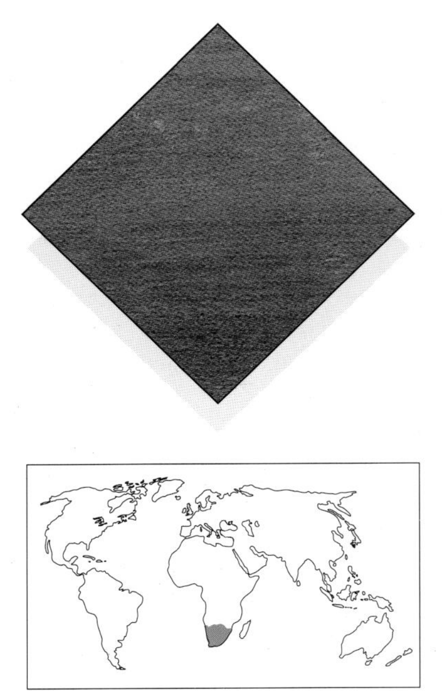

Where it grows
Stinkwood occurs in South Africa from the forested country of the Cape Peninsula, northward to Natal and eastern Transvaal. It is also known as Cape olive, Cape laurel, stinkhout and umnukane. It grows to a height of 18-24m (60-80ft) with a diameter of 1-1.5m (3-5ft).

Appearance
The heartwood varies from an even straw shade to grey-brown and very dark reddish-brown mottled with yellow, maturing to almost black. The grain varies from straight to interlocked or spiral; the surface displays exceedingly fine but pronounced medullary rays, and the texture is moderately fine and uniform. Stinkwood has an unpleasant odour when freshly worked – from where it gets its name – but this does not persist when it is dried.

Properties
The weight varies according to colour, the light type weighing 680 kg/m^3 (42 lb/ft^3) and the dark types 800 kg/m^3 (50 lb/ft^3) when seasoned. The dark wood is more difficult to dry, but the lighter type dries fairly rapidly, with little degrade. There is considerable shrinkage in service. The wood has medium bending and crushing strengths and stiffness, and high resistance to shock loads. It is not used for steam bending. The wood works fairly easily with both hand and machine tools, but has a severe blunting effect on cutting edges. Stinkwood requires pre-boring for nailing, and glues without difficulty. Smooth surfaces are obtained by scraping and sanding before polishing to an excellent finish.

Uses
This timber is highly prized in South Africa for high-class cabinets and furniture, especially Dutch designs of period furniture from the 17th, 18th and 19th centuries. It is also used for light structural work and interior joinery. Its resilient qualities make it suitable for light domestic flooring and vehicle bodies. As a turnery wood it is used for tool handles and fancy goods. There is a wide range of specialist uses – ladders, sporting goods, wheelwrighting, agricultural implements and toys. It is excellent for battery separators. Selected logs are sliced for decorative veneers for cabinets and panels.

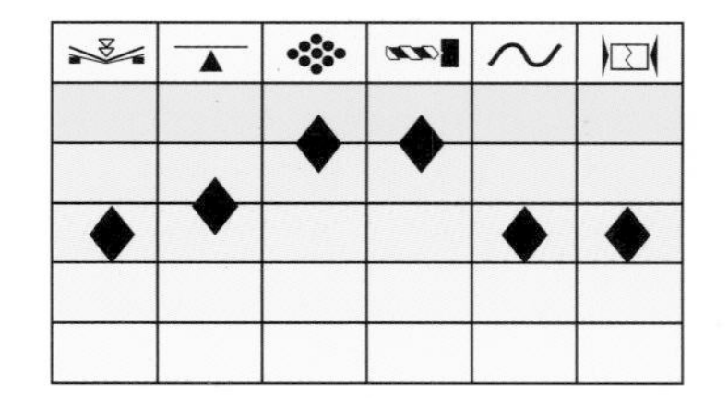

Ocotea usambarensis **Family:** *Lauraceae*
'EAST AFRICAN CAMPHORWOOD'

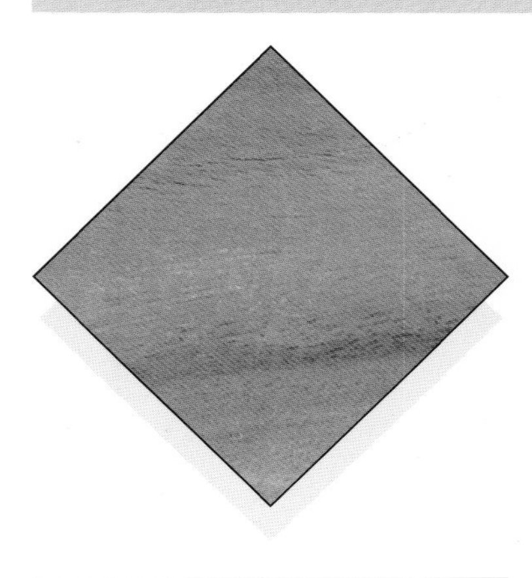

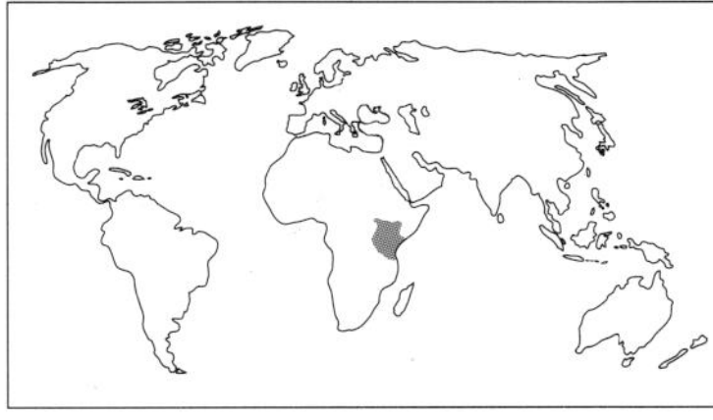

Where it grows
This tree grows in Kenya, on the southern and eastern slopes of Mount Kenya and in the Aberdare range. In Tanzania it occurs on Mount Kilimanjaro, Usambara and Upare. It is known as camphor, muzaiti, muura, munganga or mutunguru. Despite its strong scent of camphor, it should not be confused with true camphorwood (*Cinnamomium camphora*) or Borneo or Sabah camphorwood (*Dryobalanops spp*). The tree reaches a height of about 45m (150ft), with a diameter of 1.2-1.8m (4-6ft).

Appearance
The sapwood is not clearly defined from the heartwood, which is a light yellow to greenish-brown which matures into a deep brown on exposure. Many trees are ill-shaped with twisted or interlocked grain, but selected pieces produce an attractive striped figure when quartered. The texture is moderately fine.

Properties
The weight varies from 510-640 kg/m^3 (32-40 lb/ft^3), averaging about 590 kg/m^3 (37 lb/ft^3) when seasoned. The wood dries slowly if kiln dried with little degrade, but it is liable to warp and twist if air dried too rapidly. It is stable in service, has medium bending and crushing strengths, low stiffness and low resistance to shock loads. It has a moderately good steam-bending rating. It works well with both hand and machine tools, with only a slight dulling of cutting edges, but the interlocked grain affects machining operations. The timber takes nails satisfactorily, glues well, stains easily and can be brought to an excellent finish. This very durable wood is resistant to insect attack and extremely resistant to preservative treatment, but the sapwood is permeable.

Uses
It is very important to select straight-grained timber as many trees are ill-shaped with twisted grain. The distinct scent of camphor makes the wood ideal for clothes closets; it is also used for cabinets and furniture, interior and exterior joinery, light construction and vehicle building, and domestic flooring. Selected logs are sliced for decorative veneers for cabinets and panelling.

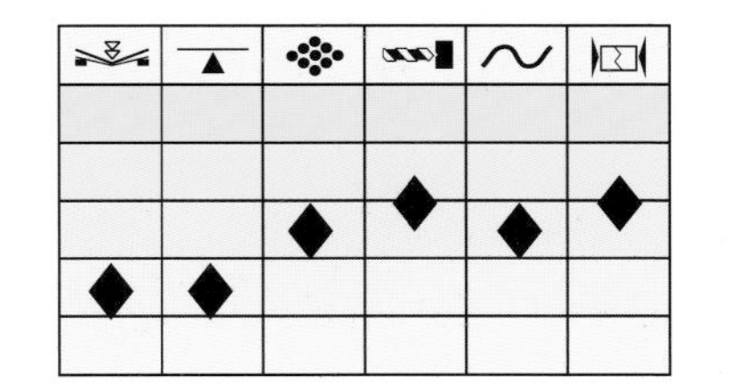

Ocotea rodiaei **Family:** *Lauraceae*
GREENHEART

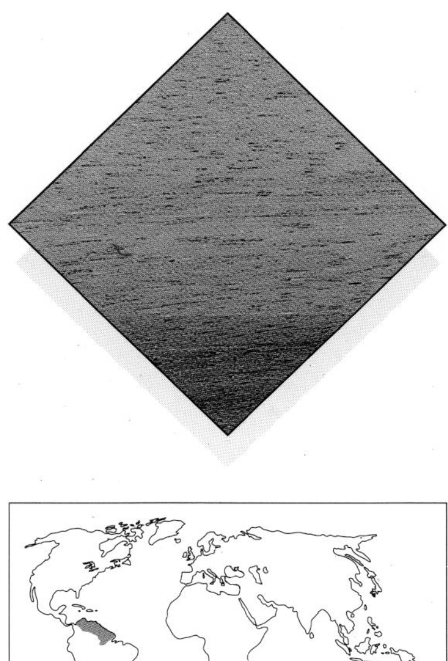

Where it grows
Greenheart is the major commercial wood of Guyana, and occurs to a limited extent in Surinam and Venezuela. It is known as Demerara greenheart in Guyana. This evergreen tree grows to a height of 21-40m (70-130ft), with a long straight cylindrical bole 15-25m (50-80ft) long and about 1.0m (3ft) in diameter.

Appearance
The pale yellow-green sapwood, 25-50mm (1-2in) wide, shades gradually into the heartwood, which varies from yellowish-green through light olive, dark olive, orange-brown, to dark brown often marked with black streaks. Local distinction between colour varieties (black, brown, yellow or white greenheart), has no bearing on the wood's properties. The grain varies from straight to interlocked, the texture is fine, uniform and lustrous.

Properties
The weight averages about 1030 kg/m^3 (64 lb/ft^3) when seasoned. It dries very slowly and with considerable degrade especially in the thicker sizes. Once dry, there is medium movement in service. This wood has exceptional strength in all categories, with a moderate steam-bending rating. It is moderately difficult and dangerous to work as poisonous splinters fly from interlocked, cross- or endgrain; the difficult grain affects many machining operations. Pre-boring is essential for screwing or nailing. A fine, smooth, lustrous surface can be obtained. Gluing results are variable; staining is rarely necessary but the wood polishes well. It is very durable, and immune to marine borers. Greenheart is extremely resistant to preservative treatment.

Uses
This is one of the world's major timbers for marine and ship construction. It is used for revetments, docks, locks, fenders, braces decking, groynes, lock gates, pier decking and handrails, jetties, piling, bridges, wharf and harbour work. In ship construction it appears as engine bearers, planking, gangways. fenders, stern posts and sheathing for whalers. Other special uses include heavy-duty factory flooring, chemical vats and filter press plates and frames. It is good for turnery of all kinds, billiard cue butts, fishing rods and the central laminae of longbows.

	◆	◆	◆	◆	◆
◆					

Olea hochstetteri **Family:** *Oleaceae*

OLIVEWOOD

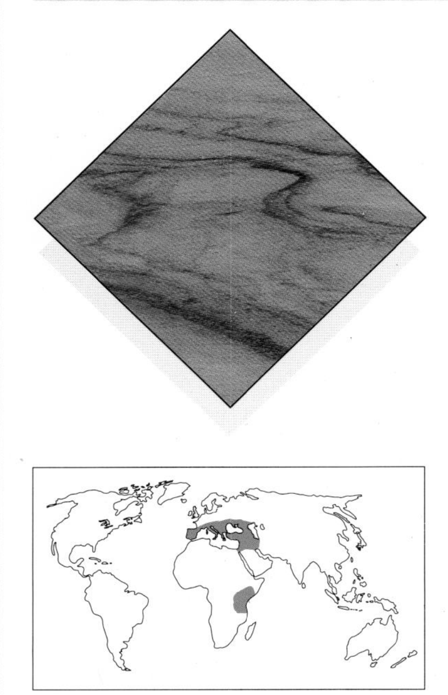

Where it grows

The olive tree *O.europaea* is grown in the Mediterranean for its edible fruit and olive oil, and is usually small and misshapen. Commercial timber comes from Kenya, Tanzania and Uganda. *O.hochstetteri* produces East African olive, also known as musheragi in Kenya, and *O.welwitschii* (Knobl.) Gilg. & Schellenb. produces loliondo in Tanzania and 'Elgon olive' in Kenya. Olivewood grows to about 25m (80ft) in height, heavily fluted and crooked, with a diameter of 0.45-0.75m (1½-2½ft).

Appearance

The sapwood is pale creamy-yellow and quite plain, but the heartwood has a pale brown background with very attractive irregular markings of mid-brown to dark brown and blackish streaks. The grain is straight to shallowly interlocked and the texture very fine and even.

Properties

O.hochstetteri weighs on average 880 kg/m³ (55 lb/ft³), and *O.welwitschii* weighs about 800 kg/m³ (50 lb/ft³) when seasoned. The wood is rather refractory and needs to be air dried slowly, especially as internal checking or honeycombing may occur in thicker pieces. It can be kiln dried successfuly. There is considerable movement in service. The wood has excellent strength in every category. The sapwood may be bent to a small radius, but because of resin exudation olivewood has only a moderate steam-bending classification. The wood is rather difficult to work as the interlocked grain affects machining. There is high resistance in cutting with a moderate blunting effect on tools. It requires pre-boring for nailing. The wood glues well, and can be brought to an excellent finish. The heartwood is moderately durable and resistant to preservative treatment, but the sapwood is permeable

Uses

This very attractive timber has a good resistance to abrasion and makes an excellent decorative flooring for public buildings. It is used for furniture, cabinets and panelling and is ideal for turnery, for tool and fancy handles and bowls. Olivewood is popular for sculpture and carving, and logs are sliced for decorative veneers.

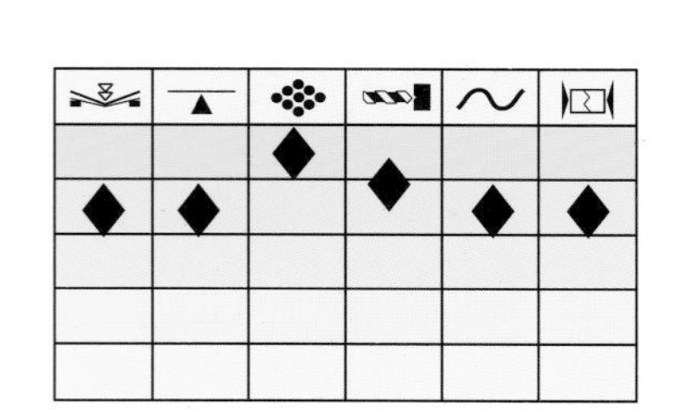

Palaquium and Payena spp. **Family:** *Sapotaceae*

NYATOH

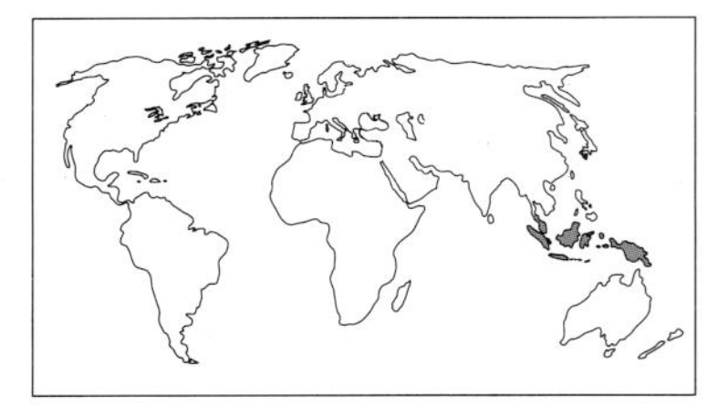

Where it grows

Certain species of the *Sapaotaceae* family which produce timbers of similar colours and properties from Malaysia, Indonesia and the south Asian islands, are collectively known as nyatoh. They are principally *P. maingayi*, *P.rostratum* and *P.xanthochymum* . Trade names of njatuh and padang are given to other species which produce light to medium weight timbers from the *Palaquium* and *Payena spp.*, chiefly Payena maingayii. Both nyatoh and padang are known as padang in the UK. Trees grow to height of 30m (100ft) with a diameter of about 1m (3ft).

Appearance

Generally the sapwood of these timbers is only slightly paler in colour and not defined from the heartwood, which is a deep pink to red-brown, often with darker streaks on quartered surfaces. The grain can be straight, slightly interlocked or slightly wavy, providing an attractive moiré or 'watered-silk' figure. The texture is moderately fine and even.

Properties

The weight of nyatoh averages 640-720 kg/m³ (40-45 lb/ft³) when seasoned. The wood dries rather slowly, with a slight tendency to end split and distort and for surface checks to develop, especially around knots. The wood is stable in service and has medium bending and crushing strength, with low stiffness and resistance to shock loads, and a moderate steam-bending rating. Some species are siliceous and can cause severe blunting of cutting edges. Pre-boring is required for nailing, but nyatoh glues well and can be brought to an excellent finish. The sapwood is liable to attack by powder post beetle and the heartwood varies from non-durable to moderately durable; it is very resistant to preservative treatment.

Uses

The name nyatoh covers a range of different woods, each varying slightly in character. Generally they are attractive timbers used for cabinet making and furniture, pattern making, interior construction, doors and shingles, high-class joinery and domestic flooring. Logs are rotary cut for plywood and corestock and sliced for decorative veneers for cabinets and panelling.

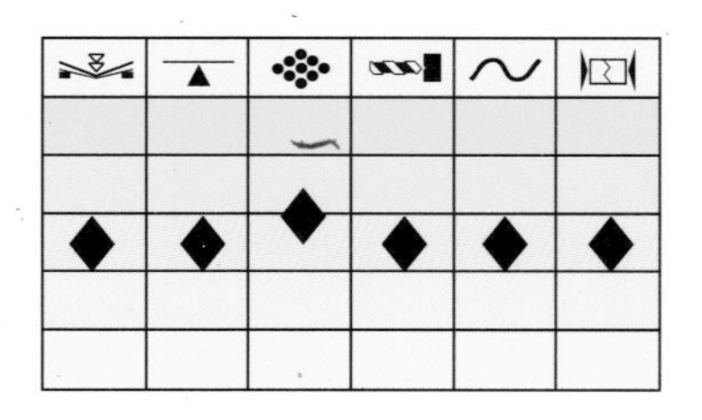

Paratecoma peroba **Family:** *Bignoniaceae*
WHITE PEROBA

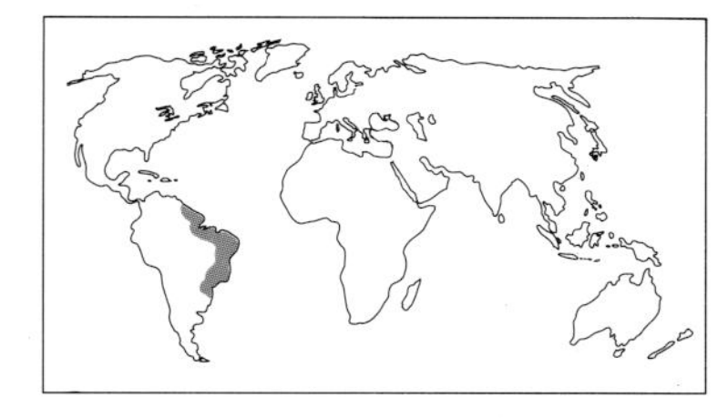

Where it grows
The tree grows in the coastal forests of Brazil. It is also known as peroba de campos, ipé peroba, peroba amarella, peroba branca and ipé claro (Brazil); and golden peroba in the UK. It should not to be confused with *Aspidosperma peroba*, which produces rosa peroba. It reaches a height of about 40m (130ft), with a straight symmetrical bole and a diameter of about 1.5m (5ft).

Appearance
The white to yellowish sapwood is clearly defined from the heartwood, which is a pale golden-olive brown, but with yellow, greenish or red shading. The grain is commonly interlocked or wavy, and the texture medium and uniform, often with a lustrous surface. The grain of quartered surfaces produces a narrow striped or roe figure.

Properties
The weight averages about 750 kg/m^3 (47 lb/ft^3) when seasoned. The wood dries easily with only negligible splitting, and distortion is not serious. There is medium movement in service. The wood has a medium bending strength, low stiffness and shock resistance, and high crushing strength. It has a moderate steam-bending classification. The material works readily with both hand and machine tools, and planes easily to a smooth, silken finish. Fine machine dust can cause skin irritation and splinters are poisonous. The wood has good holding properties, glues well and can be brought to a good finish. The very durable heartwood resists insect and fungal attacks, and is also resistant to preservative treatment.

Uses
In Brazil this very attractive timber is used extensively for high-class furniture and cabinet making. It is also used for interior and exterior joinery and in building construction; it is ideal as a pleasing heavy-duty flooring to withstand continuous traffic. Specialized uses include framing and floors for vehicle body work, decking and flooring for boat building. It is also suitable for vats for foodstuffs and chemicals. Selected logs are sliced for decorative veneers for cabinets and panelling.

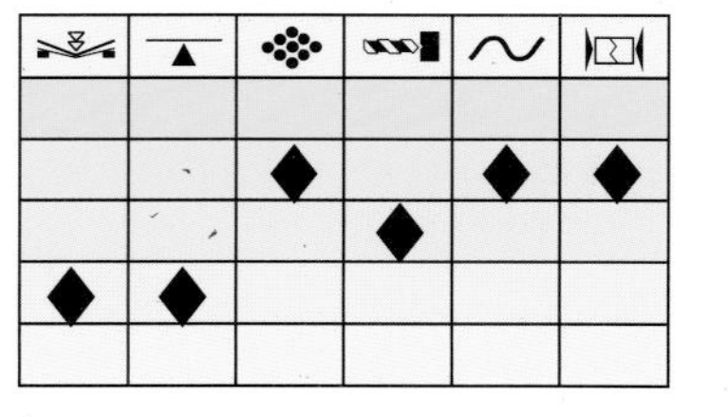

Peltogyne spp. **Family:** *Leguminosae*
PURPLEHEART

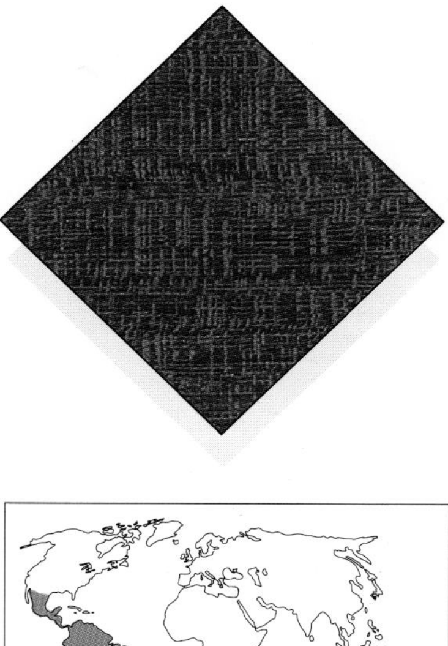

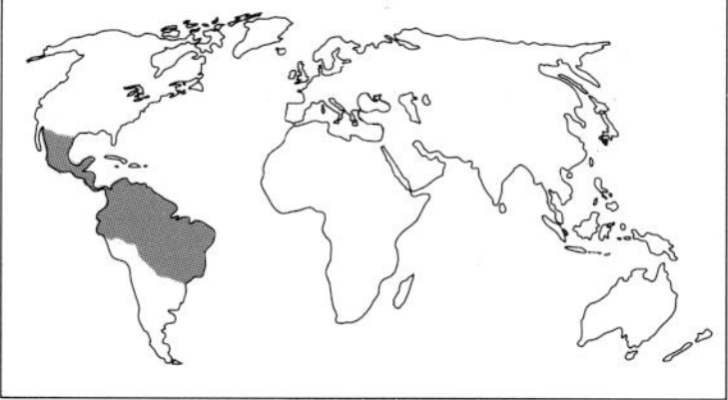

Where it grows
Purpleheart is widely distributed in tropical America from Mexico down to southern Brazil. Those of commercial importance in the Caribbean are *P.pubescens*, Benth., *P.porphyrocardia*, Griseb. and *P.venosa* (Vahl) Benth, var. *densiflora* (Spruce) Amsh. They are known as amaranth or violetwood in the USA. *P.venosa* from the Guianas is also important in the Amazonas of Brazil and other areas of South America; it is known as koroboreli, saka or sakavalli (Guyana); purperhart (Surinam); pau roxo, nazareno or morado (Venezuela); tananeo (Colombia); and amarante (Brazil). The semi-deciduous trees reach a height of 38-45m (125-150ft), with a diameter of 0.6-1.2m (2-4ft).

Appearance
The timber has a white-to-cream sapwood; heartwood, bright purple on exposure to light, then matures into a dark purplish-brown. It is generally straight grained, but sometimes wavy or interlocked with a moderate to fine, uniform texture.

Properties
The weight averages about 860 kg/m^3 (54 lb/ft^3) when seasoned. The wood dries fairly rapidly with little degrade. There is little movement in service. The timber has high bending and crushing strength and stiffness, with medium resistance to shock loads. It has a moderate steam-bending rating. It is rather difficult to work, with a moderate to severe blunting effect on tools. It requires pre-boring for nailing, but glues well and polishes easily. Spirit finishes tend to remove the purple colour.The sapwood is liable to insect attack; the heartwood is very durable and extremely resistant to preservative treatment, but the sapwood is permeable.

Uses
Purpleheart is used locally as a cabinet and furniture wood, but also for heavy outdoor constructional work – bridges, freshwater piling, dock and harbour work. It makes an attractive flooring. It is used for sculpture, carving and turnery for tool handles and small fancy items. Specialized uses include boat building, gymnasium apparatus, diving boards, skis, wheelwrighting, billiard cue butts, vats for chemicals and filter presses. It is also sliced for decorative veneers.

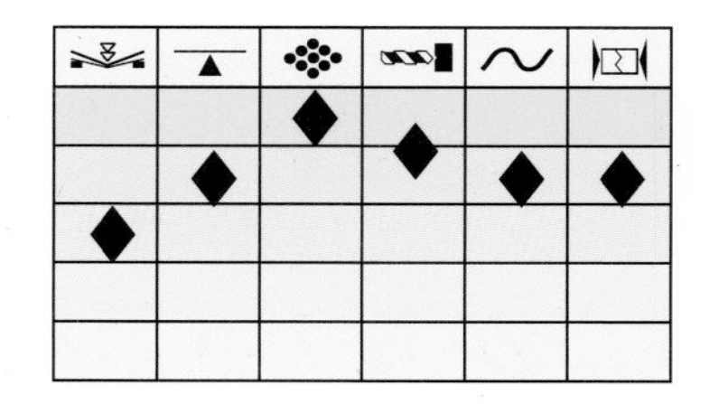

Pericopsis elata **Family:** *Leguminosae*
AFRORMOSIA

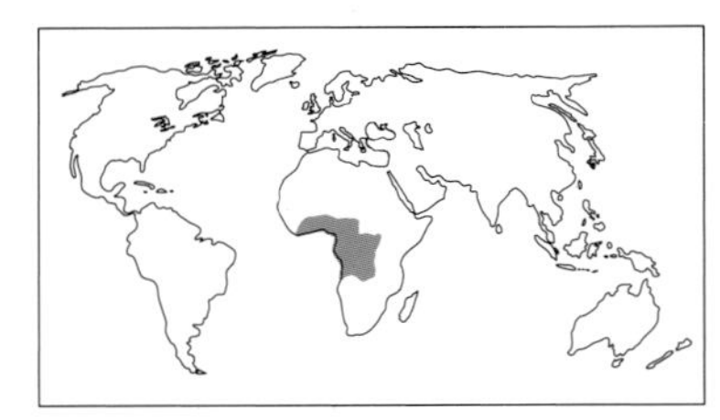

Where it grows
This tree occurs in the Ivory Coast, Ghana, Zaire and Nigeria. It is also known as assamela (Ivory Coast); kokrodua (Ghana and Ivory Coast); ayin or egbi (Nigeria); andejen (Cameroon). It reaches a height of 45m (150ft) and a diameter of about 1m (3ft).

Appearance
The creamy-buff sapwood is well defined from the heartwood, which is golden-brown when freshly felled and darkens on exposure. The grain varies from straight to interlocked, which produces a rope-striped figure on quartered surfaces. The texture is moderately fine but without the oiliness of teak. The wood is liable to blue mineral stains if in contact with iron or iron compounds in damp conditions because of its high tannin content.

Properties
The wood weighs about 690 kg/m^3 (43lb/ft^3) when seasoned, and dries slowly but well, with little degrade. There is exceptionally small movement in service. It has medium stiffness, high crushing strength and medium shock resistance with a moderate steam-bending rating. The interlocked grain can affect machining. Tipped saws should be used as there is a moderate blunting effect of tools. Pre-boring is required for nailing and screwing. Afrormosia glues well and takes a good finish. It is very durable, resistant to both fungi and termites, and extremely resistant to preservative treatment.

Uses
Afrormosia was originally used as a substitute for teak in the furniture industry for framing and fittings, edge lipping and facings for panels. Today it is used extensively in its own right where a very attractive, strong, stable and durable wood is required. It appears in high-class furniture and cabinet making, chairs, interior joinery, stairs, shop and office fitting and agricultural implements. It makes an attractive floor, and is also used for exterior joinery, boat building and marine piling. Selected logs are sliced for decorative veneers for furniture, flush doors and wall panelling.

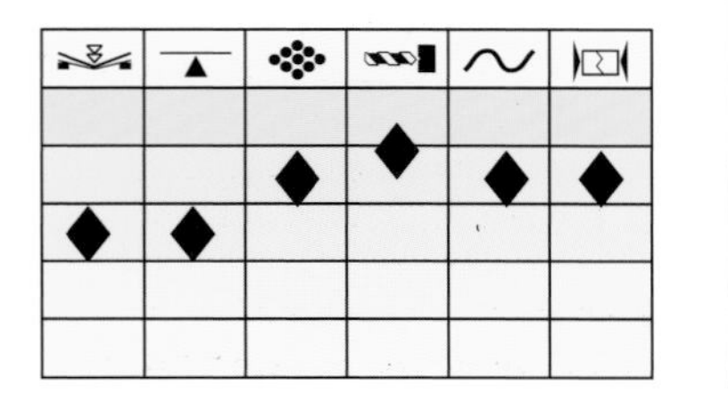

Phoebe porosa **Family:** *Lauraceae*
IMBUIA

Where it grows
Imbuia grows in southern Brazil. It is also known as imbuyia, amarela, canella imbuia and embuia (Brazil); or Brazilian walnut in the USA and UK. It grows to about 40m (130ft), with a diameter of about 2m (6-7ft).

Appearance
The beige sapwood is clearly defined from the heartwood, which varies from olive-yellow to chocolate-brown, frequently variegated. The grain may be straight but sometimes wavy or curly, with a fine to medium texture and a high lustre. It has a spicy, resinous scent and taste, most of which is lost in drying.

Properties
The weight is about 660 kg/m^3 (41 lb/ft^3) when seasoned. The wood air dries rapidly and care is needed to avoid a tendency to warp. It should be kiln dried slowly to avoid degrade. There is small movement in service. Imbuia has medium to low strength in all categories, with a very low steam-bending classification, but it is chiefly used for its decorative qualities where strength is not important. It works easily with both hand and machine tools, with only a slight blunting effect on cutting edges, and finishes very smoothly. The machining dust can be an irritant to the eyes, nose and throat. The wood has good holding properties, glues without problems, stains and polishes easily and can be brought to an excellent finish. The durable heartwood resists insect attack and is moderately resistant to preservative treatment, but the sapwood is permeable.

Uses
Finished imbuia is similar in appearance to walnut, and for many years it has been marketed as 'Brazilian walnut'. In Brazil it is considered to be one of the most valuable woods for high-class cabinets and furniture, and superior interior joinery for panelling, shop and bank fitting. It also makes a high-grade decorative flooring for light use. Among its specialized uses are sculpture and carving, turnery for handles and fancy bowls, rifle butts and gun stocks. Sliced highly decorative veneers are exported for cabinets and architectural panelling.

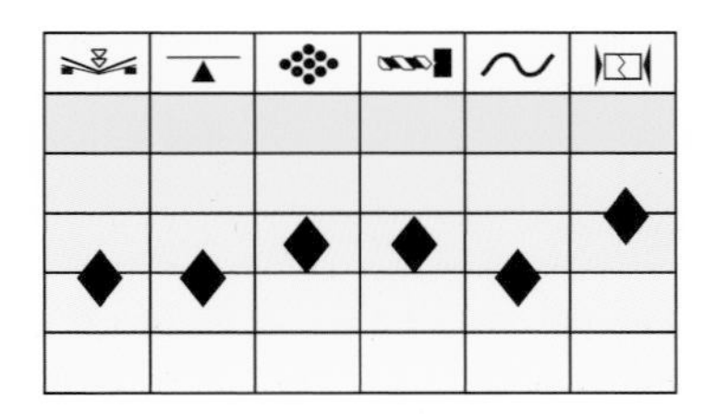

Picea abies **Family:** *Pinaceae*

SPRUCE, EUROPEAN or WHITEWOOD (S)

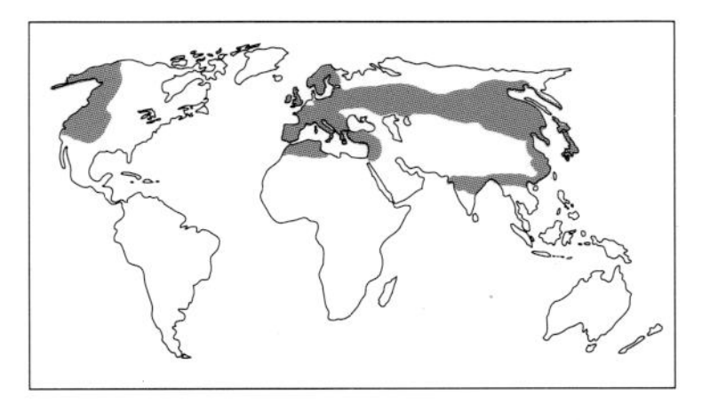

Where it grows

This wood occurs throughout Europe, with the exception of Denmark and the Netherlands, into western Russia. The tree reaches an average height of 36.6m (120ft) and a diameter of 0.76-1.2m ($2\frac{1}{2}$-4ft). In the mountains of Rumania it reaches a height of 61m (200ft) with a 1.5-1.8m (5-6ft) diameter. It is also known as white deal, common spruce or Norway spruce, and Baltic, Finnish or Russian whitewood according to the country of origin.

Appearance

The colour varies from almost white to pale yellow-brown and has a natural lustre. The annual rings are clearly defined. The wood is straight grained and has a fine texture.

Properties

Weight averages 470 kg/m^3 (29 lb/ft^3) seasoned. Spruce dries rapidly and well with some risk of distortion.It has low stiffness and resistance to shock loads, medium bending and crushing strength, and a very poor steam-bending rating. There is medium movement in service. It works easily with both hand and machine tools, holds screws and nails well, glues satisfactorily and takes stain, paint and varnishes for a good finish. The sapwood is liable to attack by the common furniture beetle; the non-durable heartwood is resistant to preservative treatment.

Uses

Norway spruce provides our traditional Christmas tree from its thinnings. The best quality spruce comes from the most northerly regions, but trees grown at the same latitude will produce different qualities according to the altitude. It is used for interior building work, carcassing, domestic flooring, general carpentry, boxes and crates. Small logs are used in the round for masts, pit props and ladder stringers. Spruce from central and eastern Europe and from Alpine areas of North America produce excellent quality 'tone-woods' for piano and other keyboard instrument soundboards, the bellies of violins, lutes, guitars and so on, because of its unsurpassed resonance qualities. It is used in the manufacture of pulp and paper, and in Germany the bark is stripped and used for tannin extraction. It is also used for plywood.

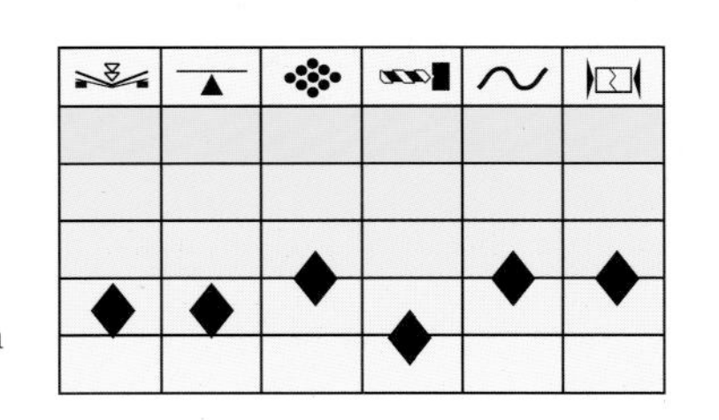

Picea sitchensis **Family:** *Pinaceae*

SITKA SPRUCE (S)

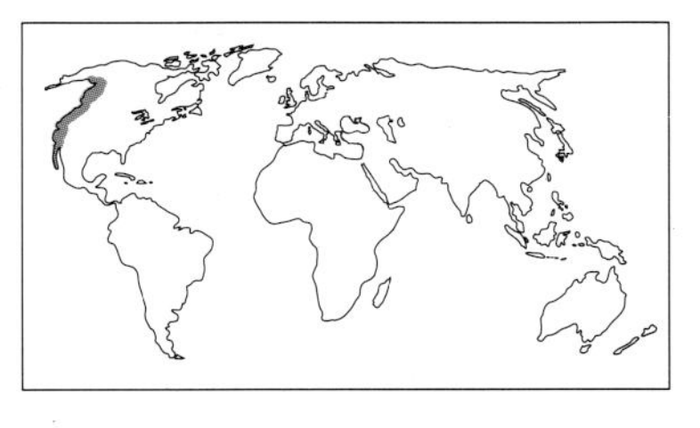

Where it grows

Sitka spruce occurs on the Pacific coast of North America and western Canada down to California. It grows to 38-53.4m (125-175ft), with a diameter of 0.9-1.8m (3-6ft); it occasionally reaches 76.3m (250ft) in height and 2.4-3.7m (8-12ft) in diameter.

Appearance

The pale pink sapwood blends into the light pinkish-brown heartwood, which is mostly straight grained, sometimes spiral. It has a fairly coarse but uniform texture and is odourless, non-resinous and non-tainting.

Properties

The wood weighs about 430 kg/m^3 (27 lb/ft^3) when seasoned. Care is needed in drying large sizes as it dries fairly rapidly and tends to twist and cup. There is medium movement in service. It has medium bending and crushing strengths, stiffness and shock resistance. Its strength-to-weight ratio is high, and it has a very good steam-bending rating. Sitka spruce works easily with both hand and machine tools, finishes cleanly, takes screws and nails without difficulty, and gives good results with various finishes when care is taken to prevent raising the grain. The heartwood is non-durable and resistant to preservative treatment.

Uses

Strength varies considerably according to the location and growth conditions, and selective grading is necessary for joists, rafters and studding for building. Split poles are used for ladder sides. In Canada sitka spruce is used for interior joinery, cooperage and boxmaking; specialized uses include boat building, oars and masts. Special grades are selected for their resonance to make soundboards for pianos, and guitar and violin fronts. Logs are rotary cut as corestock for birch and Douglas fir plywood, but sitka is seldom used for plywood faces. It is sliced for special laminates for aircraft and glider construction, sail planes and racing sculls; it is also used for box making, and is the world's most important pulp for newsprint because of its whiteness.

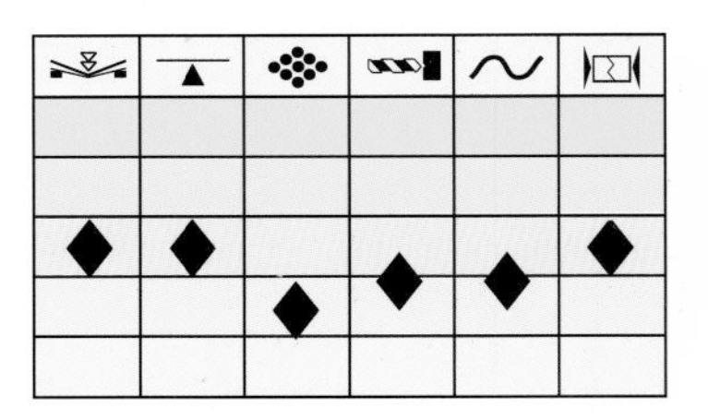

Pinus monticola **Family:** *Pinaceae*
WESTERN WHITE PINE (S)

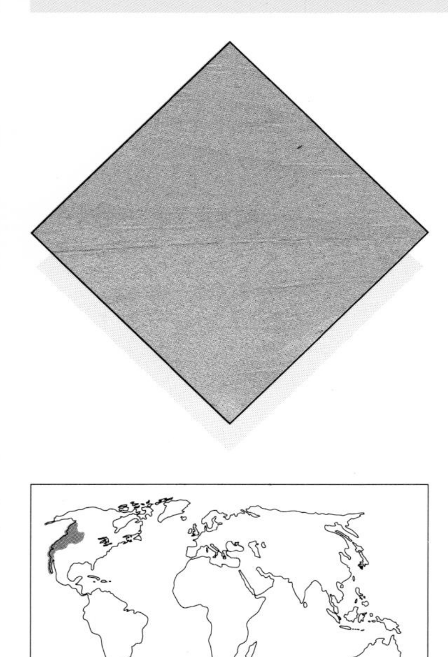

Where it grows
This tree grows in the mountain forests of western Canada and the western USA, from sea level up to over 3000m (9750ft). It occurs south down to the Kern River in California and east into northern Montana; it is most abundant in northern Idaho. It reaches 23-37m (75-120ft) with a diameter of about 1m (3ft) or more. It is also called Idaho white pine (USA). Closely related species include *P.contorta* Dougl., producing lodgepole pine known as contorta pine (UK); *P.banksiana*, Lamb. produces jack pine known as princess pine and Banksian pine in Canada and the USA.

Appearance
The sapwood is white, the heartwood is only slightly darker and varies from a pale straw colour to shades of reddish-brown. Fine brown lines caused by resin ducts appear on longitudinal surfaces. It is straight grained with an even, uniform texture. Yellow pine is always called 'white pine' in Canada and the USA, though there are differences in weight and marking.

Properties
The woods weighs about 450 kg/m^3 (28 lb/ft^3) when seasoned. It dries readily and well, with little checking or warping, and has a slightly higher shrinkage rating than yellow pine. There is little movement in service. This low-density timber has rather low strength properties, and is not suitable for steam bending. The material works easily with both hand and machine tools, takes screws and nails without difficulty, glues well and takes paint and varnish well. The wood is non-durable, liable to beetle attack and moderately resistant to preservative treatment, but the sapwood is permeable.

Uses
Western white pine is chiefly used for interior joinery for doors and windows, interior trim, fitments, shelving, light and medium building construction. Specialized uses include furniture and cabinets, boat and ship building, pattern making, drawing boards, domestic wooden ware and match splints. It is rotary cut for plywood and corestock, and selected logs are sliced for decorative panelling veneers.

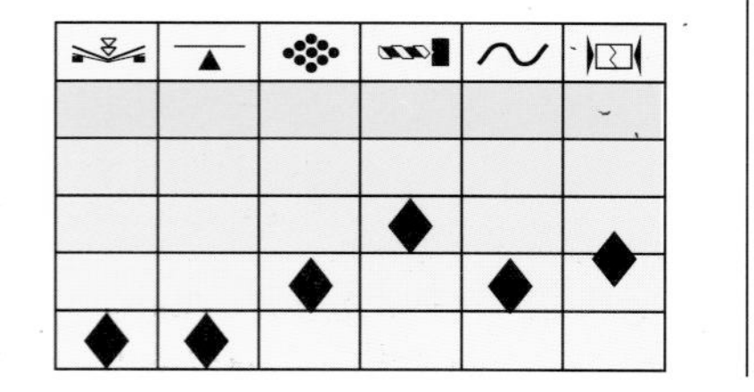

Pinus palustris **Family:** *Pinaceae*
AMERICAN PITCH PINE.(S)

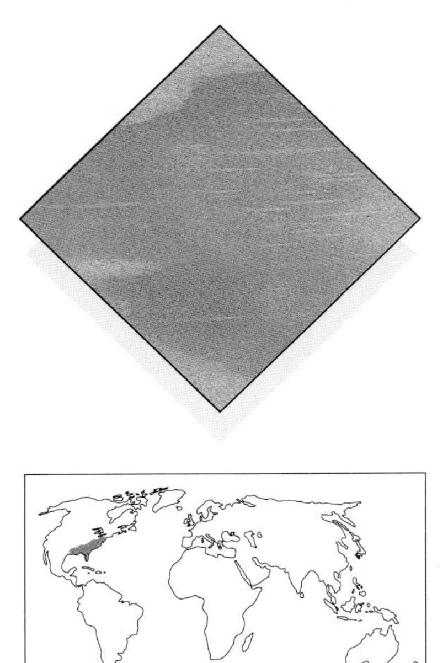

Where it grows
Pitch pine is the heaviest commercial softwood. *P.palustris* and *P.elliottii* grow through the southern US in a curve from Virginia through Florida to the Gulf. The heaviest timber is shipped as pitch pine, the lighter wood as southern pine. They grow to a height of 30.5m (100ft) and a diameter of 0.6-0.9m (2-3ft). Other names: *P.palustris*, longleaf pine, Florida longleaf, yellow pine or Georgia yellow pine; *P. elliottii*, slash pine, longleaf yellow pine or longleaf (USA); Gulf coast pitch pine or longleaf pitch pine (UK).

Appearance
The creamy-pink sapwood is quite narrow, and contrasts with the heartwood, which is yellow-red to reddish-brown, with a wide conspicuous growth ring figure especially in fast-grown timber. It is very resinous and has a coarse texture.

Properties
The weight for seasoned timber varies from about 660-690 kg/m^3 (41-43 lb/ft^3). The wood dries well with little degrade and is stable in service. It has high bending and crushing strengths and high stiffness, with medium resistance to shock loads. It is not suitable for steam bending because of the resin content. Pitch pine can be worked readily with both hand and machine tools, but resin can be troublesome in clogging cutters and saw teeth. It holds screws and nails firmly, glues without difficulty and takes paint and other finishes satisfactorily. The timber is moderately durable; sometimes beetle damage is present. It is resistant to preservative treatment but the sapwood is permeable.

Uses
The timber is used for heavy construction work, truck and railway wagons, ship building, exterior joinery, piling, dock work, bridge building, decking and chemical vats. Lower grades are used for interior joinery, general building, domestic flooring, crates and pallets. The timber is rich in resinous secretions and also produces the largest percentage of the world's rosin and turpentine.

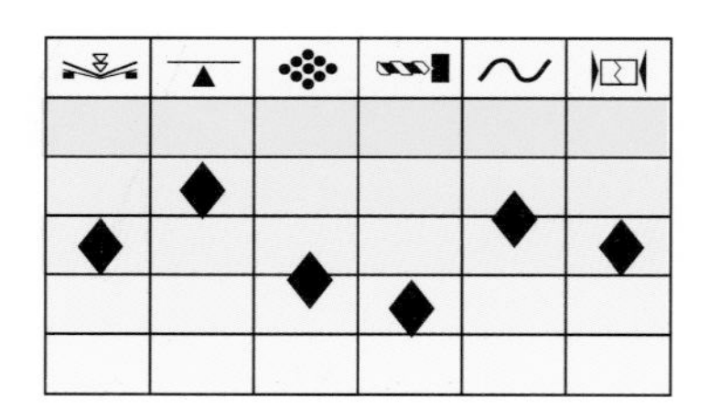

Pinus ponderosa **Family:** *Pinaceae*

PONDEROSA PINE (S)

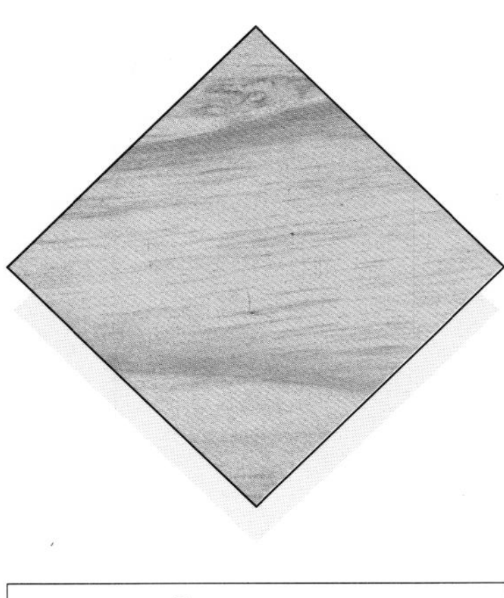

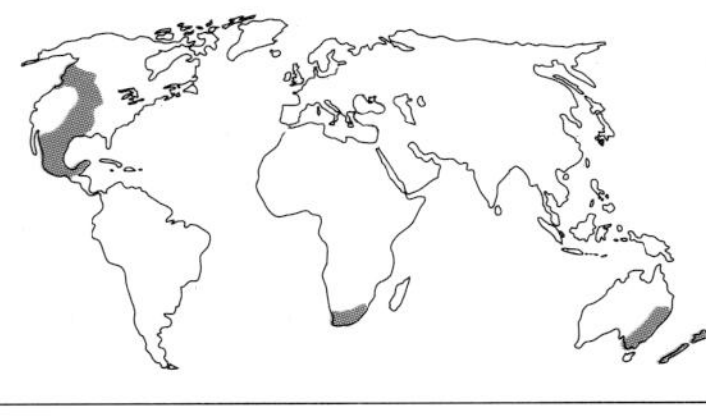

Where it grows
Ponderosa pine occurs in the drier regions of southern British Columbia and from Montana, western Nebraska and Texas into Mexico and west to the Pacific coast. The tree reaches an average height of 30.5m (100ft) with a diameter of about 0.6m (2ft), but can get up to 51.9m (170ft) with a 1.2m (4ft) diameter. It is also known as western yellow pine (USA and Australia); bird's eye pine, knotty pine, British Columbia soft pine (Canada); and Californian white pine (USA).

Appearance
Mature trees have a very thick pale yellow sapwood, soft, non-resinous and uniform in texture. The heartwood is orange to reddish-brown, with prominent dark brown resin duct lines on longitudinal surfaces. It is considerably heavier than the sapwood.

Properties
Weight is about 510 kg/m^3 (32 lb/ft^3) when seasoned. Ponderosa pine dries easily and well with little degrade, but the wide sapwood is susceptible to fungal and blue staining if the wood is not carefully piled during air drying. There is very little movement in service. It has medium bending and crushing strength, low stiffness and shock resistance, and a poor steam-bending rating. The timber works easily with both hand and machine tools but resin exudation tends to clog cutters and saws. The wood can be glued satisfactorily, takes screws and nails without difficulty, and if it is treated to remove the surface gumminess, gives good results in painting and varnishing. It is non-durable and moderately resistant to preservative treatment, but the sapwood is permeable.

Uses
The valuable sapwood is used in America for pattern making. The heartwood is used for kitchen furniture, building construction, window frames, doors, general carpentry and for packing cases, crates and pallets. When treated it is used for sleepers, poles and posts. Logs are rotary cut for veneers and sliced for knotty pine panelling.

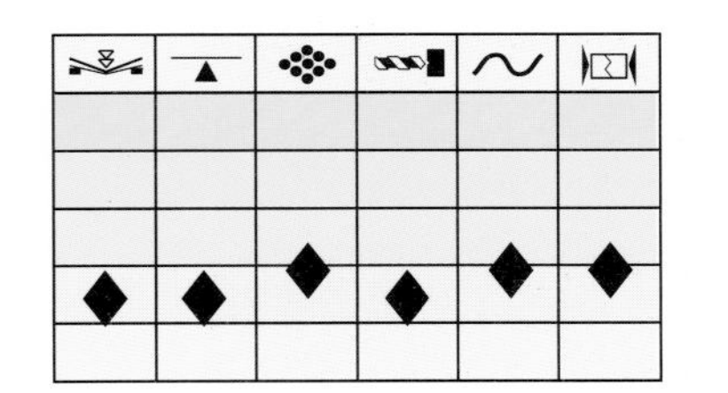

Pinus radiata **Family:** *Pinaceae*

RADIATA PINE (S)

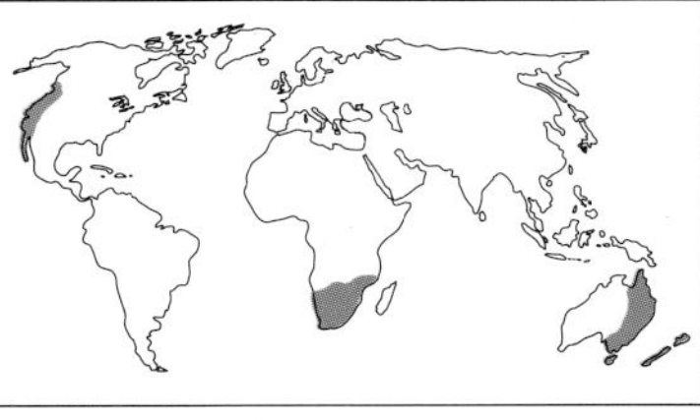

Where it grows
This is the most commonly planted softwood in the warm temperate climates of California, Australia, New Zealand and South Africa. In the southern hemisphere it grows fast, in ideal conditions to a height of 30.5m (100ft) and a diameter of 0.3-0.6m (1-2ft) in 20 years. It is also known as insignis pine and Monterey pine.

Appearance
Most commercial timber has wide pale sapwood and a rather small, pinkish-brown heartwood with inconspicuous growth rings, giving an even, uniform texture. It is knotty, and resin ducts appear as fine brown lines on longitudinal surfaces. Grain is usually straight, sometimes with some spiral.

Properties
The timber weighs about 480 kg/m^3 (30 lb/ft^3) when seasoned. The wood dries rapidly and well with little degrade, and is stable in use. It has low bending strength and stiffness with medium crushing strength and shock resistance. It is not suitable for steam bending. The timber works fairly well with both hand and machine tools. It holds screws and nails well, and gluing and finishing are satisfactory if the gumminess is treated. The wood is non-durable and there is sometimes beetle damage. The sapwood is liable to attack by the common furniture beetle but is permeable for preservation treatment.

Uses
The bulk of commercial supplies of this timber is the sapwood of young, rapidly grown plantation trees. It is extensively used for building and general structural purposes, floors and cladding, boxes, packing cases and crates; also general turnery for brush and broom handles. Specially selected dressing grades are used for furniture and interior joinery. Logs are rotary peeled for plywood and corestock, and for particleboard and fibre building boards. Selected logs are sliced for decorative veneers for panelling. In New Zealand it is an important source of paper pulp.

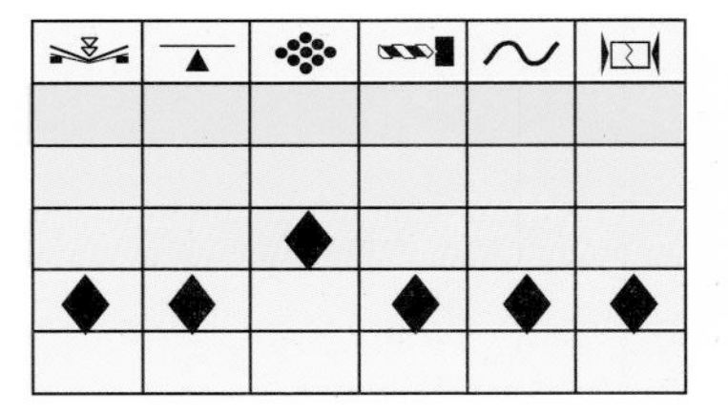

Pinus strobus **Family:** *Pinaceae*
YELLOW PINE (S)

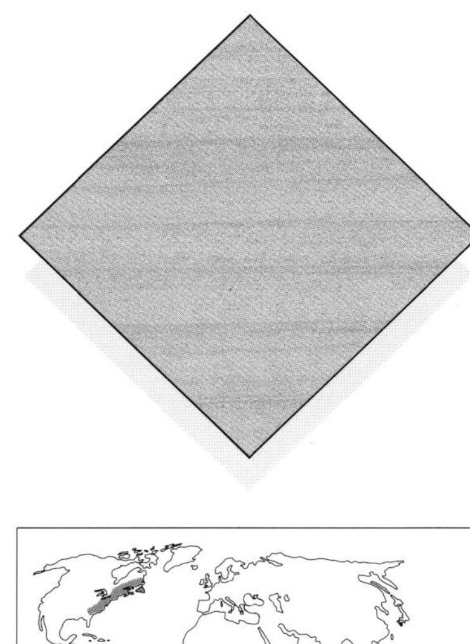

Where it grows
This species occurs from Newfoundland to the Manitoba border and south to north Georgia. It can reach a height of 45.7m (150ft) and a diameter of about 1.5m (5ft), but averages about 30.5m (100ft) high and 0.6-0.9m (2-3ft) in diameter. It is also known as white pine, eastern white, cork and soft pine (Canada and USA); northern white, northern pine (USA); Quebec yellow, Quebec pine and Weymouth pine (UK).

Appearance
The sapwood is white and the heartwood varies from a light straw brown to a light reddish-brown. It is not very resinous; the ducts appear as thin brown lines on longitudinal surfaces but the growth rings are inconspicuous. It is straight grained and the texture very fine and even.

Properties
Seasoned weight varies from 390-420 kg/m³ (24-26 lb/ft³). The wood dries fairly rapidly and well, but sap stain should be avoided when air drying. Yellow pine has extremely low shrinkage and is very stable in service. The timber is weak in all strength properties, and is not suitable for steam-bending. It works very easily with both hand and machine tools, has good screw and nail-holding properties, glues well and can be brought to an excellent finish. It is susceptible to attack by the common furniture beetle. The heartwood is non-durable and resistant to preservative treatment, but the sapwood is permeable for treatment.

Uses
Yellow pine, with its low shrinkage and extreme stability in use, it is particularly suited for engineers' pattern making for very fine detail, and drawing boards, doors and similar high-class work. It is also used for sculpture and carving, and high-class interior joinery, cabinet and furniture making, shelving and interior trim. Specialized uses include parts for stringed instruments such as guitars, organ parts, ship and boat building, and light construction. Second growth timber is much coarser in texture and usually knotty. It is used for match splints, packaging containers and wood flour.

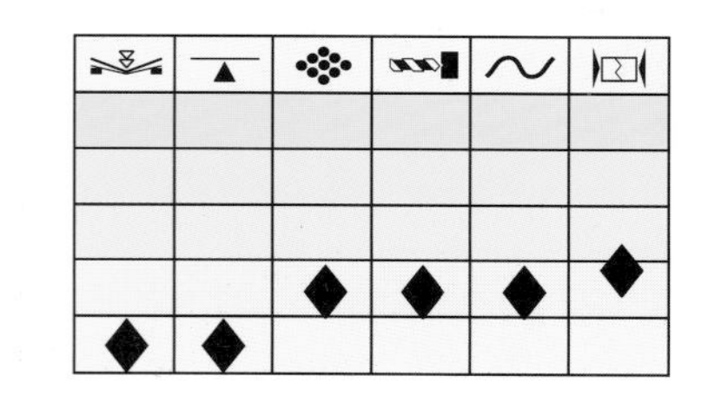

Pinus sylvestris **Family:** *Pinaceae*
REDWOOD (SCOTS PINE) (S)

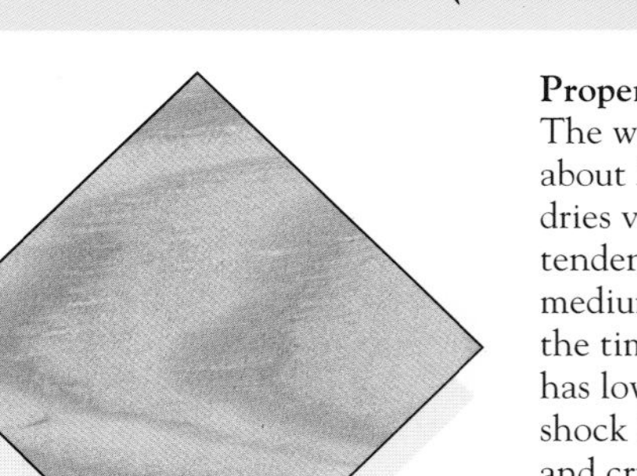

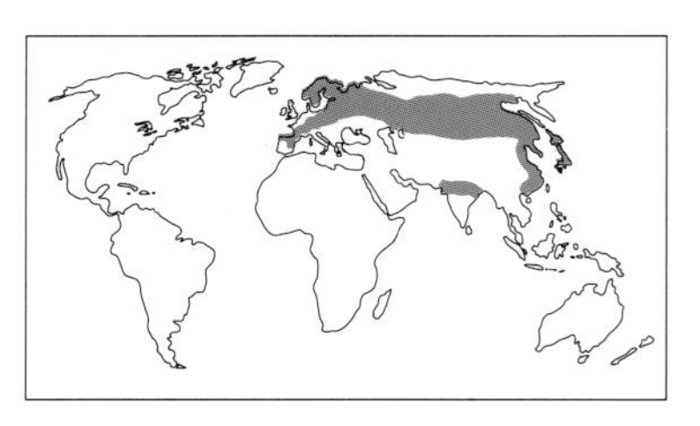

Where it grows
This common commercial softwood occurs from the Sierra Nevada in Andalusia and the mountains of west Spain, through the Maritime Alps and Pyrenees, the Caucasus and Transylvanian Alps, up into western Siberia. In good conditions the tree reaches 39.6-42.7m (130-140ft), with a diameter of 0.6-0.9m (2-3ft). In the UK, imported timber is known as redwood, red deal or simple 'red' in the north, and yellow deal or 'yellow' in the south. Homegrown timber is called Scots pine; red pine, Baltic/Finnish/Swedish/Archangel/Siberian/Polish pine; 'Norway fir' or 'Scots fir'.

Appearance
The knotty wood has a mildly resinous heartwood of pale red-brown, distinct from the paler creamy-white to yellow sapwood, with clearly marked annual rings. Texture varies from the slowly grown fine of northern Russia to the coarser and denser wood of northern Europe.

Properties
The weight of seasoned timber is about 510 kg/m³ (32 lb/ft³) and it dries very rapidly and well, with a tendency to blue sap stain. There is medium movement in service and the timber is stable in use. The wood has low stiffness and resistance to shock loads, low to medium bending and crushing strength and a very poor steam-bending rating. It works easily with both hand and machine tools. It holds nails and screws well, but gluing can be troublesome because of the resin. The wood can be stained, painted or varnished to a good finish. It is liable to attack by the common furniture beetle; it is non-durable and moderately resistant to preservative treatment, but the sapwood is permeable.

Uses
The best grades are used for furniture, interior joinery, turnery and vehicle bodies. Other grades go for building construction and carcassing. When treated, redwood is extensively used for railway sleepers, telegraph poles, piles and pit props. Logs are cut for plywood and sliced for decorative veneers. It is also used in the chemical wood-pulp industry for kraft paper.

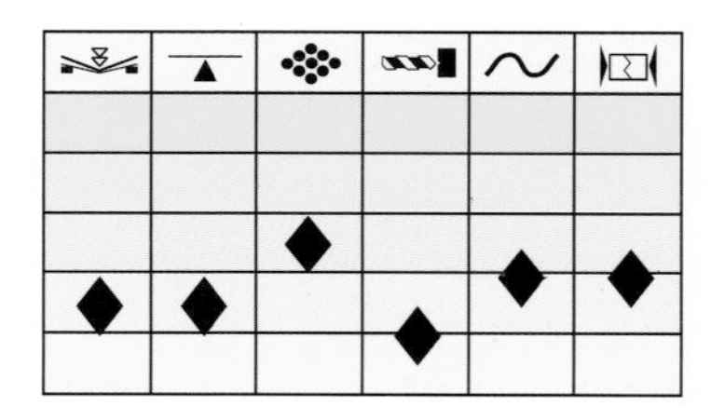

Piratinera guianensis **Family:** *Moraceae*

SNAKEWOOD

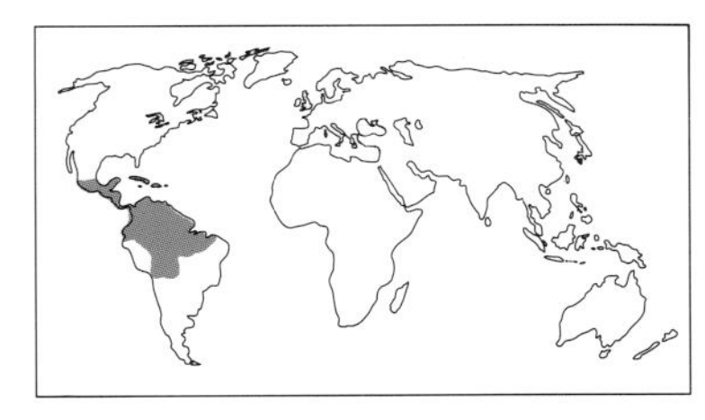

Where it grows

Many species of *Piratinera* occur in central and tropical South America, from the Amazon region of Brazil through Guyana, Venezuela, Colombia and Panama, southern Mexico and the West Indies. Commercial supplies come from Guyana, French Guiana and Surinam, with smaller quantities from Brazil, Bolivia and Trinidad. The wood is also known as letterwood (UK); amourette (France); bourra courra (Guyana); letterhout (Surinam); palo de oro (Venezuela); and leopard wood (USA). It grows to about 24m (78ft) and 0.3m-0.6m (1-3ft) in diameter.

Appearance

Snakewood is one of the most expensive woods in the world, and gets its name from the dark red to reddish-brown 'snakeskin' appearance of the heartwood. It has irregular black striped markings and dark spots like a leopard or hieroglyphic characters – hence the name letterwood. These dark spots and areas are the result of variations in the gummy deposits that fill the cell cavities. It is irregular grained, but the texture is moderately fine.

Properties

Snakewood is extremely hard and weighs 1,300 kg/m^3 (81 lb/ft^3) when air dried. The wood is difficult to dry with a tendency to degrade, and there is medium movement in service. The material is exceptionally strong in all categories, and is not suitable for steam bending. It is very difficult to work, with a severe blunting effect on tools. It needs care in gluing and finishing because of the resin, but can be polished to a smooth and beautiful finish. It is a very durable wood, immune to insect attack and extremely resistant to preservative treatment.

Uses

Snakewood is a superb turnery wood for walking-sticks, drum sticks, fishing rod butts, fancy handles for cutlery and umbrellas, fancy trinkets and brushbacks. It is used for violin bows and is the traditional wood for native archery bows. Selected logs are sliced for decorative veneers for cabinets or sawn for inlay work.

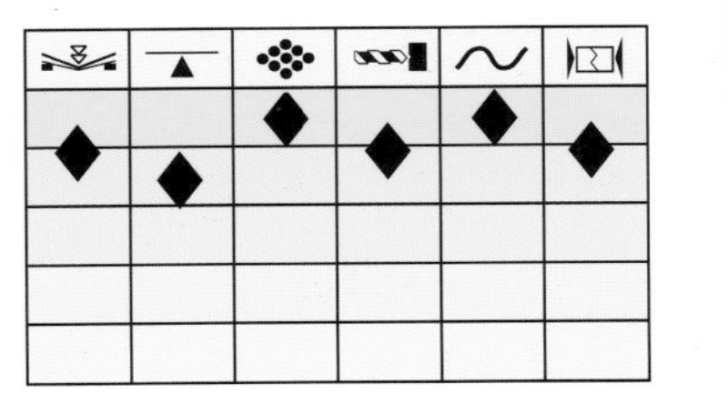

Platanus hybrida **Family:** *Platanaceae*

EUROPEAN PLANE

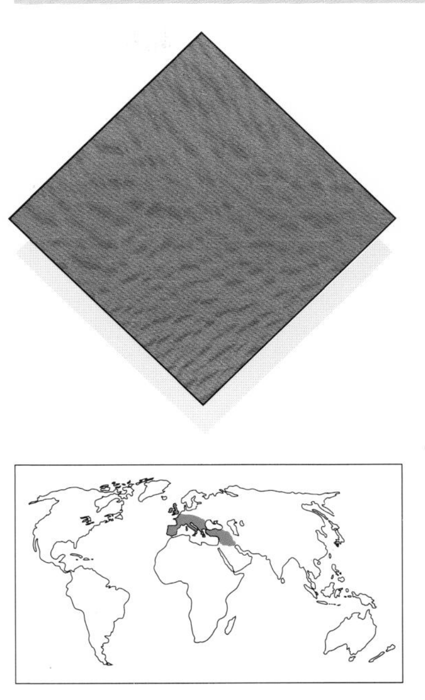

Where it grows

This tree occurs throughout Europe. The related species, *P.orientalis*, produces eastern plane tree, which occurs in south east Europe, Turkey and Iran; *P.occidentalis* produces American plane, known as 'sycamore' or buttonwood in the USA. Both have been hybridized to produce *P.hybrida*, which is found in British streets and is known as London plane or English plane. It grows to a height of 31m (100ft), with a diameter of lm (3ft).

Appearance

The sapwood is not very distinct from the heartwood, which is light reddish-brown with darker, conspicuous broad rays on quartered material. These produce a very decorative flecked figure, and such wood is sold as lacewood. The wood is straight grained, with a fine to medium texture.

Properties

Weight averages 620 kg/m^3 (39 lb/ft^3) when seasoned. It air dries fairly rapidly, is prone to splitting and has a tendency to distort. It is stable in service and has medium strength in all categories except for low stiffness, which earns it a very good steam-bending rating. The timber works well with both hand and machine tools, and has only a moderate blunting effect on the cutting edges of tools. Sharp cutters are required when planing or moulding, and there is a tendency to bind on saws. It screws and nails without difficulty, takes glue well and can be brought to an excellent finish. The sapwood is liable to attack by the common furniture beetle. The heartwood is perishable but permeable to preservation treatment.

Uses

This highly decorative wood is used for the interiors of high-class cabinets and furniture, joinery, carriage interiors, light construction, door skins and panelling; and in turnery for striking handles of tools and fancy handles. Logs are sliced for decorative veneers for cabinets and panelling. For inlay and marquetry work the veneers are treated as harewood, when the background becomes silver grey and the flecked rays retain their original colour.

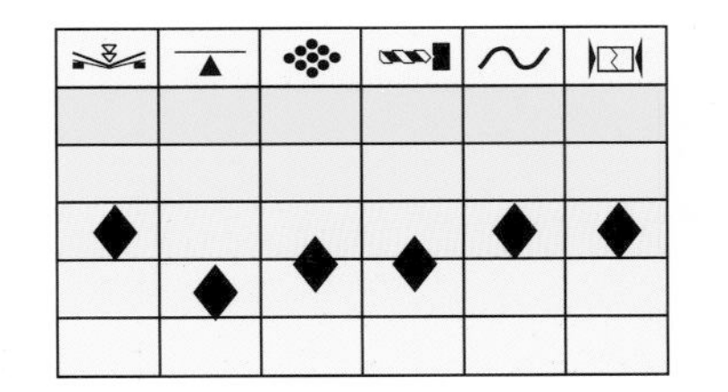

Podocarpus spp. **Family:** *Podocarpaceae*
PODO (S)

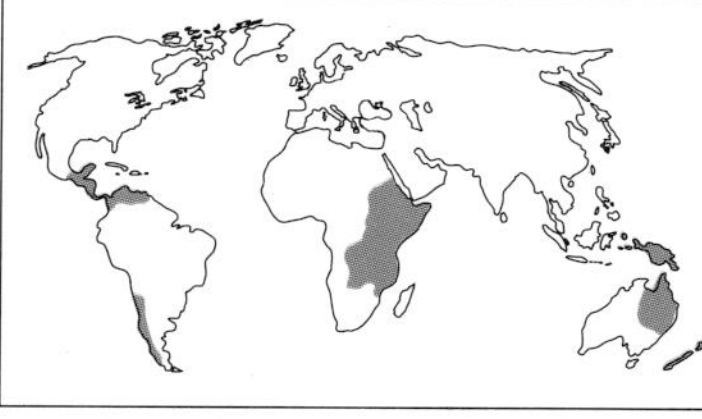

Where it grows
Podo grows in central America, East Africa and Asia. It is produced principally by *P.gracilior*, Pilg, in Kenya, Uganda, Ethiopia and Tanzania; *P.milanjianus*, Rendle is found in Kenya at higher altitudes, and also in Tanzania, Zambia and Zimbabwe. *P.usambarensis*, Pilg. occurs at lower altitudes in Kenya and Tanzania. It attains a height of 30.5m (100ft) or more, with a diameter of 0.75-1.2m (2.5-4ft). *P.gracilior*, growing to a much larger diameter, is known as yellow wood in South Africa and British Honduras; kahikatea, miro or matai in New Zealand; and manio in Chile.

Appearance
There is very slight distinction between the sapwood and the heartwood, which is light yellow-brown. The wood has no clearly defined growth rings, giving it a uniform texture. It is straight grained.

Properties
Weight is 510-630 kg/m^3 (32-39lb/ft^3) seasoned. Podo dries fairly rapidly, with a pronounced tendency to distort. There is medium movement in service. It has medium bending and crushing strength, very low stiffness, low resistance to shock loads and a moderate steam-bending rating. The wood works easily with both hand and machine tools, if care is taken as it is brittle. It requires pre-boring for nailing, but holds screws well and glues satisfactorily. It does not take stain uniformly, but can be painted or varnished well. The wood is liable to insect attack and is non-durable, but is permeable for preservation treatment.

Uses
Podo is used where durability is not of major importance for joinery and interior fittings in building construction, scaffold planks and boards, fascia boards, flooring and framing, kitchen furniture and mouldings. It is also used for turnery and, when treated, for weatherboards and boat building. *P.gracilior* is rotary cut for good-quality plywood and sliced for decorative veneers. *P.milanjianus* is used for low-grade plywood and corestock for laminated boards.

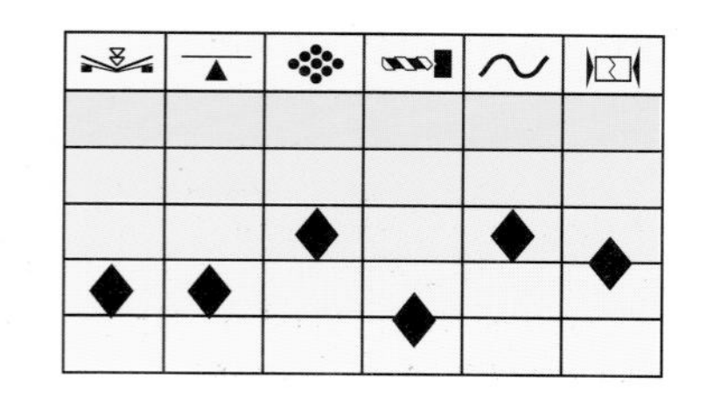

Podocarpus totara **Family:** *Podocarpaceae*
TOTARA (S)

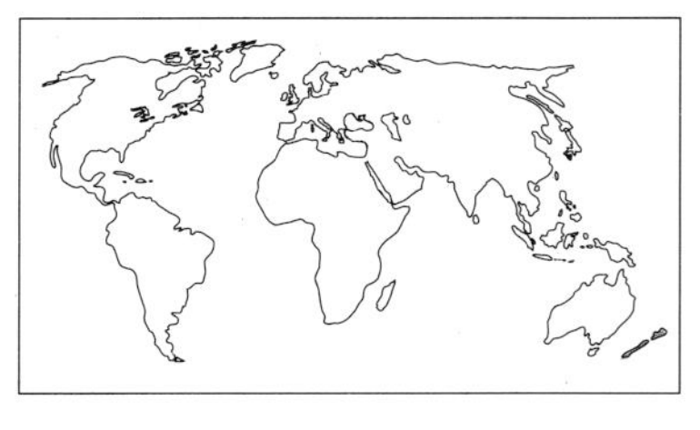

Where it grows
Totara occurs only in New Zealand and grows to an average height of 21.4m (70ft) with a diameter of 0.6-1.5m (2-5ft). It often reaches 39.6m (130ft). *P.hallii* grows to 7.3-18.3m (24-60ft).

Appearance
This is a straight-grained wood of medium reddish-brown with a fairly fine, even texture. *P.hallii* tends to be interlocked or wavy. In common with other softwoods of the southern hemisphere, growth rings are not clearly defined.

Properties
The weight when seasoned is 480 kg/m^3 (30 lb/ft^3). The wood dries fairly rapidly and well with little degrade and small movement in service. It has low bending strength and resistance to shock loads, but medium compressive strength, making it more suitable for columns and posts than beams and joists. It has a moderate steam-bending classification. The timber works easily with both hand and machine tools. *P.hallii* is more difficult to machine because of interlocked grain. Both species hold screws and nails without difficulty. Care is needed in gluing, and the resin content needs special treatment before painting to be sure of a good finish. The wood has natural durability and a high resistance to decay, but is liable to attack by the common furniture beetle. The heartwood is resistant to preservative treatment, but the sapwood is permeable.

Uses
Totara is the only softwood resistant to attack by marine borers, which makes it a very important wood in New Zealand for docks, wharf and harbour work and bridges. It has specialized uses in ship and boat building and for chemical vats. It is also widely used for flooring, cladding and shingles, and for all work in contact with the ground. It is the traditional wood for Maori carvings and canoe building.
Logs of *P.hallii* which have a more interlocked and wavy grain are sliced for decorative veneers and exported to Europe for pianos, cabinets and panelling.

					◆
		◆	◆	◆	
◆	◆				

Pometia pinnata **Family:** *Sapindaceae*

TAUN

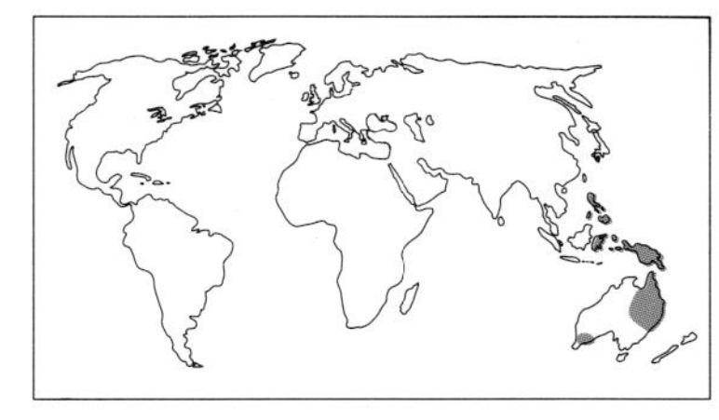

Where it grows
The tree occurs widely throughout the south Pacific and attains a height of 36-45m (118-148ft), with a diameter of 1.0m (3ft). It is also known as kasai, awa or ako (Solomons); tava (Western Samoa); ohabu (Papua New Guinea); malagai, malugay or agupanga (Philippines).

Appearance
The pink sapwood is not demarcated from the heartwood, which is a pale pinkish-brown maturing to a dull red-brown. It is usually straight grained, sometimes interlocked or wavy. Timber from Papua New Guinea and Mindanao has a dark red heartwood, a moderately coarse and uneven texture and is non-silicous and fissile; Philippines wood has a light red heartwood and a rather finer and smoother texture.

Properties
Average weight is 680-750 kg/m^3 (42-46 lb/ft^3) when seasoned. The wood needs care in drying to avoid warping and splitting. There is medium movement in service. The timber has high bending and crushing strengths, medium stiffness and resistance to shock loads, and a good steam-bending rating. It works readily with both hand and machine tools, with only a moderate blunting effect on cutting edges. It holds screws and nails without difficulty, glues satisfactorily, takes stain well and can be brought to a good finish. It is liable to blue stain if in contact with iron compounds in damp conditions. The wood is moderately durable, and resistant to preservation treatment.

Uses
Taun (or malagai) is widely used in the Philippines in constructional work for beams, joists, rafters, flooring, ceilings, interior trim and mouldings. It is also used for exterior joinery, door and window frames, boat planking and framing, wharf decking, capstan bars, masts and spars. In Australia it is used for furniture and cabinets, pianos and, as turnery wood, for bobbins, handles and fancy goods. It is rotary cut for plywood and sliced for decorative veneers.

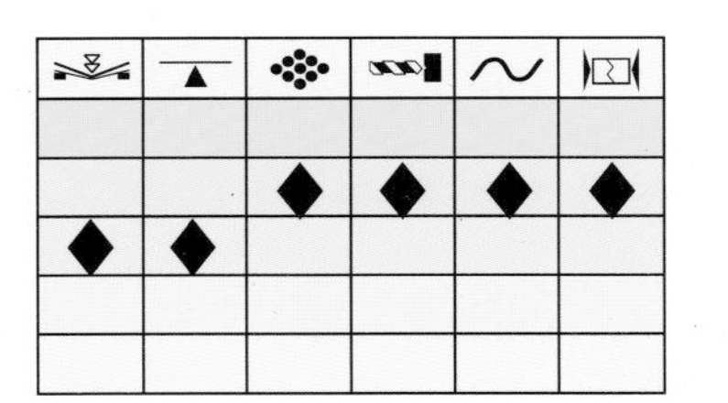

Populus tremuloides **Family:** *Salicaeae*

CANADIAN ASPEN

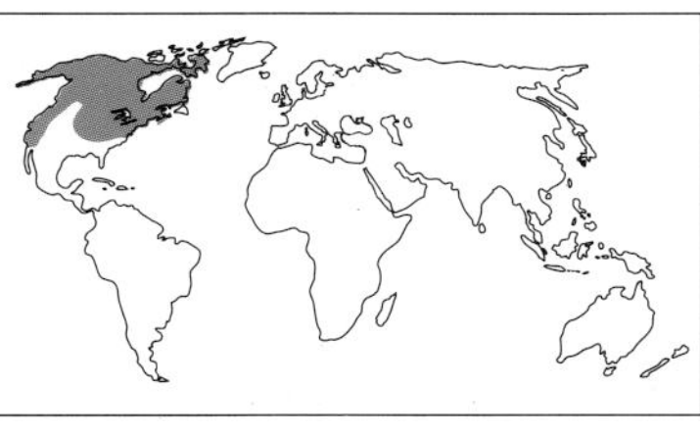

Where it grows
This tree, one of the most widely distributed in North America, occurs in the Lake States and the Rocky Mountain States of Canada, from Newfoundland and Nova Scotia to Alaska, and south along the Appalachians through New England to Minnesota. It grows to a height of 12-18m (40-60ft) and more, with a diameter of 0.2-0.3m (8-12in).

Appearance
The sapwood is not distinct from the heartwood, which is cream-white to very pale biscuit. It is mostly straight grained, occasionally wavy and inclined to be woolly. Sometimes pink, orange and golden streaks are seen; texture is fine and even.

Properties
The weight averages 450 kg/m^3 (28 lb/ft^3) when seasoned. Canadian aspen dries easily, but is inclined to warp and twist in drying, and distort unless care is taken in piling. There is small movement in service. The wood has low bending and crushing strength, low stiffness, medium resistance to shock loads, and a very poor steam-bending rating. It works easily with both hand and machine tools but tends to bind and tear on the saw. Very sharp, thin-edged tools are needed. The timber holds screws and nails without difficulty and glues well, but on woolly surfaces staining may be patchy. The surface can be painted or varnished to a good finish. The wood is non-durable and extremely resistant to preservative treatment.

Uses
This timber has a wide range of different uses; furniture interiors and fitments, brake blocks for iron wheels, vehicle bodies and the bottoms of trucks, wagons, food containers, chip and fruit baskets, boxes, packing cases, crates and pallets, wood wool and match splints. It is used in North America and Canada for veneer for matches, for chip and fruit baskets and punnets, and for wood, pulp and paper manufacture. It is also rotary cut for plywood and corestock for laminated boards and chipboard. Selected logs are sliced for highly decorative veneers for panelling.

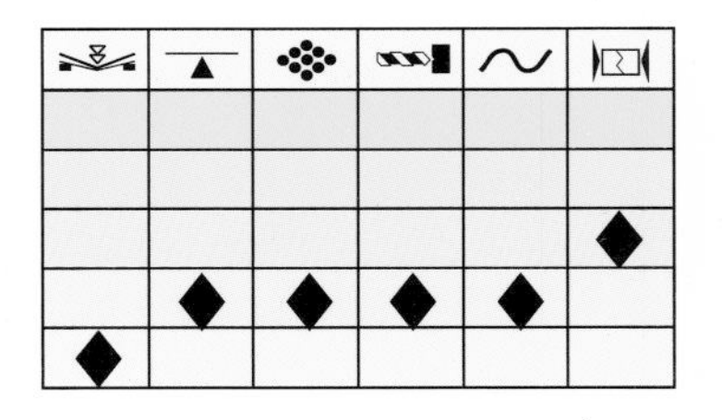

Populus spp. **Family:** *Salicaceae*
POPLAR

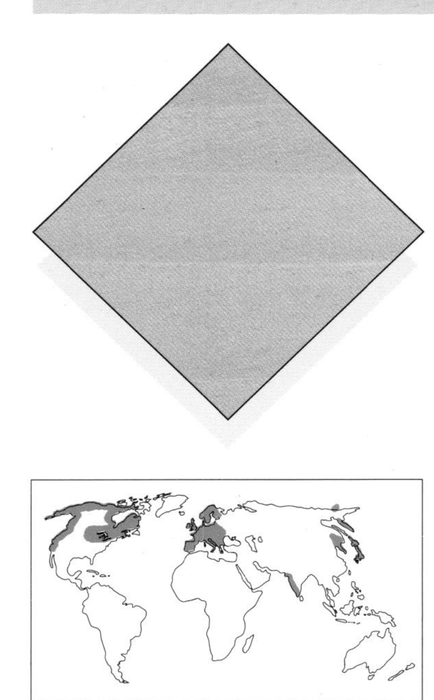

Where it grows
The principal commercial species of northern temperate regions are: *P.nigra*, L. which produces black poplar or European black poplar (Europe); *P.canadensis*, Moench. var *serotina*, which produces black Italian poplar (Europe); *P.robusta*, Schneid, which produces black poplar or robusta (Europe); *P.tremula*, L. which produces European, English, Finnish or Swedish aspen, according to the country of origin; *P.balsamifera*, L. syn. *P.tacamahaca*, Mill., which produces Canadian poplar, also known as tacamahac poplar and balsam poplar in the USA, and as balm poplar and black poplar in Canada. The black poplars reach a height of 30-35m (100-115ft); aspen grow to 18-25m (60-80ft), with a diameter of 0.9-1.2m (3-4ft) or more.

Appearance
The heartwood, not clearly defined from the sapwood, varies from cream-white to very pale straw and in some species to pale brown or pink-brown. It is usually straight grained and rather woolly, but with a fine, even texture.

Properties
The weight averages 450 kg/m^3 (28 lb/ft^3) when seasoned. The wood dries fairly rapidly and well with little degrade, and there is medium movement in service. It has low bending strength, very low stiffness and shock resistance, medium crushing strength and a very poor steam-bending rating. It works easily with hand or machine tools, but very sharp and thin-edged cutters are required. Poplar holds screws and nails well and glues easily, but staining can be patchy. The wood will paint and varnish to a satisfactory finish. Logs are liable to attack by beetles and wood boring caterpillars (*Cossidae*). The sapwood, which constitutes a large proportion of the tree, is perishable but permeable for preservation treatment.

Uses
Poplar is much less liable to splinter than softwoods and selected grades are used for interior joinery, furniture framing, toys and turnery. Logs are rotary cut for plywood and corestock and sliced for veneers. It is the main timber for match splints, wood wool, chip baskets and punnets. It also comes in for rough use for truck floors, packing cases, crates and pallets.

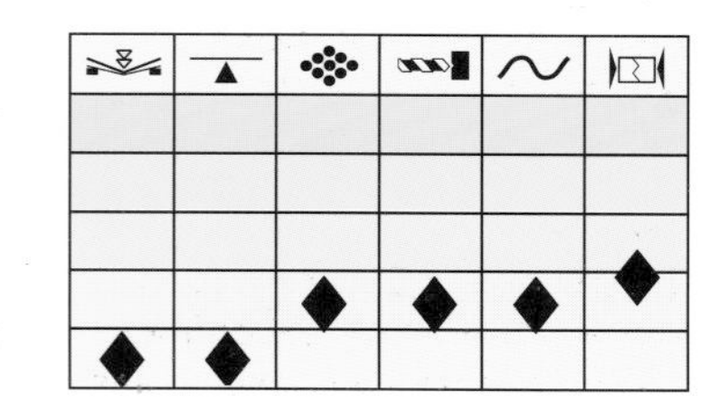

Prunus spp. **Family:** *Rosaceae*
CHERRY

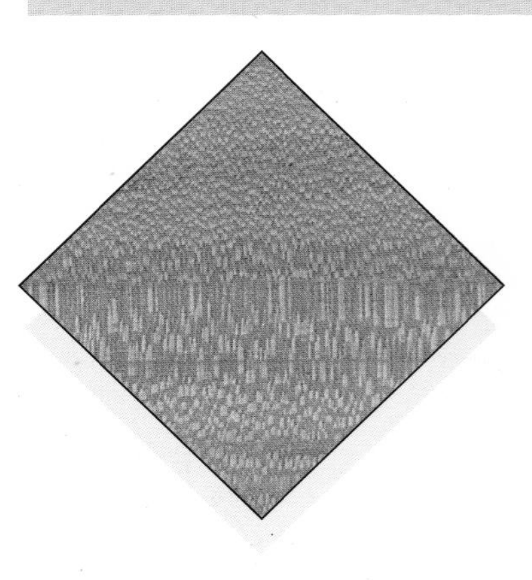

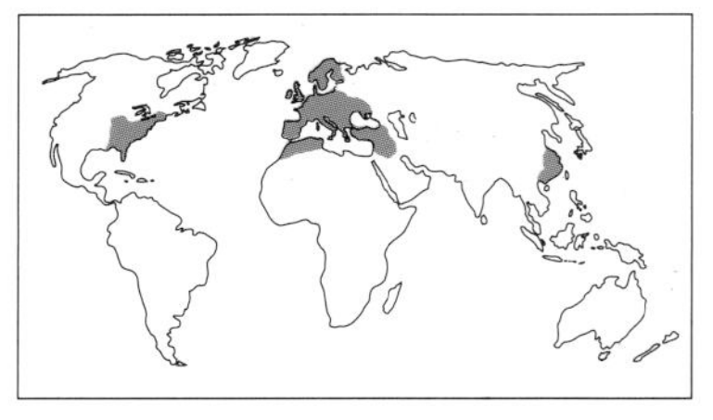

Where it grows
P.avium, L. syn. *Cerasus avium*, Moench. produces European cherry, also known as gean, mazzard, cherry or wild cherry (UK). It is native to Europe, and occurs in the mountains of north Africa. *P.serotina*, Ehrh. produces American cherry, also known as black cherry (Canada and USA) and cabinet cherry (USA), and occurs from Ontario to Florida and from the Dakotas to Texas. The tree grows to a height of 18-25m (60-80ft), with a diameter of about 0.6m (2ft). American black cherry reaches a height of 30.5m (100ft).

Appearance
The creamy-pink sapwood of *P.avium* is clearly defined from the heartwood, which is a pale pinkish-brown maturing to red-brown. *P.serotina* is a darker red-brown, with narrow brown pith flecks and small gum pockets. Both have straight grain and a fairly fine, even texture.

Properties
The weight of *P.avium* is 610 kg/m3 (38 lb/ft^3) and *P.serotina* is 580 kg/m^3 (36 lb/ft^3) when seasoned. The wood dries fairly rapidly, with a strong tendency to warp and shrink and with medium movement in service. It has medium bending and crushing strengths and resistance to shock loads, low stiffness and a very good steam-bending rating. It works well with both hand and machine tools, with a moderate blunting effect on cutting edges, but cross-grained timber tends to tear in planing. The wood holds screws and nails well, glues easily, and takes stain and polishes to an excellent finish. It is moderately durable; the sapwood is liable to attack by the common furniture beetle, but is almost immune to attack by powder post beetle. The heartwood is moderately durable and resistant to preservative treatment.

Uses
This handsome wood has a most attractive figure and colour and is used for cabinets and furniture making, carving and sculpture, and decorative turnery for domestic ware, shuttle pins, toys and parts of musical instruments. American cherry is used for pattern making, tobacco pipes, boat interiors and backing blocks for printing plates. Both types are rotary cut for plywood and sliced for decorative veneers for cabinets and panels.

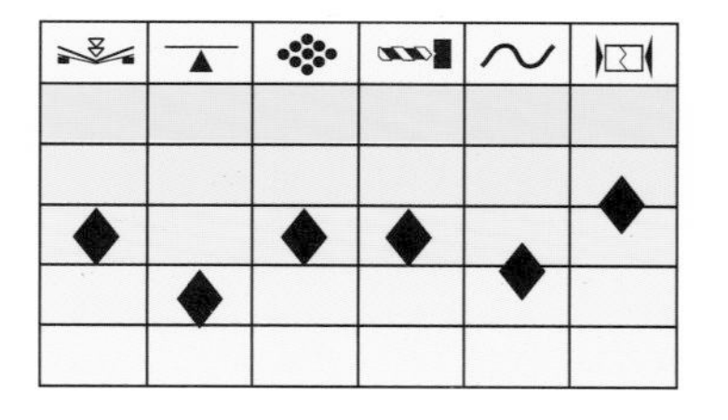

Pseudotsuga menziesii **Family:** *Pinaceae*

DOUGLAS FIR (S)

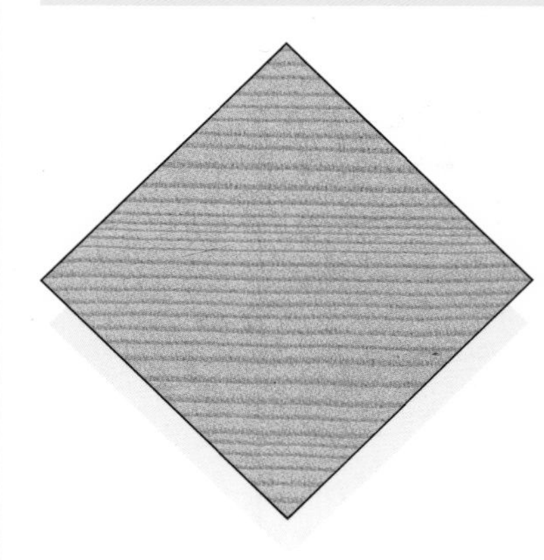

Where it grows

This softwood, not a true fir, is known as 'British Columbia pine' or 'Columbian pine' in the UK, and 'Oregon pine' in the USA. It occurs in abundance in British Columbia, Washington and Oregon, through Wyoming to southern New Mexico and west to the Pacific coast. It has been introduced to the UK, Australia and New Zealand. In Canada and America trees reach a height of 91.5m (300ft), averaging 45.7-61m (150-200ft), and 0.9-1.8m (3-6ft) in diameter. The bole is clear of branches for about two-thirds of its height, yielding a very high percentage of timber clear of knots and other defects.

Appearance

The sapwood is slightly lighter in colour than the heartwood, which is a light reddish-brown. There is a prominent growth ring figure on plain-sawn surfaces or rotary cut veneers. Grain is mostly straight, but often wavy or spiral. Texture is medium and uniform.

Properties

Weight is 530 kg/m^3 (33 lb/ft^3) seasoned. The wood dries fairly rapidly and well without much warping, but knots tend to split and loosen. Resin canals also tend to exude and show as fine brown lines on longitudinal surfaces. Douglas fir is stable in service, has high bending strength, stiffness and crushing strength, medium resistance to shock loads and a poor steam-bending rating. The wood works readily with both hand and machine tools. Cutters should be kept very sharp as there is a moderate blunting effect on tools. It is subject to beetle attack, is moderately durable and resistant to preservative treatment.

Uses

This is the world's most important source of plywood. In the solid, large baulks are used for heavy construction work, laminated arches, roof trusses, beams, interior and exterior joinery, dock and harbour work, marine piling, ship building, mining timber, railway sleepers, cooperage for vats and tanks for chemical plants, breweries and distilleries. Selected logs are sliced for decorative veneers for panelling.

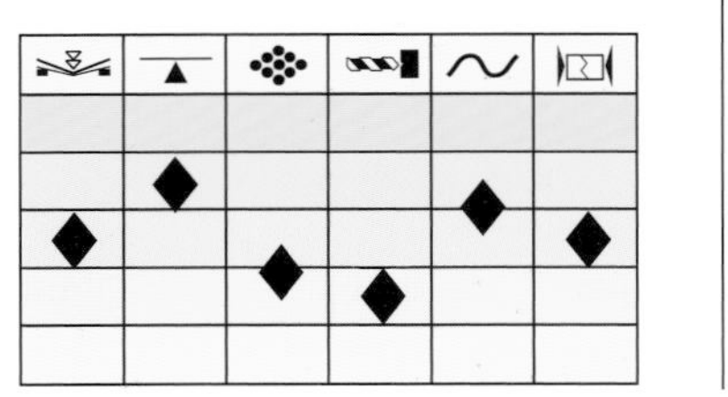

Pterocarpus angolensis **Family:** *Leguminosae*

MUNINGA

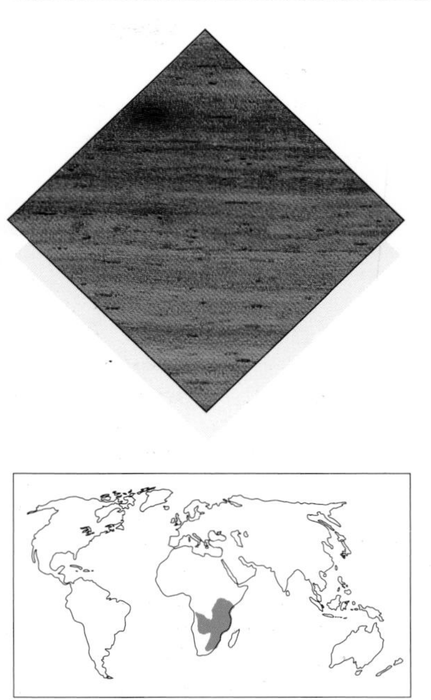

Where it grows

This wood occurs in the savannah forests of Tanzania, Zambia, Angola, Mozambique, Zimbabwe and South Africa. It grows to a height of 15m (50ft), sometimes 21m (70ft), and a diameter of 0.6m (2ft). It is also known as mninga (Tanzania); ambila (Mozambique); mukwa (Zambia and Zimbabwe); kiaat, kajat and kajatenhout (South Africa).

Appearance

The oatmeal coloured sapwood of muninga is clearly defined from the heartwood, which matures into a deep golden-brown with irregular darker chocolate-brown or dark red markings, sometimes marred with white spots or blotches. Although occasionally straight, the grain is often irregularly interlocked, which produces an attractive figure on quartered surfaces. The texture is fairly coarse and uneven.

Properties

The weight averages about 620 kg/m^3 (39 lb/ft^3) (the timber from Zimbabwe is softer and weighs only 540 kg/m^3/34 lb/ft^3), when seasoned. The wood has excellent drying properties and dries fairly slowly, especially in thicker sizes, with only a slight tendency to surface checking. It is exceptionally stable in use. This timber has medium bending strength, very low stiffness, low shock resistance, a high crushing strength and a moderate steam-bending classification. It works easily with both hand and machine tools, but with a tendency for the irregular grain to pick up when planing. It has good holding properties, glues well, and can be brought to an excellent finish. The sapwood is liable to attack by powder post beetle. The heartwood is very durable and resistant to preservative treatment.

Uses

Muninga, a very attractive wood, is excellent for turnery and is also used for carving and wood sculpture, high-class joinery and, in both solid and veneer form, for furniture, cabinets and panelling. It makes an excellent flooring with moderate resistance to wear and is suitable for domestic use, especially over under-floor heating systems.

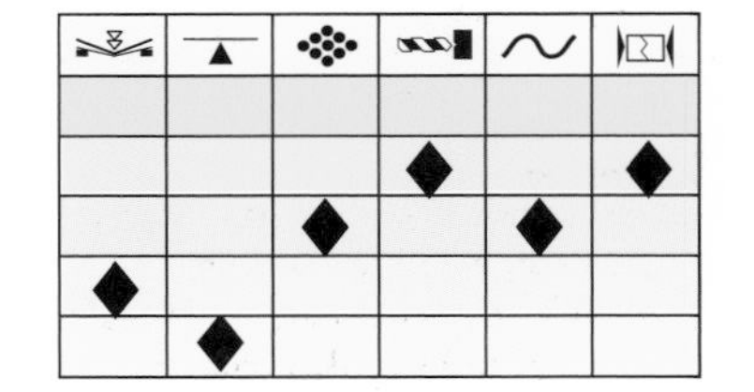

Pterocarpus dalbergoides **Family:** *Leguminosae*
ANDAMAN PADAUK

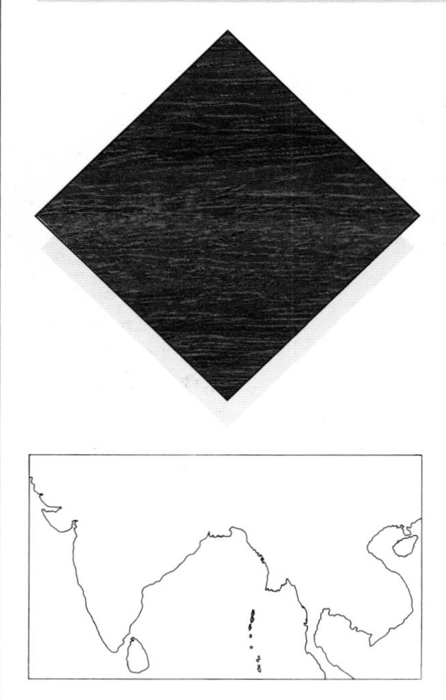

Where it grows
This tree grows only in the Andaman Islands. It is also known as Andaman redwood and vermillion wood in the USA. It grows to a height of 25-37m (80-120ft) with very large buttresses, and a diameter of 0.75-0.9m (2½-3ft) with a clean bole up to 12m (40ft).

Appearance
The narrow sapwood is fawn-grey, while the handsome heartwood varies from yellowish-pink with darker red lines to a rich, almost blood red crimson hue. It may also be reddish-purple with darker purple lines, sometimes with darker red or black streaks. Both types mature into a handsome reddish-brown. The grain is usually interlocked, with a medium to coarse texture, producing an attractive roe or curly figure on quartered surfaces.

Properties
Weight averages about 770 kg/m^3 (48 lb/ft^3) when seasoned. The wood kiln dries without undue degrade and is exceptionally stable in use. This heavy, dense timber has medium bending strength, low stiffness, low shock resistance and a high crushing strength. It is not suitable for steam-bending. There is high resistance in cutting and a moderate blunting effect on tools, especially when planing interlocked grain on quartered surfaces. It requires pre-boring for nailing, but holds screws and glues well. The wood can be brought to an excellent finish. It is very durable, noted for its high resistance to decay; it is moderately resistant to preservative treatment, but the sapwood is permeable.

Uses
This very attractive timber is used for high-class cabinets, furniture and billiard tables. It is ideal for interior joinery, especially for shop and bank fittings, and counters subject to wear. It is also popular for turnery for fancy handles and brushbacks, and for wood sculpture and carving. It is highly valued for exterior joinery and for boat building; it makes a hard-wearing and attractive domestic flooring. Logs are sliced for decorative veneers for cabinets and panelling. Note: *P.macrocarpus* produces Burma padauk, weighing 850 kg/m^3 (53 lb/ft^3), and used for similar purposes.

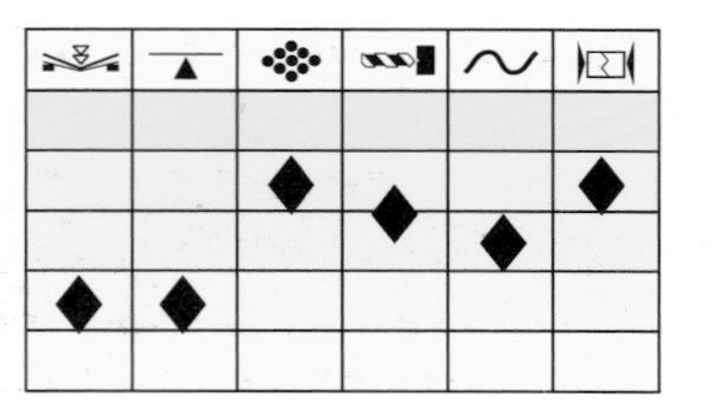

Pterocarpus soyauxii **Family:** *Leguminosae*
AFRICAN PADAUK

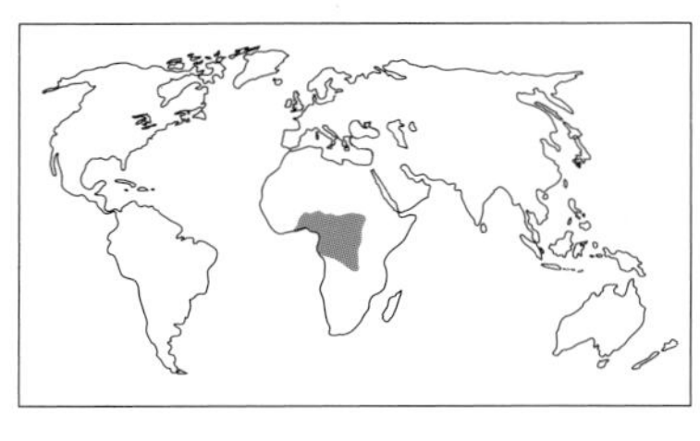

Where it grows
This tree occurs in central and tropical West Africa, and is known as camwood or barwood in the UK. It grows to a height of about 30.5m (100ft) and a diameter of 0.6-1m (2-3ft), with wide buttresses and a divided bole.

Appearance
When freshly cut, the heartwood is a very distinctive, vivid blood red maturing to dark purple-brown with red streaks. It is sharply demarcated from the straw-coloured sapwood. The grain is straight to interlocked and the texture varies from moderate to very coarse.

Properties
Padauk weighs 640-800 kg/m^3 (40-50 lb/ft^3) when seasoned and dries very well with the minimum of degrade. The wood is very dense, with high bending and crushing strengths and medium stiffness and resistance to shock loads, but it is not suitable for steam bending. It is exceptionally stable in service. The timber works well with both hand and machine tools, with only a slight blunting effect on cutting edges. It holds nails and screws without difficulty, glues easily and well and can be polished to an excellent finish. The wood is very durable and renowned for its resistance to decay. It is also moderately resistant to preservative treatment.

Uses
Padauk is world-famous as a dye wood, but it is also extensively used for high-class cabinets, furniture and interior joinery. As an excellent turnery wood it is used for knife and tool handles and fancy turnery. It is very good for carving and sculpture; other specialized uses include electrical fittings and spirit levels. It is an ideal boat-building wood. In Africa it is used for making paddles, oars and agricultural implements. Its abrasive qualities make it a good heavy-duty flooring of very attractive appearance, suitable for heavy pedestrian traffic – especially where under-floor heating is installed, as it has such good dimensional stability. Selected logs are sliced for decorative veneers for cabinets and panelling.

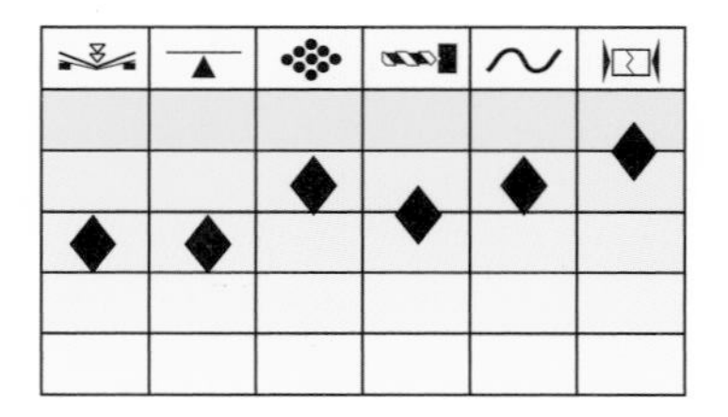

Pterocarpus indicus **Family:** *Leguminosae*
NARRA

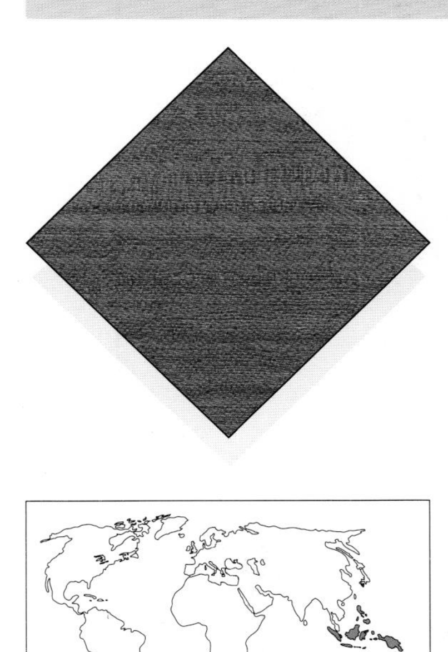

Where it grows
The tree, known as red or yellow narra in the USA, occurs throughout southern and south east Asia, producing Solomons padauk or Papua New Guinea rosewood. It is found extensively in the Philippines and is abundant in Cagayan, Mindoro, Palawan and Cotabato. It grows to a height of 40m (130ft), with a diameter of about 2m (6½ft).

Appearance
The light straw-coloured sapwood is quite clearly defined from the valuable heartwood, which varies from golden-yellow to brick red. Red narra comes from slow growing, ill-formed trees; timber from Cagayan, generally harder and heavier, is blood red. The wood produces a wide range of very attractive figures from a combination of terminal parenchyma and storied elements, and irregularities such as mottle, fiddleback, ripple and curl from wavy, interlocked, crossed and irregular grain. The texture is moderately fine.

Properties
When seasoned, the weight averages about 660 kg/m³ (41 lb/ft³), but darker wood is heavier. The wood dries fairly slowly but reasonably well, the red requiring more care than the yellow. Narra wood is very stable in use and has medium strength in all categories. Straight-grained wood works well with both hand and machine tools, with only a slight dulling effect on cutters; irregular grain requires a reduced cutting angle for best results. Narra takes screws and nails without difficulty, glues easily and polishes to an excellent finish. It is resistant to fungal and insect attack, very durable, and resistant to preservative treatment.

Uses
Narra is extensively used locally for high-class cabinets, furniture and joinery including interior trim for houses, boats and panelling. It makes a very good flooring timber. Specialized uses include sculpture and carving, cases for scientific instruments, parts for musical instruments, and sports goods. It is ideal for turnery of all kinds. It is sliced for decorative veneers for cabinets and panelling. The trade name amboyna is restricted to describe the highly valued and beautiful amboyna burr (burl) veneers.

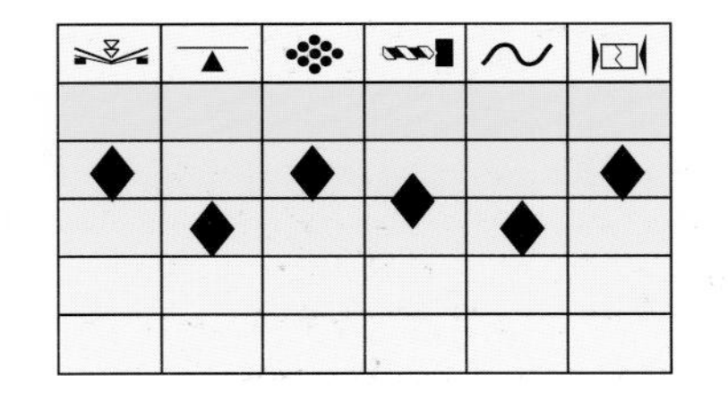

Pterygota spp. **Family:** *Sterculiaceae*
AFRICAN PTERYGOTA

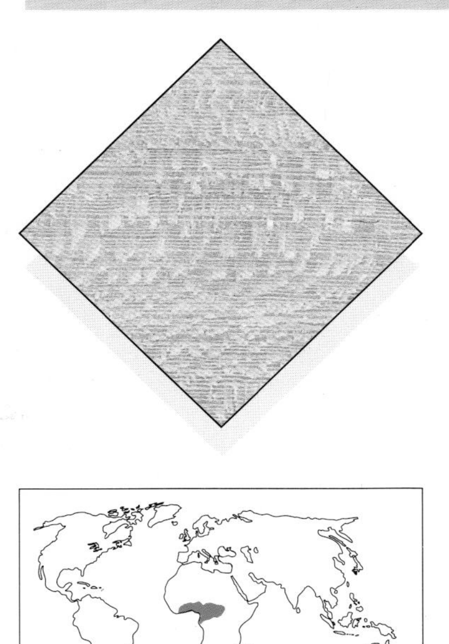

Where it grows
Two species combine to produce African Pterygota – *P.bequaertii*, De Wild., known as koto (Ivory Coast), which grows to a height of 23-30.5m (75-100ft), and *P.macrocarpa*, K. Schum, known as ware or awari in Ghana and kefe or poroposo in Nigeria, which grows to 37m (120ft). Both are found in the rain forests of Nigeria and the Cameroons. The diameter above the heavy buttresses is 0.5-1.2m (1½-4ft).

Appearance
Both sapwood and heartwood are creamy-white, with a grey tint and a shallowly interlocked grain, with small knot clusters and a moderately coarse texture. Quarter-sawn surfaces display a flecked ray figure.

Properties
The weight is 530-750 kg/m³ (33-47 lb/ft³) at 12% moisture content. The average for *P.bequaertii* is 650 kg/m³ (41 lb/ft³), and *P.macrocarpa* weighs 560 kg/m³ (35 lb/ft³) when seasoned. Both types dry fairly rapidly, and are prone to blue and grey fungal staining. Care is needed to avoid surface checking and extension of original shakes; cupping may also occur but distortion is usually small. There is medium movement in service. These medium-density woods possess medium bending and crushing strengths, low stiffness and low to medium resistance to shock loads, but have a very poor steam-bending classification. The wood works easily with both hand and machine tools provided cutting edges are kept sharp, as there is a medium blunting effect on tools. The timber glues easily, but there is a tendency for the wood to split when nailed near the edges. The grain needs filling in order to produce a good finish. This perishable species is permeable for preservative treatment.

Uses
Pterygota is used for furniture fitments, interior joinery and carpentry, also for packaging, boxes, crates and pallets. Logs are rotary cut into corestock and backing veneer for plywood, and sliced for decorative veneers, which require very careful handling as they are brittle and split easily.

Pyrus communis **Family:** *Rosaceae*
PEAR

Where it grows
There are numerous varieties of pear tree, all of which derive from the wild pear tree with harsh inedible fruit. It originated in southern Europe and western Asia; today, commercial timber comes from old orchards grown in Italy, Switzerland, France, Germany and the Tyrol. European trees grow to 9-12m (30-40ft) high, sometimes up to 18m (60ft) with a diameter of 0.3-0.6m (1-2ft), but often with poor stem form.

Appearance
The sapwood is pale yellow-apricot, and the heartwood varies from flesh tone to a pale pinkish-brown. Very minute pores and vessel lines account for an unusual uniformity and a very fine, even and smooth texture. The rays are faintly visible on quartered surfaces as tiny flecks of a deeper tone than the ground tissue. Grain is straight, sometimes irregular, which can produce a handsome mottled figure.

Properties
Weight is 700-720 kg/m^3 (43-44 lb/ft^3) when dry. Pear dries slowly, but with a definite tendency to warp and distort where irregular grain is present. The wood is strong, tough and stable in use, but as it is available only in small sizes, its strength is relatively unimportant. It is not used for steam bending. Pear machines well but is moderately hard to saw, and gives a fairly high blunting effect on cutters, which should be kept very sharp. The wood glues well, and gives excellent results with stain and polish. It is perishable, but permeable for impregnation.

Uses
Pear has excellent turning properties, and is used for fancy goods, wooden bowls, the backs and handles of brushes, umbrella handles, and measuring instruments such as set squares and T-squares. It is excellent for carving and sculpture because of the very fine grain, also for musical instruments such as recorders. It is stained black for violin fingerboards, and used for laps for polishing jewels for clocks and watches. It is sliced for decorative veneers for cabinets, panelling, marquetry and inlay work.

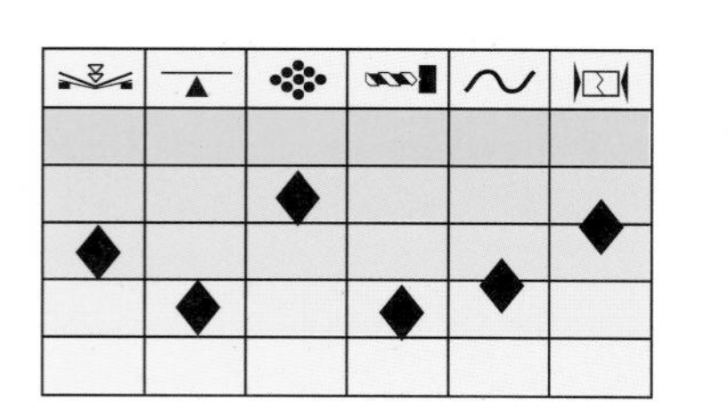

Quercus spp. **Family:** *Fagaceae*
AMERICAN RED OAK

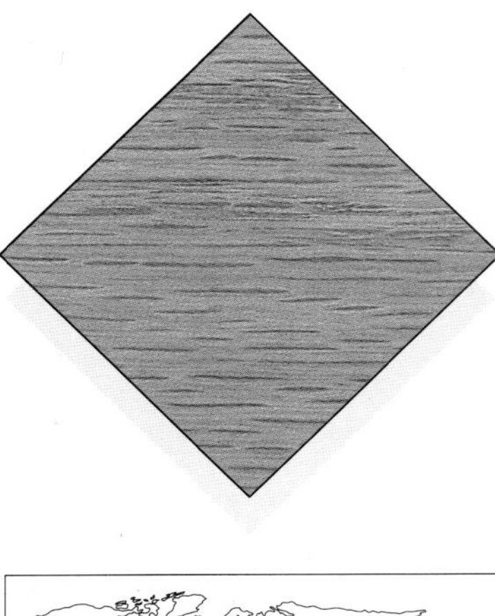

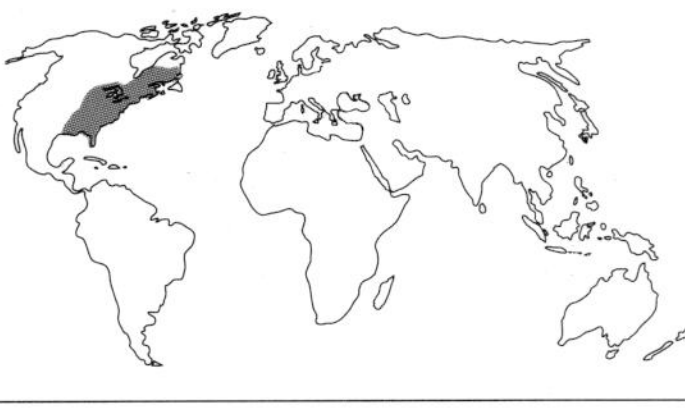

Where it grows
The red oak, growing in eastern Canada and North America, is also found in Iran as Persian oak, but is not as important commercially as the white oak. The principal species are *Q.rubra*, northern red oak (Canada and USA); and *Q.falcata* var. *falcata*, southern red oak, Spanish oak (USA). In Canada the red oak is more abundant than in the USA, and grows to a height of 18-21.5m (60-70ft), with a diameter of lm (3ft).

Appearance
The tree outwardly resembles the white oak, except that the heartwood varies from biscuit-pinkish to reddish-brown. The grain is usually straight; southern red oak is coarser textured than northern. Both species produce a less attractive figure than white oak because of the larger rays. There is a considerable variation in the quality of red oak; northern red oak grows comparatively slowly and compares favourably with northern white oak, while red oak from the southern States grows faster and produces a harder, heavier wood.

Properties
The average weight of both types is 770 kg/m^3(48 lb/ft^3) seasoned. It dries slowly, and care is needed in air and kiln drying to prevent degrade. There is medium movement in service. This dense wood has medium bending strength and stiffness, high shock resistance and crushing strength, and a very good steam-bending classification. It usually offers a moderate blunting effect on cutters, which should be kept sharp. It requires pre-boring; gluing results are variable, but red oak takes stain well and polishes to a good finish. The wood is non-durable, moderately resistant to preservative treatment, and unsuited for exterior work.

Uses
Red oak is too porous for tight cooperage purposes and its lack of durability and drying problems limit its use. However it is good for domestic flooring, furniture fitments, interior joinery and vehicle construction. Logs are rotary cut for plywood manufacture and sliced for decorative veneers. Persian red oak *(Q.castaneaefolia)* is impermeable and used for barrel staves.

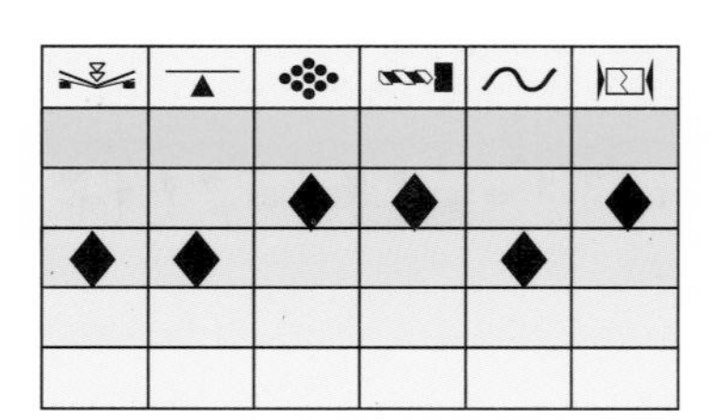

Quercus spp. **Family:** *Fagaceae*
OAK

Where it grows
The genus *Quercus* produces the true oaks and has more than 200 different species. Most white oaks occur in the temperate regions of the northern hemisphere; in warmer climates they grow in the montane forests. The white oaks occur in Europe, Asia Minor, North Africa, the eastern USA, south eastern Canada and Japan. The principal species producing European oak are: *Q.petraea* (sessile oak, durmast oak in the UK, and English/French/Polish/Slavonian oak according to origin); also *Q.robur* producing pedunculate oak (UK). American white oaks are *Q.alba* (white oak – USA), *Q.prinus* (chestnut oak), *Q.lyrata* (overcup oak) and *Q.michauxii* (swamp chestnut oak). Japanese oak is from *Q.mongolica*. White oaks are all similar in character and grow to 18-30m (60-100ft) and 1.2-1.8m (4-6ft) in diameter.

Appearance
The sapwood is lighter than the heartwood, which is light tan or yellow-brown, usually straight grained, but often irregular or cross-grained. It has a characteristic silver-grained figure on quartered surfaces because of broad rays and a moderately coarse texture.

Properties
Weight is 720-750 kg/m^3 (45-47 lb/ft^3). Volhynian, Slavonian and Japanese oaks weigh 660-672 kg/m^3 (41-42 lb/ft^3) when seasoned. The wood air dries very slowly with a tendency to split and check. These dense woods have high strength, low stiffness and resistance to shock loads, and a very good steam-bending rating. They are corrosive to metals, and liable to blue stain in damp conditions. Machining is generally satisfactory and they can be brought to an excellent finish. The wood is durable but liable to beetle attack; it is extremely resistant to preservative treatment though the sapwood is permeable.

Uses
One of the world's most popular timbers, light oaks are ideal for furniture and cabinet making. English oak is used for boat building, dock and harbour work, vehicle bodywork, high-class interior and exterior joinery and flooring. It is excellent for ecclesiastical sculpture and carving, and also tight cooperage for whisky, sherry and brandy casks. It is sliced for decorative veneers.

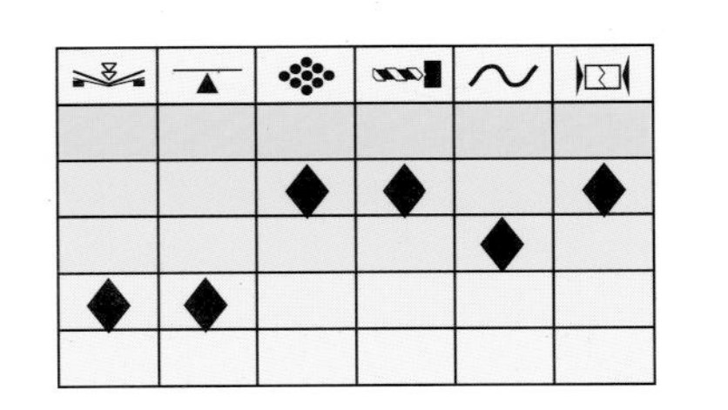

Salix spp. **Family:** *Salicaceae*
WILLOW

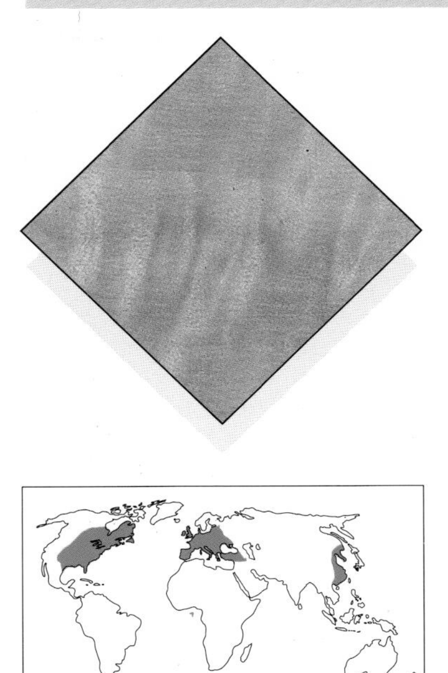

Where it grows
The main commercial species of willow occur in Europe, western Asia, north Africa and the USA. *S.alba* produces white willow and common willow (UK); *S.fragilis* produces crack willow (Europe and north Asia); *S.alba* var. *coerulea* produces cricket-bat willow and close-bark willow; *S.nigra*, black willow (USA). The 'weeping' willow thrives in wet acidic soil near streams and rivers. Many species are either pollarded or cut very low to induce long slender shoots known as osiers, used for basket and wicker work. Willows grow to a height of 21-27m (70-90ft) with a diameter of 0.9-1.2m (3-4ft).

Appearance
The timber has a white sapwood and a creamy-white heartwood with a pink tinge. It is typically straight grained, with a fine, even texture.

Properties
Weight averages 450 kg/m^3(28 lb/ft^3) seasoned. Cricket-bat willow weighs 320-420 kg/m^3 (21-26 lb/ft^3) dry. Willow dries well and fairly rapidly, degrade is minimal but it often retains pockets of moisture and special care is needed in checking the moisture content. Crack willow splits badly in conversion. The timber is stable in use. Willow has low bending and crushing strength, very low stiffness and resistance to shock loads, and a poor steam-bending rating. It works easily by both hand and machine, with a slight blunting of tools; sharp cutters are needed to avoid woolliness. The wood takes nails and glues well, and can be stained and brought to an excellent finish. Willow is perishable and liable to attack by powder post and common furniture beetles. It is resistant to preservation treatment but the sapwood is permeable.

Uses
Selected butts of cricket-bat willow are used for cleft cricket-bat blades. Other willows are used for artificial limbs, clogs, flooring, brake blocks in colliery winding gear, toys, sieve frames, flower trugs, vehicle bottoms, boxes and crates. Osiers are grown for wickerwork, fruit baskets and punnets. Pollarded shoots are used for stakes and wattle hurdles. Willow is sliced for attractive moiré, mottled figured veneers for architectural panelling, and treated as harewood for marquetry work.

Scottellia spp. **Family:** *Flacourtiaceae*
ODOKO

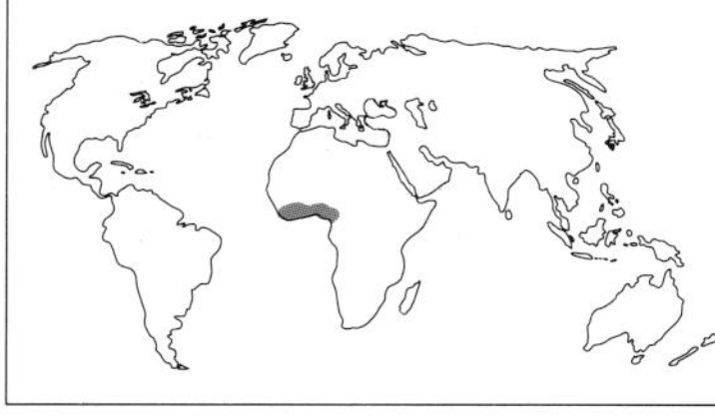

Where it grows
Several species of *Scottellia* occur in west Africa. *S.coriacea*, A.Chév. produces odoko and occurs from Liberia to southern Nigeria. Related species include *S.chevalieri*, producing akossika à grandes feuilles, and *S.kamerunensis*, which produces akossica à petites feuilles. Trees grow to about 30m (100ft) in height, with a diameter of 0.3-0.6m (1-2ft).

Appearance
The sapwood is not clearly defined from the heartwood, which is pale yellow to biscuit coloured with darker zones. Grain is generally straight, but sometimes interlocked, showing an attractive silver grain figure on quartered surfaces because of the broad rays. The texture is fine, even and lustrous.

Properties
Weight averages 620 kg/m^3 (39 lb/ft^3) when seasoned. The wood air dries fairly rapidly with a tendency to check and split, and it is prone to staining. In kilning hair shakes may develop and existing shakes extend, but warping is not serious. These woods have medium movement in service and possess medium bending strength, stiffness,and resistance to shock loads with a high crushing strength and a poor steam-bending classification. The material is fairly easy to work by both hand and machine, with only a slight blunting effect on tools. Brittleness can cause chipping or flaking on quartered stock. The wood requires pre-boring for nailing, but glues well and can be brought to an excellent finish. These species are non-durable and prone to discoloration by staining fungi, but are permeable for preservative treatment.

Use
Odoko is used for domestic woodware, furniture and fitments, interior joinery and light construction. It is used for carving and turnery, brushbacks and shoe heels. It makes an excellent flooring timber with high resistance to wear; locally in Nigeria it is used for cutting boards, models for casts, wooden spoons, bowls and rollers. Rotary cut logs produce veneers used for corestock, utility plywood and decorative panelling.

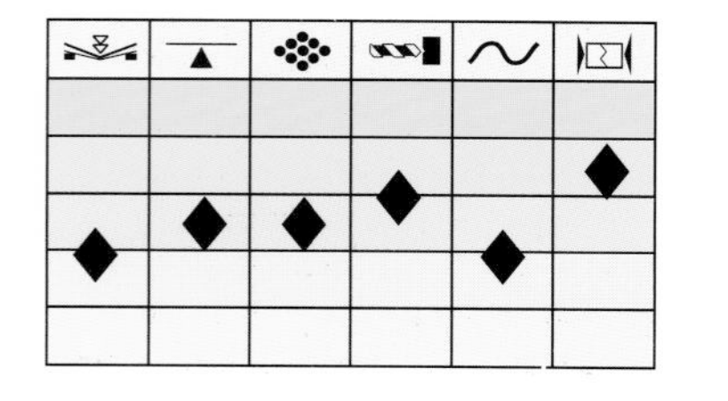

Sequoia sempervirens **Family:** *Taxodiaceae*
SEQUOIA (S)

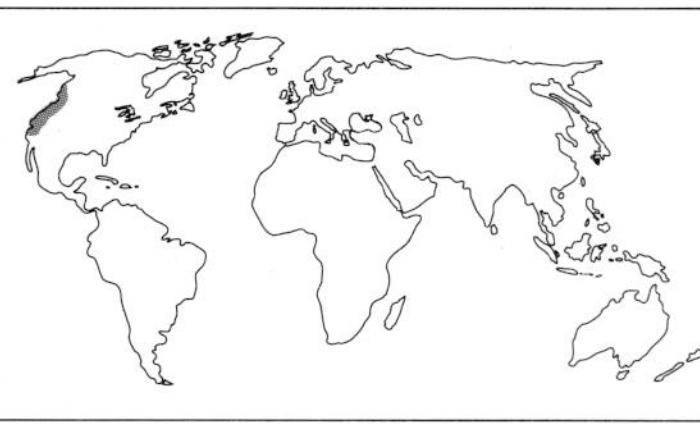

Where it grows
The greatest living organism on earth is the *Sequoia gigantea*, Decne. or the 'Big tree' of California. A protected species whose dimensions are unequalled, its wood is of no commercial value. Its cousin *S.sempervirens*, Endl., produces the sequoia, known as the Californian redwood in the UK and USA. It occurs in southern Oregon, and northern California. This massive tree reaches 61-104m (200-340ft), with a diameter of 3-4.6m (10-15ft).

Appearance
The sapwood is white and the heartwood a dull reddish-brown, with a distinct growth-ring figure. It is straight grained, and the texture varies from fine to coarse.

Properties
Average weight is 420 kg/m^3 (26 lb/ft^3) when seasoned. The wood air dries fairly rapidly and well with little degrade, and is stable in use. It has low bending and crushing strength, low resistance to shock loads, very low stiffness, and a poor steam-bending rating. It works easily with both hand and machine tools, but it is prone to splintering. Sharp tools are needed to reduce chip-bruising. It takes nails well, but stains with alkaline adhesives, which should be avoided. Sequoia provides a good finish. It is invaluable for use in exposed situations for its durability, and is resistant to preservative treatment.

Uses
The excellent durability of sequoia makes it ideal for wooden pipes, flumes, tanks, vats, silos and slats in water-cooling towers. It is also used for coffins. It is extensively used for posts, interior or exterior joinery, organ pipes, and for exterior cladding and shingles, windows and doors of buildings. The cinnamon-coloured very thick bark is used in the manufacture of fibre board. Logs are rotary cut for plywood faces, and selected logs are sliced for highly valued veneers. The extremely attractive burrs (burls) are marketed as Vavona burr.

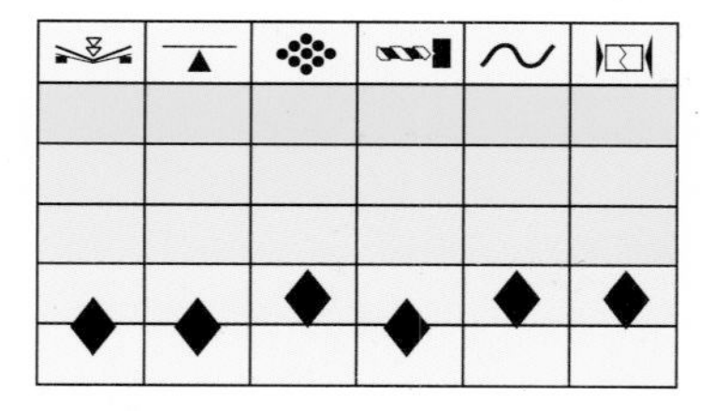

Shorea spp. **Family:** *Dipterocarpaceae*
LIGHT/DARK RED MERANTI/SERAYA/LAUAN

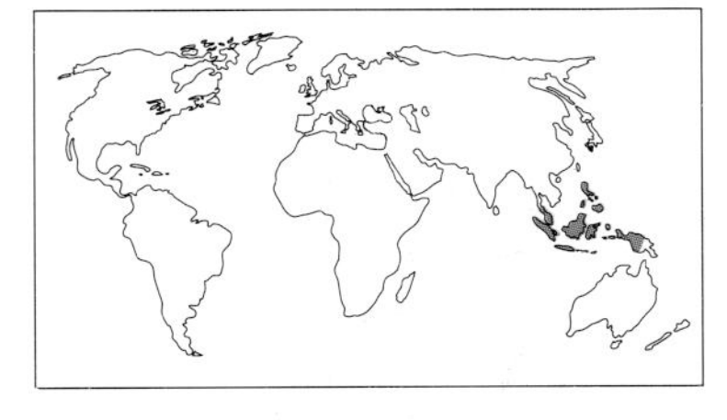

Where it grows
A large number of species of the genus *Shorea* occur in south east Asia, which produce meranti, seraya or lauan. Meranti is from Malaya, Sarawak and Indonesia; seraya from Sabah, and lauan from the Philippines. These timbers vary in colour and density and are grouped as follows: light red meranti, light red seraya and white lauan; dark red meranti, dark red seraya and red lauan. These trees reach 60-70m (200-225ft) in height and 1-1.5m (3-5ft) in diameter.

Appearance
Because of the number of species, these details are very general; in the first group the colour is pale pink to red, and in the second it is medium to dark red-brown with white resin streaks. Both have interlocked grain and a rather coarse texture.

Properties
The light red timbers of the first group have an average weight of 550 kg/m³ (34 lb/ft³), and the dark red woods weigh about 670 kg/m³ (42 lb/ft³) on average when seasoned. Drying is usually fairly rapid, without serious degrade. Both timbers are stable in use, but the first light red group is much weaker than the darker timbers. Both types have medium bending and crushing strengths, low stiffness and shock resistance and a poor steam-bending rating. They work well with both hand and machine tools, hold screws and nails satisfactorily, can be glued easily and produce a good finish when filled. The light red timbers are non-durable, and the dark red group are moderately durable and resistant to impregnation.

Uses
The light red timbers are used for interior joinery, light structural work, domestic flooring, cheap furniture and interior framing. The dark red second group is used for similar purposes, plus exterior joinery, cladding, shop fitting and boat building. Logs of both groups are used for plywood manufacture and sliced for decorative veneers for cabinets and panelling.

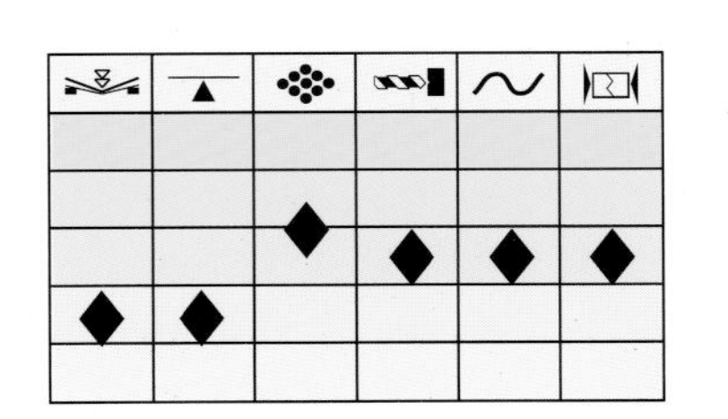

Shorea spp. **Family:** *Dipterocarpaceae*
WHITE/YELLOW MERANTI, YELLOW SERAYA

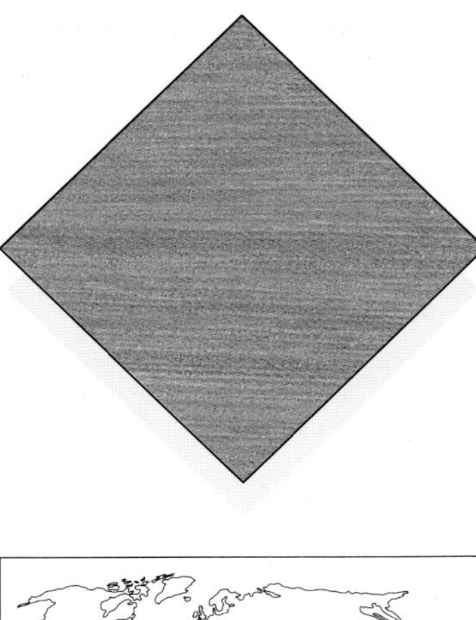

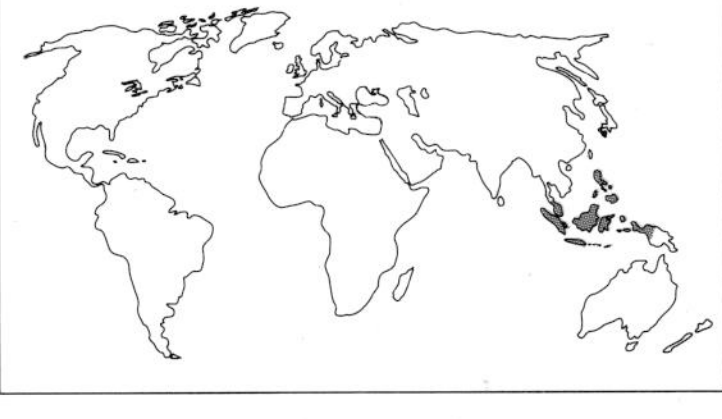

Where it grows
The name meranti is applied to timbers from Malaya, Sarawak and Indonesia; seraya to timber from Sabah. However, white meranti is not the equivalent of white seraya or white lauan. Yellow meranti of Malaysia is also known as meranti damar hitam, and yellow meranti of Brunei and Sarawak as lun or lun kuning. Yellow seraya of Sabah is known as seraya kacha and selangan kacha. White meranti, or melapi, occurs in West Malaysia, Sarawak, Brunei and Sabah, and is also known as lun or lun puteh in Sarawak and as melapi in Sabah. Trees grow to 60m (200ft) tall and 1.0-1.5m (3-5ft) diameter.

Appearance
Yellow meranti and yellow seraya sapwood is lighter in colour and distinct from the heartwood, which is light yellow-brown and matures to a dull yellow-brown. They have shallowly interlocked grain and moderately coarse texture. The sapwood of white meranti is well defined, and the heartwood matures to a much lighter golden-brown. The texture is moderately coarse but even.

Properties
Yellow meranti/seraya weighs on average 480-670 kg/m³ (30-42 lb/ft³) and white meranti weighs 660 kg/m³ (41 lb/ft³) when seasoned. It dries slowly but well, apart from a tendency to cup, and brittleheart is sometimes present. It has low bending strength and shock resistance, medium crushing strength, very low stiffness and a moderate steam-bending rating. White meranti/seraya dries without serious degrade, has medium strength in all categories and a very poor steam-bending rating. The timbers are stable in use. Yellow meranti works satisfactorily, but the silica in white meranti has a severe blunting effect on tools, making tipped saws a necessity. Both hold nails and screws well, glue easily, and when filled provide a good finish. They are moderately durable and the sapwood is permeable.

Uses
Both groups are used for light construction, interior joinery, furniture and flooring. White meranti is also used for exterior joinery, ship and boat planking, shop fitting and carriage framing. Logs are cut for plywood and veneers.

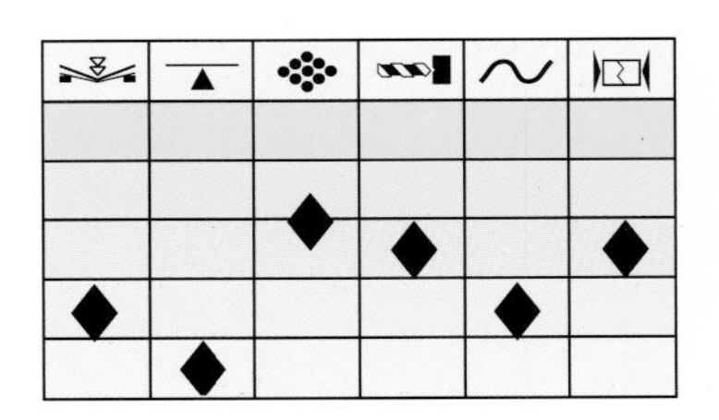

Swietenia spp. **Family:** *Meliaceae*
AMERICAN MAHOGANY

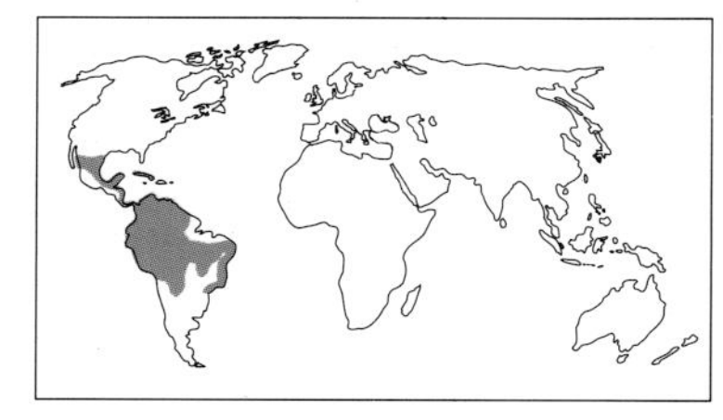

Where it grows
Cuban or Spanish mahognay (*S.mahoganii*, Jacq.) was brought to Europe by the Spanish in the late 16th century. Since the 18th century it has been the most cherished cabinet wood in the world and is now rare. Today, commercial supplies are of *S.macrophylla*, King, which occurs from southern Mexico south along the Atlantic coast from Belize to Panama; in Colombia and Venezuela, Peru, Bolivia and Brazil. The timber is named after its country of origin. Forest trees grow to 45m (147ft) in height, but plantation trees average 30m (100ft) with a diameter of 1.2–1.8m (4–6ft).

Appearance
The sapwood is yellowish-white, and the heartwood varies from pale red to dark red-brown in heavier timber. It is mostly straight grained, but pieces with interlocked or irregular grain produce a highly valued beautiful figure on quartered surfaces. It has a moderately fine to medium and uniform texture.

Properties
The weight averages 540 kg/m³ (34 lb/ft³) when seasoned. The wood can be air or kiln dried rapidly and well without warping or checking, but tension wood and gelatinous fibres can result in high longitudinal shrinkage. The timber is stable in use. It has low bending strength, very low stiffness and shock resistance, medium crushing strength, and a moderate steam-bending rating. This is one of the best woods to use with either hand or machine tools, and sharp cutting edges will overcome woolliness. It holds nails and screws well, glues well and gives an excellent finish. Liable to insect attack, this timber is durable and extremely resistant to preservative treatment.

Uses
Mahogany goes into high-class cabinets and reproduction furniture, chairs, panelling, interior joinery, domestic flooring, exterior joinery, boat building, pianos and burial caskets. It is excellent for carving, engravers' blocks and engineers' patterns, moulds and dies. It is cut for plywood and sliced to produce a wide range of fiddleback, blister, roe, striped, curl or mottled figures in veneer for cabinets and panelling.

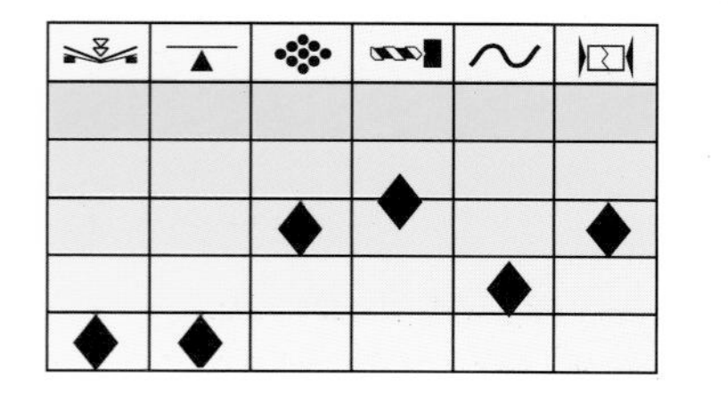

Taxus baccata **Family:** *Taxaceae*
YEW(S)

Where it grows
The common yew is widely distributed through Algeria, Asia Minor, the Caucasus, northern Iran, the Himalayas and Burma, and throughout Europe. It grows to a height of 12-15m (40-50ft) with a short twisted or fluted bole, and often consists of several vertical shoots which have fused together to form multiple stems.

Appearance
The heartwood varies from orange-brown streaked with darker purple, to purplish-brown with darker mauve or brown patches, and clusters of in-growing bark. The irregular growth pattern produces wood of varying ring widths, which combine with narrow widths of dense latewood to give a highly decorative appearance.

Properties
Yew is among the heaviest and most durable of softwood timbers, and weighs 670 kg/m³ (42 lb/ft³) when seasoned. The wood dries fairly rapidly and well, with little degrade if care is taken to avoid shakes developing or existing shakes from opening. Distortion is negligible and it is stable in use. This hard, compact and elastic wood has medium bending and crushing strength, with low stiffness and resistance to shock loads. Straight-grained air-dried yew is one of the best softwoods for steam bending, even though it is inclined to check during drying. It works well in most hand and machine operations, but when irregular, curly or cross grain is present, it tears easily. Nailing requires pre-boring and the oiliness of the wood sometimes interferes with gluing, but it stains satisfactorily and provides an excellent finish. Yew is durable, but not immune from attack by the common furniture beetle; it is resistant to preservative treatment.

Uses
For many centuries, yew was prized for the English archer's longbow. It is an excellent turnery and carving wood, and is used for reproduction furniture making, interior and exterior joinery, garden furniture, fences and gate posts. It is the traditional wood for Windsor chair bentwood parts. It is sliced for highly decorative veneers and burrs (burls).

Tectona grandis **Family:** *Verbenaceae*
TEAK

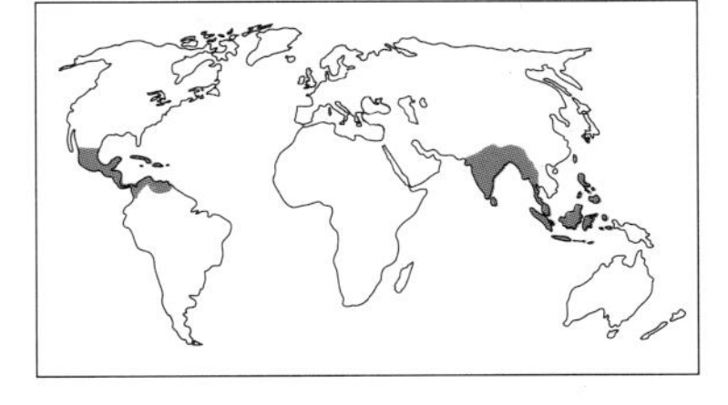

Where it grows

Teak is indigenous to Burma and grows extensively throughout India, in Thailand, Indonesia and Java. It has been introduced into Malaysia, Borneo and the Philippines, tropical Africa and Central America. In favourable locations it grows to 40-45m (130-150ft), with a diameter of 1.8-2.4m (6-8ft), but averages 9-11m (30-35ft) with a diameter of about 0.9-1.5m (3-5ft).

Appearance

True Burma teak has a narrow, pale yellow-brown sapwood and a dark golden-brown heartwood, darkening on exposure to mid or dark brown. Other types have a rich brown background colour with dark chocolate-brown markings. The grain is mostly straight in Burma teak, and wavy in Indian teak from Malabar. The texture is coarse and uneven and it feels oily to the touch.

Properties

Weight is 610-690 kg/m^3 (38-43lb/ft^3), averaging 640 kg/m^3 (40 lb/ft^3) when seasoned. The wood dries rather slowly, and there is small movement in service. Teak has medium bending strength, low stiffness and shock resistance, high crushing strength, and a moderate steam-bending classification. It works reasonably well with both hand and machine tools and has a moderately severe blunting effect on cutting edges, which must be kept sharp. Machine dust can be a severe irritant. Pre-boring is required for nailing, and it glues well and can be brought to an excellent finish. The timber is very durable.

Uses

Teak enjoys a well-deserved reputation for its strength and durability, stability in fluctuating atmospheres and its excellent decorative appearance. Its vast number of uses include furniture and cabinet making, decking for ship and boat building, deck houses, handrails, bulwarks, hatches, hulls, planking, oars and masts. It is also used for high-class joinery for doors, staircases and panelling, and externally for dock and harbour work, bridges, sea defences, and garden furniture. It makes a very attractive flooring. Good chemical resistance enables it to be used for laboratory benches, fume ducts and vats. It is cut for all grades of plywood and decorative veneers.

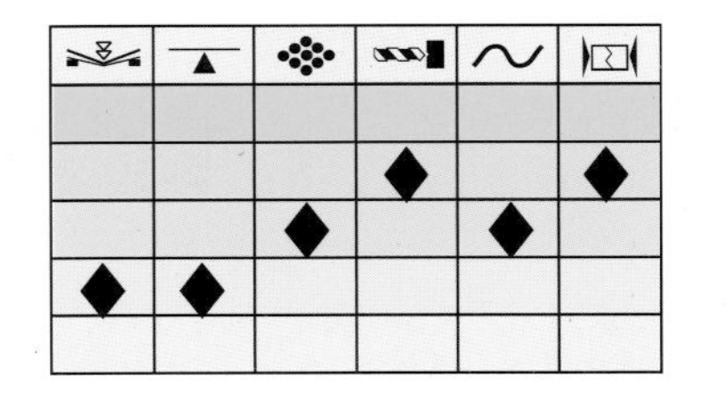

Terminalia spp. **Family:** *Combretaceae*
'INDIAN LAUREL'

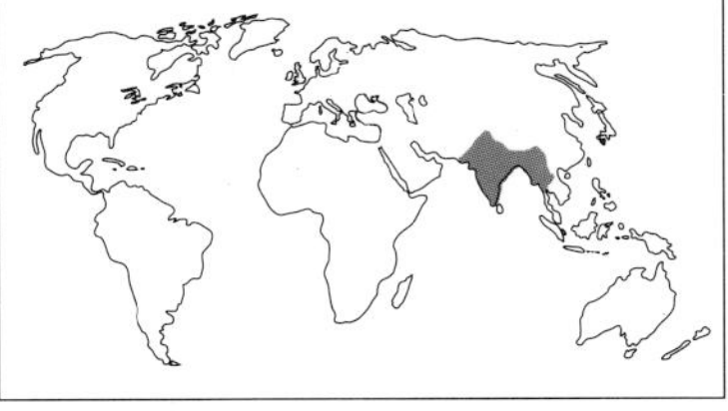

Where it grows

The timber is derived from *T.alata*, Roth., *T.coriacea*, W & A., and *T.crenulata*, Roth. It is not a true laurel. It occurs in India, West Pakistan, Bangladesh and Burma and is known as taukkyan (Burma) and asna, mutti or sain (India). It reaches a height of 30m (100ft) with a straight bole of 12-15m (40-50ft) and a diameter of lm (3ft).

Appearance

The sapwood is reddish-white and sharply defined from the heartwood, which varies from light brown with few markings or finely streaked with darker lines, to dark brown banded with irregular darker brown streaks. The grain is fairly straight to irregular and coarse textured.

Properties

The weight averages about 850 kg/m^3 (53 lb/ft^3) when seasoned. Indian laurel is a highly refractory timber to dry, prone to surface checking, warping and splitting; it must be dried slowly to avoid degrade. There is small movement in service. It is very dense, has medium bending strength, shock resistance and stiffness, high crushing strength and a poor steam-bending classification. It is rather difficult to work with hand tools and moderately hard to machine, especially when interlocked grain is present. It is difficult to nail but holds screws well and glues firmly. The wood requires filling for a good finish, which is best achieved with oil or wax. The sapwood is subject to attack by the powder post beetle but the heartwood is moderately durable and resistant to preservative treatment. The sapwood is permeable for treatment.

Uses

This is one of India's most valuable woods, used extensively for furniture and cabinet making, high-class interior joinery for panelling, doors and staircases. It is an excellent turnery wood for tool handles, brushbacks and police batons. It is used in India for boat building and, when treated, for posts and pit props, harbour and dock work. Logs are sliced into flat-cut or quartered highly decorative veneers for cabinets and panelling.

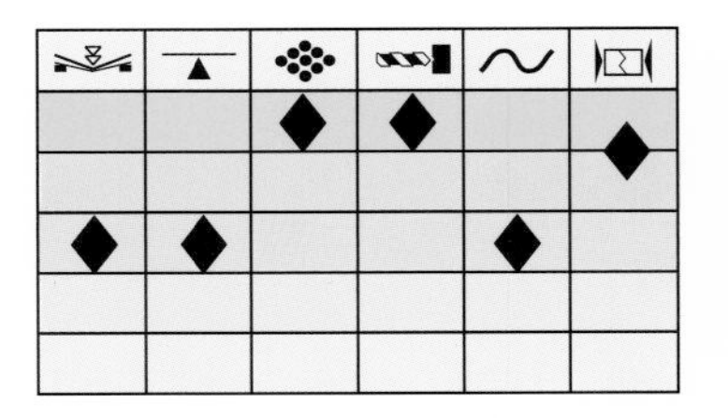

Terminalia bialata **Family:** *Combretaceae*
INDIAN SILVER GREYWOOD

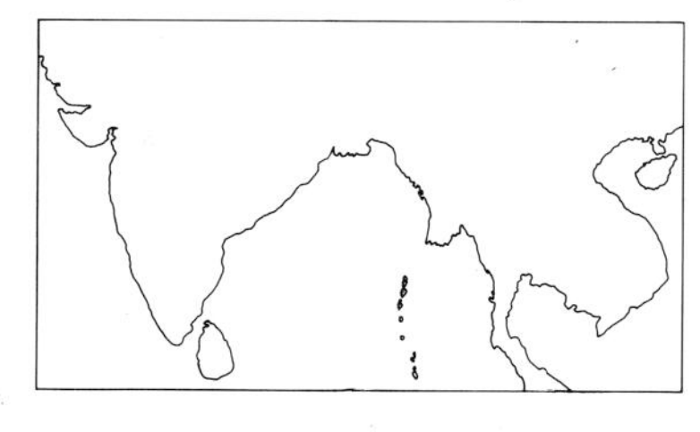

Where it grows
The species *T.bialata* produces a pale wood known as white chuglam and the darker, highly figured Indian silver greywood. Both occur on the Andaman Islands. The trees attain a height of 30-50m (100-160ft), with a diameter of 0.8-1.5m (2½-5ft).

Appearance
The sapwood, which may comprise the whole tree, is a uniform greyish-yellow – this is white chuglam. In other logs there is a false heartwood of grey to smoky yellow-brown or sometimes olive to nut brown, banded with darker brown streaks. This highly ornamental marbled effect is described as Indian silver greywood. Grain in both is usually straight, sometimes wavy. The texture is medium.

Properties
Weight averages 670 kg/m^3 (42 lb/ft^3) when seasoned. The wood air dries very well, with some tendency for end splits and checks to develop. Conversion while the timber is green and prompt storage under cover for air drying will prevent appreciable degrade. Kiln drying is trouble free. There is small movement in service. This dense timber has medium strength in all categories, but a poor steam-bending classification. It works easily with hand and machine tools, and produces a smooth finish on straight-grained surfaces. It holds nails and screws well, and can be glued, stained and polished to a good finish. The figured timber is moderately durable. The sapwood is liable to attack by powder post beetle. Indian silver greywood is extremely resistant, and white chuglam moderately resistant to preservative treatment.

Uses
The highly figured Indian silver greywood is used for all kinds of decorative work for furniture, cabinet work and high-class joinery, including decorative panelling in public buildings, railway coaches and ocean liners. The plainer white chuglam is used in India for flooring and interior joinery including staircases; mathematical instruments, boat fittings, plain furniture, edge lippings and tea chests. Selected logs of Indian silver greywood are sliced for highly decorative veneers.

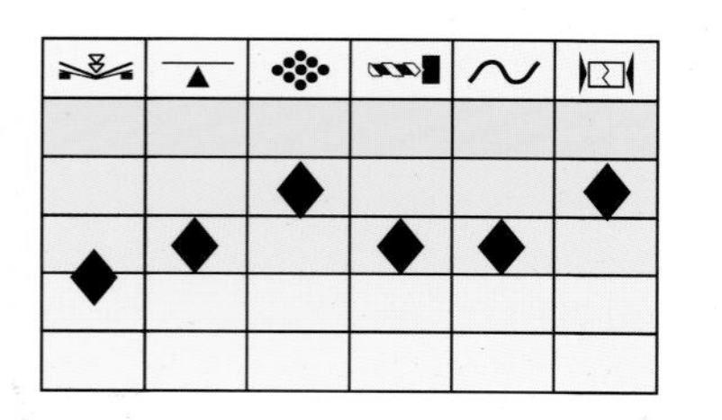

Terminalia ivorensis **Family:** *Combretaceae*
IDIGBO

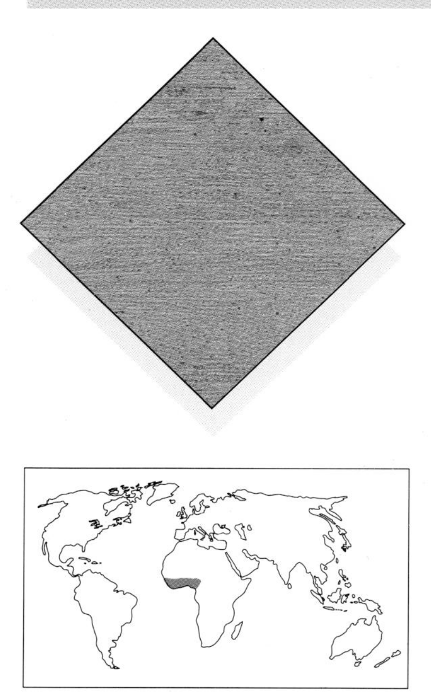

Where it grows
The tree occurs in equatorial Guinea, Sierra Leone, Liberia, Ivory Coast, Ghana and southern Nigeria. It grows to 45m (150ft) and a diameter of 0.9-1.2m (3-4ft), with a clean cylindrical bole for 21m (70ft) above the broad, blunt buttresses. It is known as framiré (Ivory Coast) and emeri (Ghana).

Appearance
There is little distinction between the sapwood and heartwood, which is plain pale yellow to light pinkish-brown. Grain is straight to slightly irregular, with a medium to fairly coarse uneven texture.

Properties
The weight of idigbo is variable due to lightweight brittleheart, especially in large over-mature logs. Weight averages 540 kg/m^3 (34 lb/ft^3) when seasoned, and dries rapidly and well with little distortion. There is small movement in service. The wood has medium density, low bending strength, very low stiffness, medium crushing strength, very low shock resistance and a very poor steam-bending classification. The wood works easily with both hand and machine tools with little dulling effect on cutting edges, but is prone to pick up the grain on quartered surfaces. It has good screw- and nail-holding properties, glues well, takes stain readily, and, when filled, provides a good finish. The timber contains a natural yellow dye which may leach out and stain fabrics in moist conditions; also its high tannin content will cause blue mineral stains if it comes into contact with iron or iron compounds in the damp. It is slightly acidic and will corrode ferrous metals. The sapwood is liable to attack by powder post beetle and the heartwood is durable and extremely resistant to preservative treatment.

Uses
Idigbo is very useful for furniture and fine interior and exterior joinery such as window and door frames, cladding, shingles, carpentry and building construction. It makes a good domestic flooring. Logs are rotary cut for plywood and sliced for decorative panelling veneers.

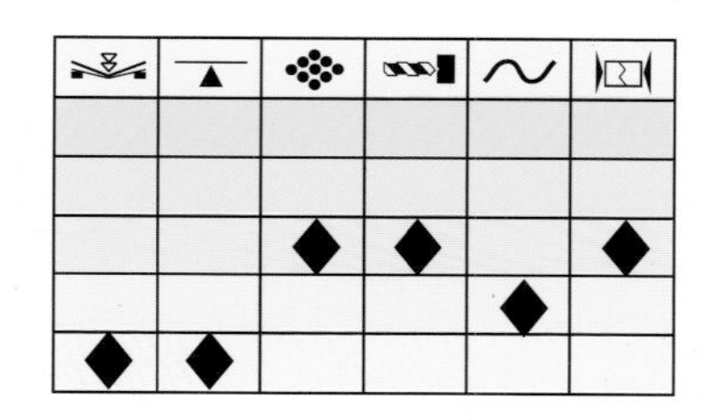

Terminalia superba **Family:** *Combretaceae*
AFARA(LIMBA)

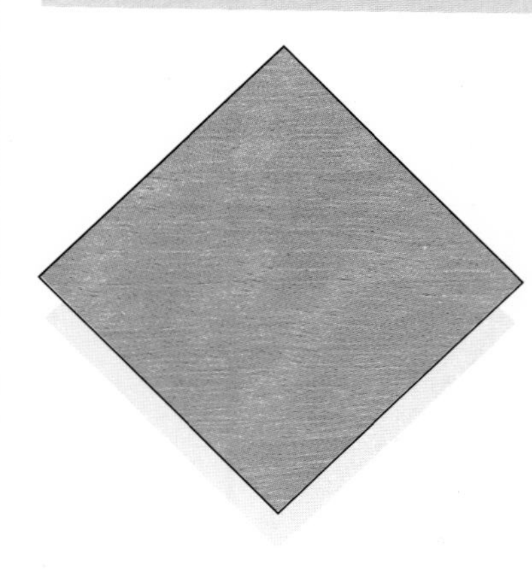

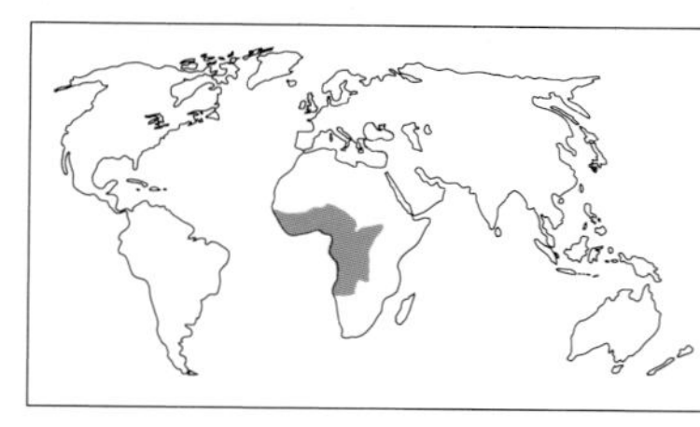

Where it grows
The pale yellow-brown to straw-coloured heartwood is known as light afara, or light limba. Heartwood with grey-black streaked markings is known as dark afara, dark limba or limba bariolé, and as korina in the USA. This tree occurs throughout West Africa, from the Cameroons to Sierra Leone. It reaches a height of 45m (150ft) and about 1.5m (5ft) in diameter. Above the buttress the bole is straight and cylindrical for 27m (90ft).

Appearance
The sapwood and heartwood are uniformly pale straw to light yellow-brown, but some logs have an irregular dark heart with grey-brown or almost black streaks and markings. The timber is straight and close grained, but often wavy. The texture is moderately coarse but even.

Properties
The weight is 480-640 kg/m^3 (30-40 lb/ft^3), averaging 550 kg/m^3 (34 lb/ft^3). This medium density timber has low bending strength and stiffness but medium crushing strength. Care is needed in air drying as there is a tendency for the heart or 'brash' wood to split and shake, but kilning is rapid with little or no degrade and there is only small movement in service. Afara works well with hand and machine tools but a low cutter angle is required when planing irregular grain to prevent tearing. It glues well, provides an excellent finish when filled, and requires pilot holes for nailing or screwing. The wood is non-durable, and the sapwood is liable to attack by the powder post beetle and blue sap stain. The heartwood has moderate resistance to preservative treatment.

Uses
The light-coloured wood is widely used in furniture production and interior joinery, shop and office fitting and coffins. The blackish streaked heartwood produces very attractive face veneers for panelling and furniture, flush doors and marquetry. It is also a good turnery wood. Both are used for plywood manufacture. The greyish-brown type is used for corestock and light construction work such as school equipment.

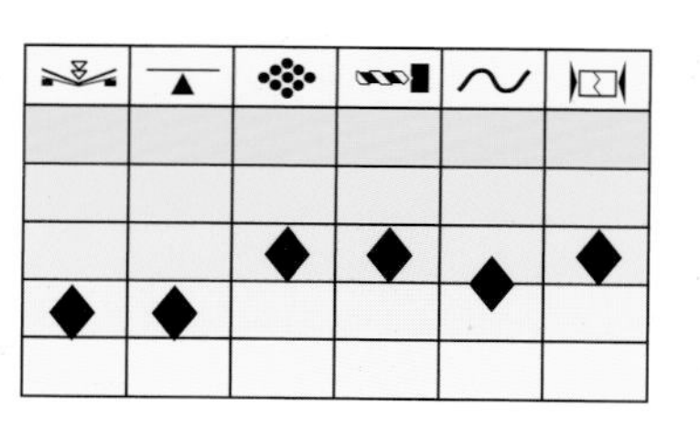

Thuya plicata **Family:** *Cupressaceae*
'WESTERN RED CEDAR' (S)

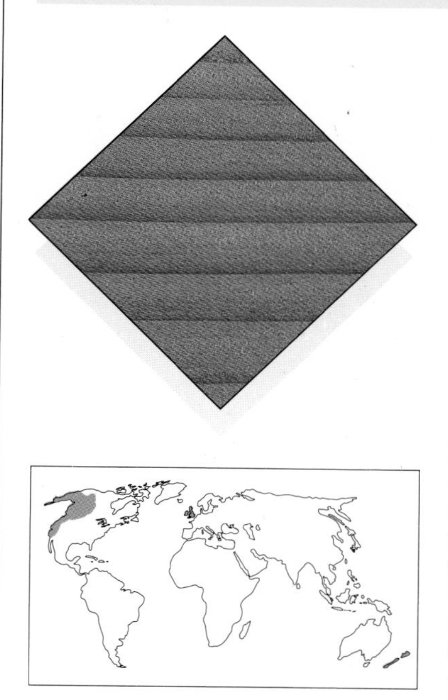

Where it grows
The tree is known as British Columbia red cedar in the UK, giant arbor vitae in the USA and red cedar in Canada. It occurs from Alaska south to California, and east from British Columbia to Washington, Idaho, Montana and the northern Rockies. It has been planted in the UK and New Zealand. It grows to a height of 46-76m (150-250ft) with a diameter of 0.9-2.5m (3-8ft). It is not a true cedar.

Appearance
The sapwood is white, in contrast with the heartwood, which varies from dark chocolate-brown in the centre to a salmon pink outer zone which matures to a uniform reddish-brown. Once dry and exposed, the timber weathers to silver grey, which makes it a particularly attractive prospect for shingles, weatherboard and timber buildings. The wood is non-resinous, straight grained and with a prominent growth-ring figure. It has coarse texture and is rather brittle.

Properties
Weight is about 370 kg/m^3 (23 lb/ft^3) when seasoned. Thin sizes dry readily with little degrade, but thicker stock requires careful drying. There is very small shrinkage in changing atmospheres, and stability in service. The wood has low strength in all categories and a very poor steam-bending classification. It works easily with both hand and machine tools, with little dulling effect on tools. Cutters should be kept very sharp. It has fairly good nailing properties but galvanized or copper nails should be used, as its acidic properties cause corrosion of metals and black stains in the wood in damp conditions. The wood can be glued easily and nailed satisfactorily, and takes stains of the finest tint without fading. It can be polished to an excellent finish. Sapwood is liable to attack by powder post beetle; the heartwood is durable and resistant to preservative treatment.

Uses
This softwood is used in the solid extensively for greenhouses and sheds, shingles, exterior weatherboarding and vertical cladding. It also goes into bee-hives, and is used in the round for poles and fences.

Tieghemella heckelii **Family:** *Sapotaceae*
MAKORÉ

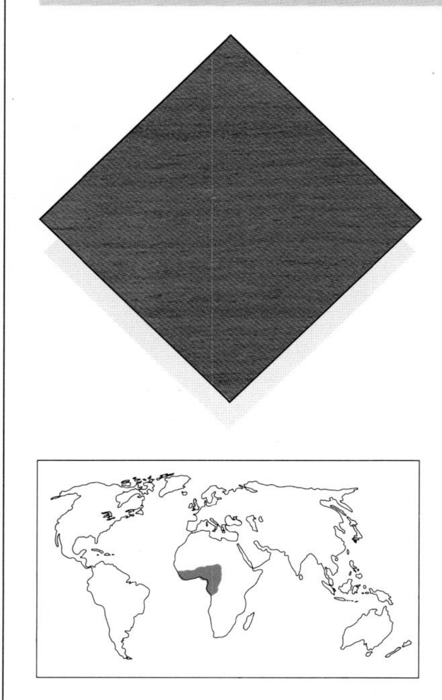

Where it grows
This tree occurs in Sierra Leone, Nigeria, the Ivory Coast, Ghana and Liberia. It is known as agamokwe (Nigeria); baku and abaku (Ghana). A similar timber from Cameroon and Gabon from *T.africana* is known as douka. These trees attain a height of 37-45m (120-150ft), with a diameter of about 1.2m (4ft).

Appearance
The heartwood varies from pale blood red to reddish-brown, and the sapwood is slightly lighter. Some logs have irregular veins of darker colour. Most timber is straight grained but selected logs provide a broken stripe or mottle and a very decorative moiré or 'watered-silk' appearance, with a high natural lustre. The texture is much finer and more even than mahogany.

Properties
Average weight is 620 kg/m^3 (39 lb/ft^3) when seasoned, and makoré dries at a moderate rate with little degrade. There is small movement in service. The timber has medium bending and crushing strength, low stiffness and shock resistance, and a moderate steam-bending rating. The silica content causes severe blunting of cutting edges, and in dry wood, tungsten carbide-tipped saws are required. Machine dust causes an irritant to eyes, nose and throat, and efficient dust extraction is essential. The wood tends to split in nailing, but takes glue well and stains and polishes to an excellent finish. It tends to blue stain if in contact with iron or iron compounds in damp conditions. The sapwood is liable to attack by powder post beetle. The heartwood is very durable and extremely resistant to preservation treatment.

Uses
Makoré is used for doors, table legs, chairs, fittings, superior interior joinery and panelling, and for exterior joinery for cladding, doors, sills and thresholds. Specialized uses include framing for vehicles and carriages, boat building, laboratory benches and textile rollers. It is an excellent turnery wood and makes a good flooring. Logs are peeled for veneers for marine quality plywood, and sliced for highly decorative veneers.

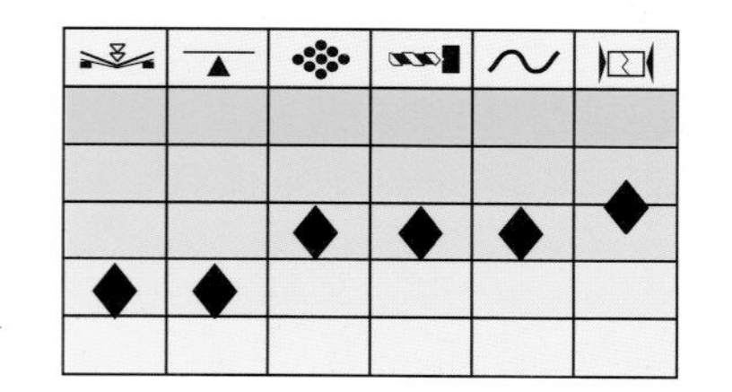

Tilia spp. **Family:** *Tiliaceae*
LIME (BASSWOOD)

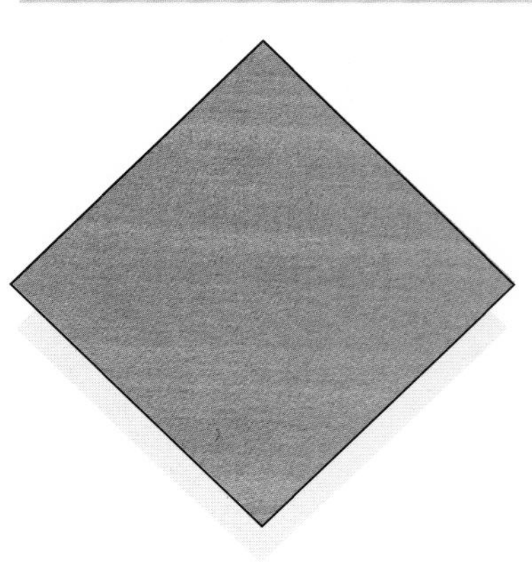

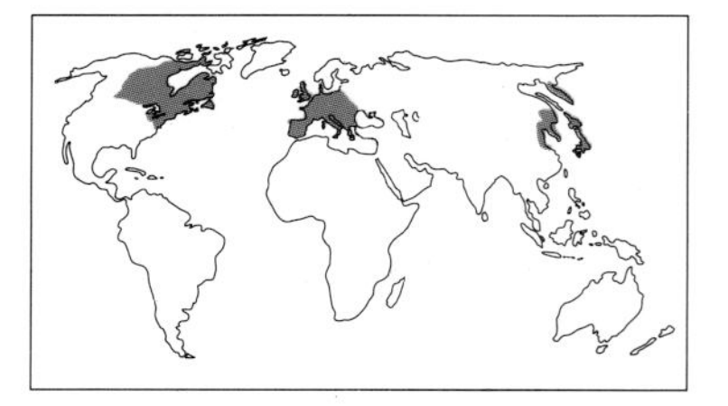

Where it grows
Several species of the lime or linden tree, *T.vulgaris*, grow throughout Europe and eastern Asia. The closely related species *T.americana* produces basswood in Canada and the eastern USA, where it is known as American lime; *T.japonica*, Japanese lime, is known as Japanese basswood in the UK. Trees attain an average height of 20-30m (65-100ft), and a diameter up to 1.2m (4ft).

Appearance
There is no distinction between the sapwood and heartwood, which is creamy-white when the tree is felled and matures to a pale brown when dried. It has a straight grain and a fine uniform texture, and is soft, weak, odourless and taint free.

Properties
The weight of European lime is 540 kg/m^3 (34 lb/ft^3). American basswood and Japanese lime are 420 kg/m^3 (26 lb/ft^3) when seasoned. They dry fairly rapidly with little degrade and only a slight tendency to distort. There is medium movement in service. Lime has low to medium bending and crushing strength, low stiffness and shock resistance, and a poor steam-bending rating. It works easily with both hand and machine tools, but needs thin-edged sharp tools for a smooth finish. The wood nails and glues and stains and polishes satisfactorily for a good finish. The sapwood is liable to attack by the common furniture beetle. The heartwood is perishable but permeable to preservative treatment.

Uses
Lime resists splitting in any cutting direction and is ideal for carving, cutting boards for leather work, pattern making and so on. Hatblock manufacturers regard lime as an alternative to alder. Other specialized uses include artificial limbs, piano sounding boards and harps. It is a good turnery wood for broom handles and bobbins. It is used for bee-hive frames, flat paintbrush handles, cask bungs, toys, clogs, dairy and food containers. It is rotary cut for corestock and plywood. Selected logs are sliced for decorative veneers for marquetry and architectural panelling.

		◆		◆	◆
◆	◆		◆		

Triplochiton scleroxylon **Family:** *Triplochitonaceae*

OBECHE (WAWA)

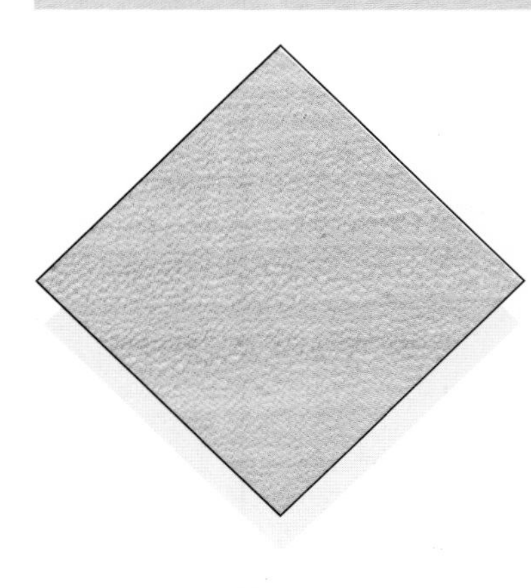

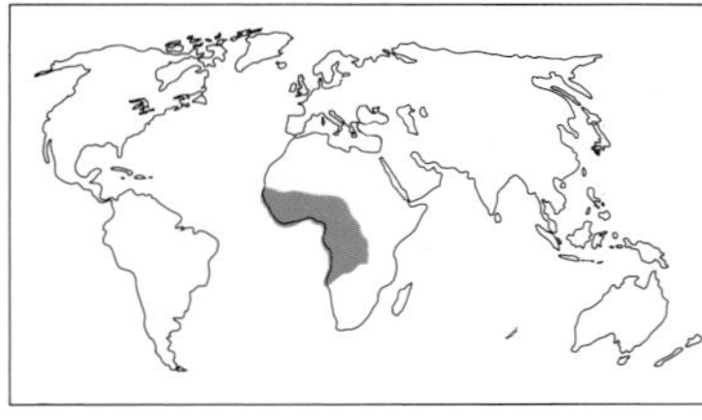

Where it grows
The tree occurs throughout West Africa. In Nigeria it is also known as arere, in Ghana as wawa, in the Ivory Coast as samba and in the Cameroons and Zaire as ayous. It attains a height of 45-55m (150-180ft), with a diameter of 0.9-1.5m (3-5ft), and a clean cylindrical bole free from branches up to 25m (80ft).

Appearance
There is little distinction between the sapwood and heartwood, which is creamy-yellow to pale straw. The grain is interlocked, producing a striped appearance on quartered stock, and the texture is moderately coarse but even. The wood has a natural lustre.

Properties
Obeche weighs 380 kg/m^3 (24 lb/ft^3) when dry. It dries very rapidly and easily with no tendency to split or for shakes to extend, but slight distortion may occur. There is small movement in service. The wood has low bending and crushing strength, very low stiffness and shock resistance, and a moderate to poor steam-bending classification. It works easily with both hand and machine tools with only a slight blunting effect. Sharp cutters with a reduced sharpness angle are recommended. The wood nails easily but has poor holding qualities. It glues well, but requires a light filling to obtain a good finish. This non-durable timber tends to blue stain if in contact with iron compounds in moist conditions. The sapwood is liable to attack by the powder post beetle. The heartwood is perishable and resistant to preservative treatment, and the sapwood is permeable.

Uses
Where durability and strength are relatively unimportant, obeche is used for whitewood furniture and fitments, interior rails, drawer slides and linings, cabinet framing, interior joinery, mouldings, sliderless soundboards for organs, and in model making. It is rotary cut for constructional veneer for corestock and backing veneer for plywood, and it is sliced for decorative striped veneers. The blue-stained obeche is particularly sought by marquetry craftsmen.

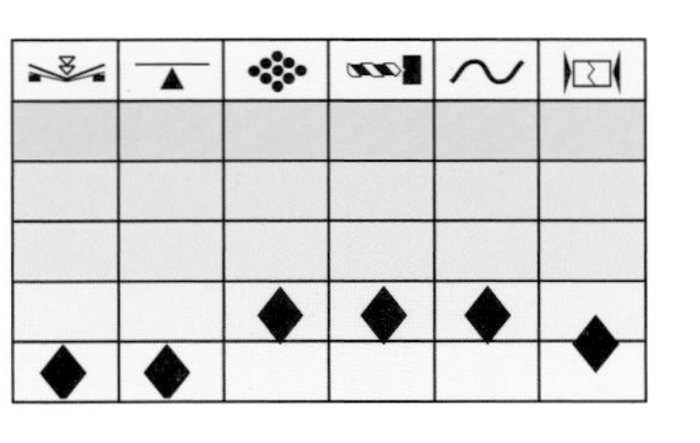

Tsuga heterophylla **Family:** *Pinaceae*

WESTERN HEMLOCK (S)

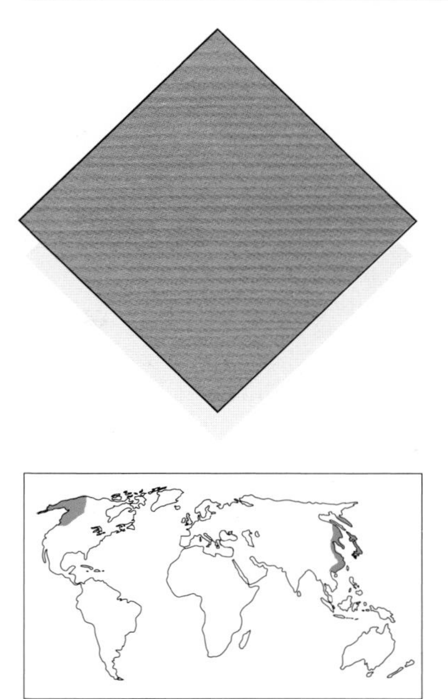

Where it grows
This softwood tree occurs in Alaska, British Columbia, northern Washington, Idaho and the western slopes of the Cascades. It has also been planted in Britain. Known as Pacific hemlock and British Columbian hemlock in the USA, it grows to a height of 61m (200ft) with a diameter of 1.8-2.4m (6-8ft).

Appearance
The heartwood is cream coloured with a very pale brown cast. Darker late-wood bands often produce a well-marked growth-ring figure with purplish lines. It is straight grained with a fairly even texture, and is somewhat lustrous.

Properties
Hemlock weighs 490 kg/m^3 (30 lb/ft^3) when seasoned. The initially high moisture content of this wood demands careful drying to avoid surface checking and ensure uniform drying in thick stock. Distortion is minimal. There is small movement in service. The wood has medium bending and crushing strength, low hardness and stiffness, and a moderate steam-bending rating. It works readily with both hand and machine tools with little dulling of cutting edges. It can be glued, stained, painted or varnished to a good finish. It should be pre-bored for nailing near the ends of dry boards. Damage by *Siricid* wood wasps is sometimes present. The sapwood of seasoned timber is liable to attack by the common furniture beetle. Dark brown or black resinous scars ('black check') are also sometimes found, caused by fly larvae (*Syrphidae*). The timber is non-durable or resistant to decay. The heartwood is resistant to preservative treatment.

Uses
Western hemlock, regarded as one of the most valuable North American timbers, is exported to all parts of the world in large baulk dimensions for use in general building construction, joists, rafters, studding, interior and exterior joinery, doors, floors, fitments, suspended ceilings and vehicle bodywork. It is widely used for wood turnery, broom handles, and for packaging cases, crates and pallets. It is also treated for railway sleepers. Logs are cut into veneers for plywood and decorative panelling. The timber is also used for quality newsprint pulp.

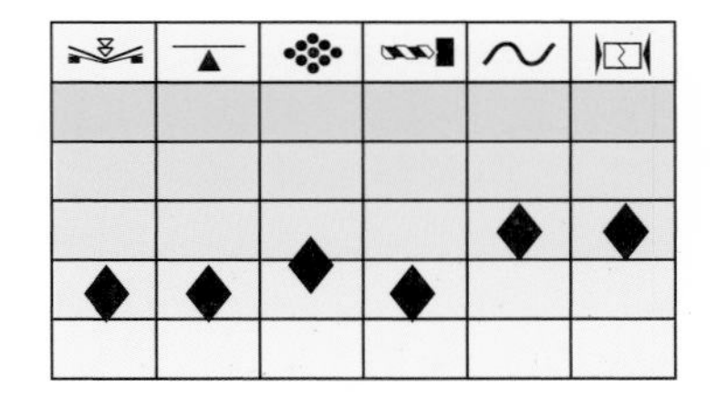

Ulmus spp. **Family:** *Ulmaceae*
EUROPEAN ELM

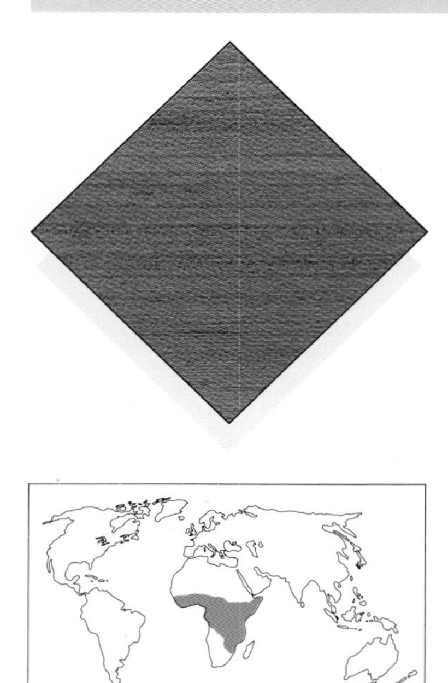

Wych elm has a greenish tinge. Continental elms are usually more straight grained.

Properties
European and Japanese elms (*U.laciniata*) weigh about 550 kg/m^3 (34 lb/ft^3); wych elm 670/kg/m^3 (42 lb/ft^3) when seasoned. The wood dries fairly rapidly, with a strong tendency to distort. The wood has low bending and crushing strength, with very low stiffness and resistance to shock loads. All elms have a very good steam-bending rating. It can be difficult to work, tending to pick up in planing and bind on the saw, but it takes nails well, glues satisfactorily and provides a good finish. The sapwood is liable to attack by powder post beetle and common furniture beetle. Elm is non-durable, moderately resistant to preservative treatment, and the sapwood is permeable.

Where it grows
Trees of the *Ulmus* genus occur widely throughout the temperate climes of Europe, western Asia, North America and Japan. English elm (*U.procera*) is known as red elm and nave elm; smooth-leaved elm,(*U.carpinifolia*) is the common elm of Europe, known as French elm or Flemish elm; there is Dutch elm (*U.hollandica*); and wych elm (*U. glabra*) is known as Scotch elm, mountain elm or white elm. European elms reach an average height of 38-45m (120-150ft), with a diameter usually of 1-1.5m (3-5ft); wych elm grows to 30-38m (100-125ft) tall and up to 1.5m (5ft) in diameter.

Appearance
The heartwood is usually a dull brown, often with a reddish tint, and is clearly defined from the paler sapwood. The heartwood has distinct and irregular growth rings, giving the wood a rather coarse texture. It is cross-grained and of irregular growth, which provides some very attractive figure.

Uses
The Rialto in Venice stands on elm piles. Elm is used for boat and ship building, dock and harbour work, weatherboards, gymnasium equipment, agricultural implements, vehicle bodywork, ladder rungs, coffins and the seats of Windsor chairs. It makes an attractive flooring; it is used for meat chopping blocks, and as a turnery wood for bowls. It is sliced for highly decorative veneers, and elm burrs (burls) go for cabinets and panelling.

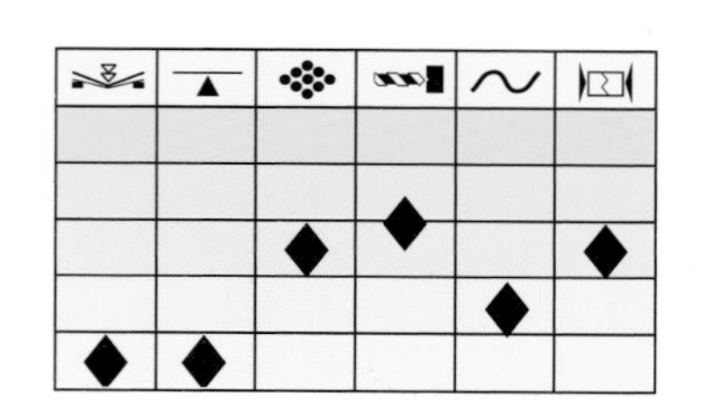

Ulmus thomasii **Family:** *Ulmaceae*
AMERICAN ELM

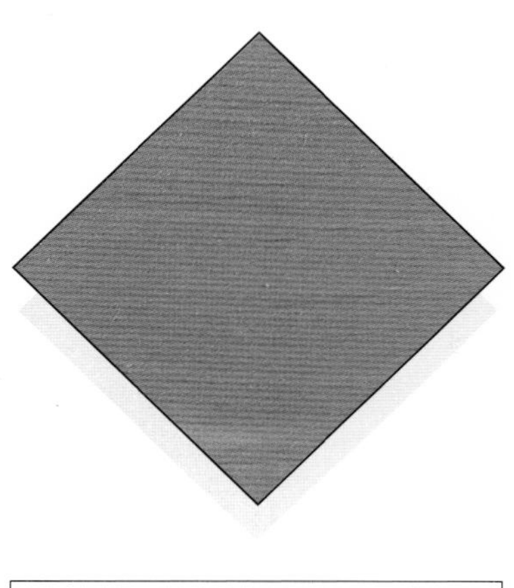

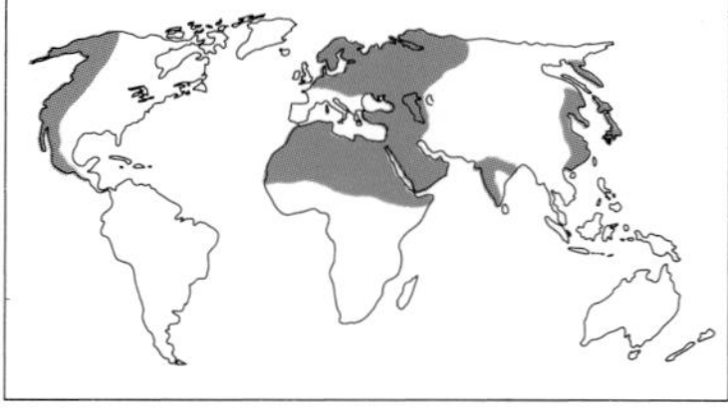

Where it grows
The *Ulmus* genus has five species growing in eastern Canada and the USA. *U.americana*, L. produces white elm in eastern Canada as far west as Saskatchewan, and in eastern and central parts of the USA; it is also known as water elm, swamp elm and American elm. *U.fulva*, Michx. produces slippery elm, also known as soft elm, red elm and slippery barked elm, which occurs in the St Lawrence River valley; and *U.thomasii*, Sarg. produces rock elm also known as cork elm, hickory elm and cork bark elm, which occurs in southern Quebec and Ontario, extending into the USA. Elms average 15-25m (50-80ft) in height, with a diameter of 0.3-1.2m (1-4ft).

Appearance
The heartwood is medium to light reddish-brown, and the sapwood is slightly paler. Rock elm is straight grained with a moderately fine texture. White elm is sometimes interlocked and the texture coarse and rather woolly.

Properties
White elm weighs 580 kg/m^3 (36 lb/ft^3); slippery elm is a little heavier. Rock elm weighs 620-780 kg/m^3 (39-49 lb/ft^3) when seasoned. The wood dries readily with minimum shrinkage. American elms have medium bending and crushing strength and very low stiffness. White elm has high resistance and rock elm very high resistance to shock loads. All have very good steam-bending ratings. The timbers work fairly easily with only moderate blunting effect on tools. They take nails without splitting, glue easily and provide an excellent finish. The sapwood is liable to attack by powder post beetle but is resistant to fungus. Elm is non-durable and moderately resistant to preservative treatment.

Uses
American elm is used in boat and ship building for stern posts, ribs, general framing, gunwales, bilge stringers, keels, rubbing strips and components that are completely submerged in water; also for underwater parts in dock and harbour work. Other uses are wheel hubs, blades of ice hockey sticks, agricultural implements, chair rockers, gymnasium equipment, bent work for vehicle bodies and ladder rungs. The wood is excellent for turnery, and is sliced for decorative veneers and burrs (burls).

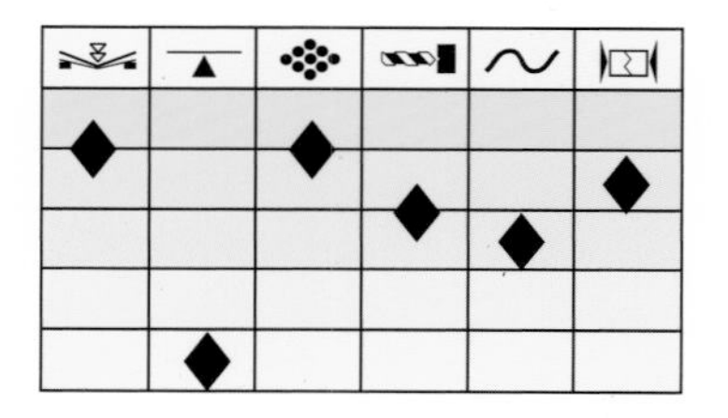

HOW WOODS ARE NAMED

Names in common usage are often misleading, as they allude to some prominent characteristic of the tree. For example, a very heavy tree may be called 'ironwood', but there are more than 80 species, of completely different families, all known as ironwood. Hornbeam is one example.

Sometimes the common name of a tree is quite simply untrue. Black Italian poplar, is not black but white – nor Italian. It grows in Britain. Indian silver greywood likewise is not from India, nor is it silver or grey.

Vernacular names are those by which the wood is known in its country of origin, and are naturally in the same language. Trade names are sometimes given to a wood by traders who seek to glamorize an otherwise ordinary species such as ayan by calling it Nigerian satinwood; afara is sold as korina in the USA.

There is also a remarkable difference in naming woods between nations in the English-speaking world. For example in Australia, oak, ash and elm are entirely different species from those recognized in Europe and the USA:

BRITAIN	**USA**	***LATIN***
Lacewood	Lacewood	*Cardwellia sublimis*
Silky oak	Plane Sycamore	*Platanus acerifolia*
Sycamore	Maple	*Acer pseudoplatanus*
Maple	Sugar maple	*Acer saccharum*

The International Code of Botanical Nomenclature is the standarised code of Latin names to enable people who work, study or play with wood in all parts of the world, whatever their native tongue, to identify a species correctly.

The first name is assigned to a genus and the second to a specific epithet to indicate the particular species within the genus. Thus we can see, for example, that Parana pine (*Araucaria angustifolia*) is not a pine at all; it belongs neither to the Pinus genus, nor even to the family (*Pinaceae*).

Botanically every tree has a classification from which the generic and specific names are taken, as in this example of American walnut:

Kingdom .... Vegetable
Division .... Angiospermae
Class .... Dicotyledoneae
Order .... Juglandales
Family .... Juglandaceae
Genus .... Juglans
Species .... Nigra.

In the Directory of the Encyclopedia, we have provided the family and genus of each of the 150 woods for positive identification.

Henry Moore, Composition, 1933, Walnut wood.
This composition fully exploits the flowing grain and
sensuous qualities of the wood.

THE WORLD OF WOOD

The myriad uses, facts and fables connected with the world's best-loved trees are often intriguing and, sometimes, amusing. From the richness and beauty of 18th century furniture in the 'Age of Walnut' to the humble wassail bowl carved from lignum vitae, the unique properties of each wood have been admired and exploited through the centuries, and provide a revealing insight intc the multitude of ways in which man has benefitted from this extraordinarily versatile natural material.

OAK

There is no tree more closely associated with the history and patriotic fervour of a nation, no tree considered to have served a nation so well and so consistently as the oak for Britain. Soil and climate promote different quality timbers, and although the common oak grows straighter and more predictably in Europe and Asia, it is in Britain that it achieves the utmost strength, durability, and variety. From the Latin name (*Quercus robor*) we even derive our word 'robust'.

THE GREAT OAK TRADITION

Two recent events in England have brought the values of the timber back to national consciousness. In July 1984 the roof of York Minster was struck by lightning and partially destroyed, leading to extensive rebuilding in the best medieval tradition.

The spanning of the vaults, and the logistics of getting the massive timbers into place even with modern equipment, have heightened not only the admiration for the original craftsmen but also for the suitability of the timber for projects of this kind. No other material was considered to be as proven or reliable.

Roofs of medieval barns exploited this versatility even more, in that every crooked oak branch could find use as a triangulation brace, and every offcut could be cleft into a roofing batten or peg for fixing slates. Medieval hammerbeam roofs, whose geometric construction spans vast spaces without the need for tie beams, even incorporated fine moulding and sumptuous carved angels into structural magnificence.

The other event was the lifting of the 'Mary Rose', the flag ship of HenryVIII, which sank off Southsea in July 1545. A wave of popular sentiment swept through England as the remains lifted from the water on a salvage cradle; the oak hull settled, effortlessly puncturing the steel supporting frame. This seemed entirely appropriate to a race who has adopted the oak as a national symbol, principally for the role it played in building the ships of the Royal Navy from the 16th to the 19th century. As the song goes:

'Hearts of oak are our ships,
Hearts of oak are our men.
We always are ready,
Steady, boys, steady.'

Oak is not generally spoken of in the same breath as flexibility – it epitomizes unbending strength – so David Garton's elegant chairs and table in steam-bent oak (LEFT) are quite a surprise. The timber's high tannic acid content renders it unique to the tastebuds; oak casks like the ones at bottlers Tollot Père et Fils in Beaune, France (RIGHT) lend a smoky flavour to maturing alcohol. The row of half-timbered medieval houses (FAR RIGHT) in Miltenberg, South Germany, illustrate how a tree's natural twists would be put to advantage in building not only dwellings but ships too.

This jingoism has obscured the fact that the demands of shipbuilding wiped out the country's natural forests (one large warship might consume as many as 3,000 trees) and has led to more historical conflicts over keeping the Baltic trade routes open than a nation ought to feel proud about. The Dutch and Portuguese soon discovered that teak was superior, but sentiment dimmed the vision of the British naval procurers.

PROPERTIES, PRODUCTS AND USES

For lock gates (before the advent of greenheart), cruck frame roofing, timber frame houses and the spokes of wheels, oak has been employed with lasting success.

For furniture, oak was the prime timber until walnut and mahogany arrived. Periodically it has come back into favour, notably during the Arts and Crafts revival (1880-1910), and especially recently in the form of American white oak. Although the planks are narrower, the trees are straighter, and the wastage in converting the timber is a fraction of that of English oak, making the American species ideal for mass-production.

In cooperage (the making of barrels) oak has not just been used for its strength and structure, but also because of the qualities it lends to the contents. Brandy, sherry, wine and beer have conventionally been stored in oak, the storage period being considered an essential part of the making process. This has much to do with the tannic acid in the wood.

Tannin is also used to arrest the natural decay of hides and skins, and lends its name to the 'tanning' of leather. Acorns are another by-product, high in both starch and tannin, which apart from being prized as pig food, have even served as a substitute for coffee.

OAK MYTH AND MAGIC

It is hardly suprising with its size,majestic appearance, and application that the oak should become a holy tree. Druids venerated not only the tree but also the mistletoe sometimes found in oak boughs.

For the Romans it was the king of trees,dedicated to Jupiter whose weapons of punishment were, according to Shakespeare, 'oak-cleaving thunderbolts'. In fact it has been confirmed by research that lightning strikes oak more frequently than other trees, which may explain the persistence of this pagan tradition.

Nordic myths ascribe the tree to Thor, the God of Thunder. Objects made from wood thus struck were considered most blessed, which should have given some comfort to the Archbishop of York, as he surveyed the charred remains of his Minster.

YEW

From the earliest times, the evergreen leaves of yew trees have been held to symbolize everlasting life, a belief held by the Egyptians 1,000 years before the Greeks, who themselves held the tree sacred to Hecate, Queen of the Underworld. A Christian development of this pagan trend has encouraged the planting of yew trees in country churchyards.

There is a popular myth that yew was compulsorily planted in English churchyards to provide timber for the famous longbow, whose reputation and place in history were assured in the battles of the Hundred Years' War. But many churchyard trees are over 800

OPPOSITE PAGE: the magnificent 40ft-high yew hedge planted in 1720 round the courtyard of Cirencester Park, Gloucestershire. The timber's mellow colour and curly grain are displayed in Alan Dixon's luscious carved fruit (RIGHT); the bowl, the nearest apple and one pear are of yew. Alan Peters sets off the angular lines of his low table (ABOVE) with its complex Japanese dovetails, against the grain's organic flow.

years old, which would indicate that they were never removed for this early version of the arms race, and that their presence was more symbolic than military. It is true that yew is poisonous to cattle and churchyards were among the few areas where grazing was prohibited.

CHARACTERISTICS AND USES

If ever there was a wood whose characteristics contradicted the normal distinction between 'hardwoods' and 'softwoods', it is yew. Softwoods come from cone-bearing trees with evergreen needle-like leaves; hardwoods come from broad-leaved deciduous trees. Whereas lime is one of the softest hardwoods (and certainly softer than yew), yew, one of the hardest woods of all, still classifies as a 'softwood'.

The trunks tend to be deeply fluted and twisted, which makes conversion into planks uneconomic and impractical, because the sizes would never be satisfactory. Equally, the irregular growth makes the timber unpredictable and volatile to work. This explains the desirability of yew for the creation of smaller objects such as snuff and pill boxes, plates and spoons, and even the working parts of windmills.

A striking exception to this is the staircase at Llanivangel Court in Wales where large timber sections were used from what must have been some remarkable trees. Although the veneers have found some favour with reproduction furniture makers, there is a denser variety in the form of Irish yew, which was found to be better suited for work in the solid, such as ornate carving. Yew is also one of the few softwoods that can be steam-bent effectively, and as a result has been conventionally used in the curved work of country chairs.

THE LONGBOW

The most celebrated use for yew was in the manufacture of the longbow, a powerful, devastatingly effective weapon which changed the strategy of medieval wars.

At the Battle of Crécy (1346), the first battle at which the prowess of the weapon was tested, the French army outnumbered the English by three to one. But for every bolt released by the mechanical crossbow, a longbow could fire eight arrows. It could be reloaded every four seconds. The effect of 7,000 archers on the charging French cavalry was demoralizing in the extreme.

Although it is true that the timber needed to be selected with care, there is a popular misconception that only one bow could be hewn from one trunk. The myth has developed, it seems, as a dramatized response to the fact that English yew was rarely found in long straight lengths. The straightest, most desirable timbers actually came from Spain, which led to a number of serious economic conflicts of interest between trading partners who were frequently at war with each other.

ASH

Problems with warts? Rub them with a piece of bacon, cut a slice in the bark of an ash tree and slip in the bacon, and your warts will transfer to the tree – or so medieval herb doctors would have had you believe. The warts would reappear on the tree itself as burrs, so went the tales. Similarly, a 'sure cure' for children with rickets was to pass the body of the ailing child three times at sunrise through an ash sapling which has been split longitudinally. The sapling would then be tightly bound, and if it was seen to heal, so the child also would be cured.

A distillation of young ash shoots is reputed to be effective for earache, unsteady hands and certain snake bites. Slavic tradition has it that the shadows of ash leaves in the wind induce fear in snakes, so travellers may rest untroubled beneath the shade of the tree.

A more potent legend still concerns Yggdrasil, the sacred ash of Norse mythology. Its roots are anchored in the underworld, watered by streams of wisdom and destiny. Its trunk is supported by the earth, and its crown touches the apex of the heavens. Mythical animals lodge in it, and every day the gods cross the bridge between heaven and earth – the rainbow – to hold court under the tree's shade.

CHARACTERISTICS AND USES

It is the porous bands of spring-growth timber interspersed with the denser, harder bands of summer growth that lend the flexibility, strength and reliability to ash, almost as if its growth were a natural lamination process. As a result, the timber has proved invaluable for military use – lances, spears, pikes and javelins were made from ash since before the Bronze Age to the late 18th century.

The pliability and capacity to absorb shock has made ash ideal for all kinds of transport; from skis to eskimo sledges, from the frames of canoes to the skeletons of the famous Mosquito aeroplanes (still in service 10 years after the Second World War), from the rims of cart-wheels to the early-19th century 'boneshaker' bicycles. Long after the days of horse-drawn transport, many motor-car companies were still employing former carriage builders in the construction of ash vehicle frames, and not just from benevolent concern about their employment. The material with which these skilled workers had such a wealth of experience was also suitable for use in contemporary designs of sports cars.

The shock-absorbency of ash, and its ability to withstand repeated impact without fracturing, has made ash ideal for sports equipment and the handles of tools. Many a blister must have erupted on the hands of chain-gang labourers wielding ash-handled tools – before hickory, which is also used for baseball bats, (from southern Canada and the eastern United States) was found to be a better substitute for handles.

Used in furniture, ash is best with a stained finish because the broad grain tends to pick up grime, but it is supremely suited to steam bending, taking the least time of all woods in a steam kettle to

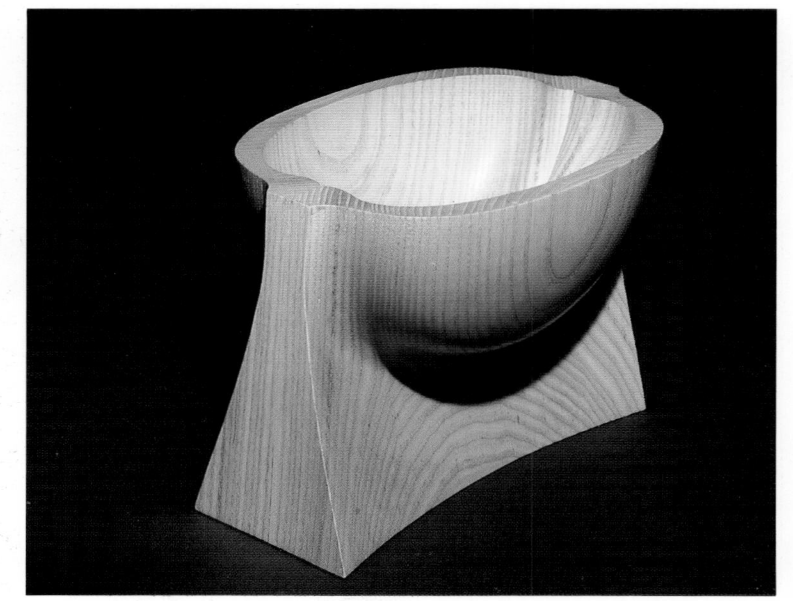

Odin, the 'All-Father' of Norse mythology, hangs from Yggdrasil, the Tree of the World – an ash – in this drawing by Jan Salerno (BELOW FAR LEFT). Strong grain figure and three-dimensional form combine in Alan Dixon's bowl, while Nic Pryke's desk (RIGHT) employs a rare burr ash veneer on pedestal and leg. (BELOW LEFT), traditional coachbuilding skills in ash still abide at the Morgan sports car factory, Worcestershire, England; (BELOW RIGHT), a chairmaker exploits the flexibility of the steamed timber to bend it on a mould for the Windsor bow shape.

be rendered supple. Once bent, it also keeps its shape reliably. There is a long tradition of making ladder-back steam-bent chairs which is still alive both in England and America, and this has found interesting development in the work of some modern furniture designers, such as Wales-based David Colwell. Andrew Whateley, who was chief cabinetmaker at the Dorset workshops of John Makepeace during a period of production of some important work, has developed a steam-bent, plank-backed ash chair with double curvature, which is as technically innovative as it is visually exciting and practically comfortable.

MAHOGANY

The first description of the qualities of mahogany came as early as 1595, from the carpenter on Sir Walter Raleigh's ship during an exploratory voyage to South America. The Spaniards began to use it regularly for ship repairs from the early 17th century.

The first recorded purchase in England of 'Jamaica' wood for domestic purposes appears in the accounts for Hampton Court Palace for the year 1661, and the first recorded piece of furniture in Great Britain, a butcher's chair now in Trinity Hall Museum in Aberdeen, dates from the same year.

But it was not until 1721, when the English Parliament repealed the heavy duty on timber imported from the colonies – imposed in the 17th century to preserve supplies for the Royal Navy – that mahogany became a popular timber. This popularity was immediately distorted by the shortage of walnut (the most favoured furniture wood at the time), a result of a devastating frost which wiped out a large proportion of European walnut trees in 1704; it is, therefore, hardly suprising a comparable substitute would be in such immediate demand.

Crisp carving is easily worked in mahogany (ABOVE), as is the elaborate decorative detail of furniture such as the 1760 desk (LEFT) by John Townsend of Rhode Island. Sixteenth century Spanish explorers in South America (RIGHT) found it ideal for reparing their ships, but hundreds of years of greedy logging has eliminated the wood from Cuba and now threatens other countries' supply. Mercier's 1733 painting of Frederick Prince of Wales and his sisters (FAR RIGHT) shows a mahogany music stand that must have been very new - Chippendale himself only started to use the wood in the early 1730s.

There was no better wood to turn to than mahogany, with its crisp, strong texture, its colour, its fine working qualities and exceptionally wide planks.

THE LUXURY WOOD - AND THE TRUE COST

The first mention of mahogany in furniture advertisements comes from South Carolina in 1730 – around the time that Thomas Chippendale founded the company that was to make such an impact on furniture design on both sides of the Atlantic for the next century. His designs most fully exploited the properties of the wood: light, strong and versatile, it is as suited to the thin sections of lattice-work in music stands and chair backs as it is to the broad expansive surfaces of dining tables. Auction room catalogues still refer to a dining table as a 'mahogany'.

The capacity to take fine mouldings was especially appreciated by Neo-classical architects, such as Robert Adam, who became purveyors of furnishing taste after the middle of the 18th century.

Early names for mahogany include Baywood (after the Bay of Honduras) and Havana wood. Indeed, the Cuban variety is the most prized, but the indiscriminate felling of Cuban woods over the last three centuries has virtually destroyed the country's principal asset. This sad story serves as an object lesson for other countries tempted to neglect proper forest conservation and management. The early Jamaican merchants were amongst the most short-sighted; they not only sold off their own trees, but imported Cuban mahogany to sell on to the voracious European and North American trade.

Being impervious to worm and rot, mahogany was ideal for shipping, as Sir Walter Raleigh's carpenter correctly surmised. But the early use as decking was quickly superseded by fashioning it for elegant fittings, and this is still the case. A favourite subject for Victorian photographers was the Royal Families of Europe aboard their yachts; every hatch-cover, bulk head and chronometer case is shining mahogany, in a tradition very much alive today.

The wood's natural colour, a pinky red, was not considered the most attractive (the light tones were considered a bad match for Robert Adam's wallpaper designs), so it became customary to stain it dark with a solution of potash (potassium bichromate). This also, of course, allowed for the disguise of shoddy workmanship, as it has done from the earliest applications to the present day. In fact the 'mahogany style' pieces that inhabit the reproduction furniture market are rarely made from mahogany at all, but from cheaper woods like agba, idigbo, ramin and iroko which are easily sprayed to resemble mahogany.

THE DEALS

Larch, spruce and pine have been collectively known as 'deal' since the 14th century. It is a confusing word, more precisely held to describe the converted timbers of coniferous softwoods. Sheraton (1803) defined deal as 'fir or pine timber being cut to thin portions'; early classification specified these portions to be 2-4in thick by 9-11in wide, but regional preferences vary.

PINE

Pines are associated with Attis, the young Greek shepherd chosen by Cybele as her priest. She imposed on him a vow of chastity, which he broke; in the course of his frenzied, delirious guilt, he 'unmanned' himself, and began to bleed to death under a pine tree. Cybele took pity, and turned him into a pine, beginning a colourful (if paradoxical) association, in classical and more recent folklore, of pines with both enforced celibacy and fertility.

Scots pine is found in varying abundance across Northern Europe as far as the vast forests of Siberia. It is widely employed as a 'nurse' for oak trees – the conical growth when young affords protection from the wind while allowing sunlight to get to the seedling oaks. In the United States, pines are categorized as 'hard' (Longleaf, Loblolly, and Ponderosa) and soft (Stone pines, Foxtail, and Nut).

Like maple, pines are as valued for their by-products as for the timber. 'Naval stores' – pitch, tar, resin, and turpentine – are all derived from pines by making sloping cuts in the bark which are reopened each morning in spring; the oozing resin is collected in a cup,then distilled to separate the solvent turpentine.The solid crystalline form, rosin, is used by violinists to prime their bows for getting more tone from their strings, and by French rock climbers to get more grip from the soles of their boots.

LARCH

Larch, indigenous to the Swiss Alps, was well established all over Europe by the 16th century. It was first introduced to Britain in 1737 by the Duke of Atholl, who planted five at Dunkeld, and a dozen at Blair Castle, but the carrier, John Menzies, dropped a few samples along the road from London. Any old examples are said to be

Pine suggests northern temperate forests, but the wood occurs as far south as the Mediterranean, and on any high ground. Prolific and fast growth means easily accessible timber is put to a variety of local uses, like building fishing boats on England's Northumbrian coast (BELOW LEFT) or ornately fretworked izba – peasant dwellings – in northern Siberia (BELOW). Human spirituality finds expression through the neighbourly pine; the East Siberian Buryats decorate their sacred pine groves (BELOW CENTRE) while an Alaskan totem pole (RIGHT) speaks of the creatures of an individual's dreams or of ancestral mythology.

descended from these trees. Eight still survive, and the Duke, impressed by the alacrity with which the larches established themselves, planted several million between 1750 and 1780, with an entrepreneurial eye for the naval market which was growing rapidly carrying exports all around the world..

Larch is popular in part for the early yield of marketable timber, but it was also found that the falling autumn needles (it is one of the few conifers that is not evergreen) improve the quality of the soil, giving a better chance to more demanding species of spruce and pine which would be interplanted with it.

SPRUCE

Norway spruce was well established in Britain by 1548, and by 1800 was the main component of forestry. It is a fecund tree,with the everlasting greenery held to be a symbol of hope and immortality in Norse legends, adapted more recently for use as Christmas trees.

The twin associations of the Norway spruce are embodied in the annual gift of the Norwegian monarchy, who present a large spruce as a Christmas tree to the British people, to be erected in London's famous Trafalgar Square in gratitude for the help given in the Second World War.

A string of telephone poles stretch into the Oregon distance (LEFT). Coniferous softwoods are unbeatable where straightness, strength and economy are needed. Ships' masts often use larch for the same reasons; the Chinese developed a method of hardening the wood by burying it before fitting their junks (ABOVE) with it. Spruce's workability and resonance puts it at a premium for the soundboards of musical instruments; this piano (RIGHT) is by the old-established London maker Broadwood. All the coniferous softwoods contribute to America's characteristic architecture, such as this Pennsylvania house (BELOW).

Spruce has exceptionally resonant qualities, and is favoured for use as soundboards for musical instruments. Indeed there has been much research into why the violins of Stradivarius are of so much higher quality than those of his rivals. It has been suggested that the spruce which he used was floated down the mountain rivers to his workplace in Cremona, where the waters are known to be heavily silted.The combination of the silt deposit and the long and total immersion may have lent peculiar properties, so far unimitated by any other maker.

ROSEWOOD

The name rosewood has nothing to do with garden roses, though it does derive from the supposedly fragrant scent of the timber. The Chinese writer Cao Zhao, in *The Essential Criteria of Antiquities* (1388) wrote: 'Its fragrance much resembles that of the truth-bringing incense'. Quite what truths are brought remain a mystery to those who work with it; the dust is amongst the most irritating, debilitating, allergy-provoking of any timber.

Collectively rosewood denotes various species of *Dalbergia*; in particular Rio or Brazil rosewood, Indian or Bombay rosewood, and the Chinese huang-hua-li. Close cousins include tulipwood (or pinkwood in the USA), kingwood, and cocobolo, the diverse colours of all of which were used in the elaborate marquetry decoration of the furniture and joinery of 18th-century Europe. The wood is seldom seen in the solid, except in oriental furniture. African blackwood, as its name implies, is the only one of the family to be almost black. Like ebony, it is often used for woodwind instruments – oboes, clarinets and the chanters of bagpipes.

The Chinese first spotted the outstanding qualities for furniture. It was a perfect colour, being neither too showy nor too subdued.They especially valued timber with the jagged, contrasting figure poetically described by one ancient commentator as the 'devil's face grain'. Rosewood's strength and weight lent themselves admirably to the faultless workmanship characteristic of the Chinese, with their command of line, curve and cubic proportion.

Nowhere have joinery techniques reached such sophistication as in classical Chinese cabinet-making, though it is unclear whether this is due more to ethical pretensions or to pragmatism; since the oils in rosewood make it extremely difficult to glue.

A wood that has inspired cabinetmakers through the ages and across the world, rosewood first came to the West in the late 18th century. The decorative writing desk with brass details (FAR LEFT)*, made in the reign of King George III of England, contrasts tellingly with the unadorned lines of Luke Hughes' cigar humidor* (BELOW)*, inspired both in its proportion and use of visually exciting 'devil's-face grain' by early Chinese aesthetic restraint. Rosewood's hardness is popular with individualist craftsmen like Colorado clockmaker Wayne Westphale* (LEFT)*, who uses it for the gears and accents in his creations; at an auction of rosewood logs in Kadur, South India* (RIGHT)*, policemen take both an official and personal interest.*

If today's furniture-makers do not know the the precaution of wiping both surfaces to be bonded with benzene dry-cleaning fluid for reliable adhesion, they are likely to spend time and effort researching obscure and expensive epoxies. Another unlikely workshop technique especially associated with the wood is for enhancement of the colour; the effect of painting nitric acid on the surface and then burning it off with a blowtorch has a welcome timelessness.

In India, experts agree, there was little by way of a furniture tradition until the Europeans arrived. The Portuguese were the first to import Bombay rosewood from Goa, but they soon began to send Western furniture prototypes to the East for copying, a tradition speedily imitated by the British and Dutch. The result was a marked lack of distinctive indigenous styles.

The Portuguese also introduced the more prized Brazil rosewood from their other colonies in South America, so they may rightly be considered pioneers in the rosewood trade. As usual, it was the establishment of trade patterns which led to changes in furniture styles. In both America and England the introduction of rosewood made a dramatic impact on early 19th-century interiors; the emphasis of simple lines and elegant proportions was much enhanced by the rich figure.

Dense, expensive woods are mostly used for decorative smaller items, and rosewood is no exception. For knobs, drawer pulls, and handles of furniture, cutlery and carpenters' tools, rosewood is ideal; but lest anyone mistake the daintiness of such items for lack of strength, they should appreciate that the Chinese used it for the spokes of the wheels of their military chariots.

HAZEL AND WILLOW

These are country woods, used more as a regenerating crop than as planks of converted timber since they rarely grow to a sufficient height, and to which there is more folklore attributed than to almost any other woods.

The Greeks believed the weeping willow trees to represent mourning and forsaken love, a theme to which Shakespeare returned several times: Dido stood with a willow in her hand waiting for Aeneas; Ophelia drowned herself for love of Hamlet where a willow grew 'aslant a brook'; and in *Othello*, a moment used by Verdi in his opera *Otello*, of a maid called Barbara, he wrote:

"She was in love, and he she lov'd proved mad
And did forsake her; she had a song of 'willow'
An old thing 'twas, but it expressed her fortune,
And she died singing it.'

MEDICINE, DIVINING AND WATER

As the Latin name (*salix*) implies, from willow can be derived salicylic acid, still the principal ingredient of asprin, and certainly known to the Romans for its power to relieve pain and reduce fever. It has recently been considered helpful in countering heart disease.

To willow is even attributed good fortune in child birth. During the Festival of Green George, the chief celebration of spring in Transylvania and Rumania, young willows are cut down and adorned with garlands and leaves, then set up in the ground. Pregnant women place a garment under the branches. If next morning they find a leaf of the tree lying on the garment, they may be assured of an easy delivery. Less quaintly perhaps, witches are deemed to fly on broomsticks of hazel, and love-smitten witches are said to disappear into the hollows of willows, reappearing as hissing cats.

The association with water has had other implications, and not just in water divining (identifying underground sources ofwater), which is best performed with hazel or willow twigs. Indeed both species favour damp and wet soil, such as the banks of rivers, ponds and lakes; the roots enable the tree to keep river banks together. England has some of the oldest such trees in the Fens near Cambridge, but even in Tibet willows are planted along the banks of the massive rivers susceptible to flash floods that descend from the Northern Himalayas.

Once submerged, willow is suprisingly durable.The Dutch have pefected the art of stabilizing their sea defences with 'polder'mats, huge mattresses of bound willow saplings which are then weighted down with stones. The foundations of England's ancient Salisbury Cathedral, being built in a former marsh, were laid on similar mats.

DISTINCTIVE PROPERTIES AND PRODUCTS

Hazel is thin, and best kept that way by cutting back every seven years; it is soft and pliable but not specially durable. Occasionally the butts are used for veneers, but the chief uses are for spars (the bent twigs that fasten thatched reed and straw to the roofs of country cottages), sheep-hurdles, cask hoops, basket work, and the groundwork for 'wattle and daub' plastering (strengthening and binding a plastered surface with a basket-like structure of hazel woven between load-bearing timbers).

Willow is light, soft, resilient, tough, flexible and easily worked. Cricketers have immortalized these properties by using willow for the making of bats. Until the advent of plastic and fibreglass, artificial limbs were often made of willow, since it could be whittled into the correct shape.

Known of old as 'sallywood', willow is recorded by John Evelyn in the 17th century as useful for "boxes, such as apothecaries and goldsmiths use", but it is in the form of osiers(literally, tough flexible twigs) for weaving wickerwork that willow has been most used for furniture and baskets. To this end, the trees are pollarded after the tree is established – cut back hard to a point from which they continue to shoot.

Willow means the sound of ball on bat (OPPOSITE PAGE) for the many countries of the world that have inherited the tradition of cricket. The trees are cultivated by some old-established firms in England and harvested young (15-30 years) for bats; willows such as these near the English university town of Oxford (FAR LEFT) are also cut back hard, or pollarded, to stem branch growth and produce osiers, used in traditional basketmaking (BELOW LEFT). Other long-standing country crafts dependent on the flexibility and durability of trees such as willow and hazel include thatching (ABOVE LEFT) where reeds are held down by spars or staples made of hazel. David Drew's willow chair (RIGHT), is modelled on the croquet seat of the country house.

EBONY

It is a name that conjures up exotic images: medieval merchant ships laden with pirated cargoes; harems and silks; bribes and despotism. A name that appears in the yarns of travellers in the same breath with onyx, lapis lazuli, and rare treasures.

Believed to be an antidote to poison, ebony was popular with the ancients for use in drinking vessels – such was their perennial anxiety. The Greek historian Herodotus records that Ethiopia paid an annual tribute of 200 ebony logs to the Persian Empire. Most supplies came from India and Sri Lanka, so it was not readily available in Europe until the 17th century. Sheraton wrote (rather dismissively) of it in 1803: 'a foreign wood lately introduced. . . . '

Because of its cost and the difficulty with which it was worked (it is exceptionally hard), ebony first became popular in the form of veneers, thin slices cut mechanically which could be glued onto cheaper, more workable woods such as oak or pine. This was especially so in France, where the art of veneering was most extensively developed; there cabinet-making became known as *ébénisterie*, and the French word for a cabinetmaker is *ébéniste*.

VARIETIES AND APPEARANCE

Ebony is a collective trade name given to all species of *Diospyros* which have a predominantly black heartwood, by contrast, say, to the North American white ebony, known in the trade as persimmon. Coromandel, or calamander (*D. marmorata* – appropriately translated as 'marbled ebony') refers to variegated ebony with grey or brown mottling, and lent the name to the heavy black screens associated with China and Japan.

Macassar ebony from Indonesia is more variegated still, like a tiger skin with inverted stripes (the black predominates). It became popular in the 1920s in extravagant widths for Art Deco furnishings, and is still used where the sumptuousness of the material is required to compensate for inadequate design.

There is nothing quite like the 75ft-long ebony floor of Henry VIII's Banqueting Hall at Leeds Castle in Kent, England (LEFT). Although this exceptionally hard and expensive timber is reputed to be used for railway sleepers in inaccessible Brazil, where local needs outweigh the difficult economics of export, such extravagance would be laughable in the West.

African craftsmen like this bowl maker in Malawi (BELOW) have always valued ebony's impermeability and durability - and now its attractiveness to tourists - while graceful woodturnings and sophisticated furniture by British designers Alan Dixon, macassar ebony vases, (LEFT) and Martin Grierson, chest of drawers, (RIGHT) exploit the variety of colour in what many think of as a 'black' wood.

EBONIZING AND THE IVORIES

Nineteenth-century craftsmen found it hard to resist simulating ebony's deep black lustre on cheaper timbers. Sheraton betrayed the early art: 'pear tree and other close grainwoods have sometimes passed for ebony by staining them black. This some do by a few washes of a hot decoction of galls and when dry, adding writing ink, polishing it with a stiff brush and a little hot wax'. There was a time when Victorian interior decorators would ebonize everything from chair frames to door cases, but the method continues today, in a more perfect form, in the finishing of the cases of grand pianos.

The tradition of using ebony for the black notes of keyboard instruments began with the use of rosewood on early harpsichords, for contrast with the ivory keys. The decorative effect of opposing the two materials is also apparent in the mounts for Chinese porcelain and ivory carvings, but it became laden with heavier symbolism (the forces of good and evil) in the religious carvings of medieval Europe. A typical example might be a virgin carved in walrus-tusk ivory, overcoming a turbulent ebony underworld.

Fingerboards, tuning pegs, and the tailpieces of stringed instruments – all the points of stress – are still of ebony, but it was for its clearer, crisper tone that it was appreciated for Baroque woodwind instruments. Ebony is also the source of the compelling sound of the castanet. More recently, however, African blackwood from the rosewood family has been preferred.

BEECH

Because of its close grain and structural consistency, beech is easy to shape and has thus always been popular for the stocks and bodies of craftsmen's tools, like these pieces of hatmaker's equipment (RIGHT). Eighteenth and 19th-century carpenters would have a large collection of beech-bodied moulding planes, each one shaped along with its blade to follow a particular profile for furniture decoration, skirting boards or picture rails. The large curved 'T'-shaped tool in the picture incorporates a sliding device for tensioning the hat rims.

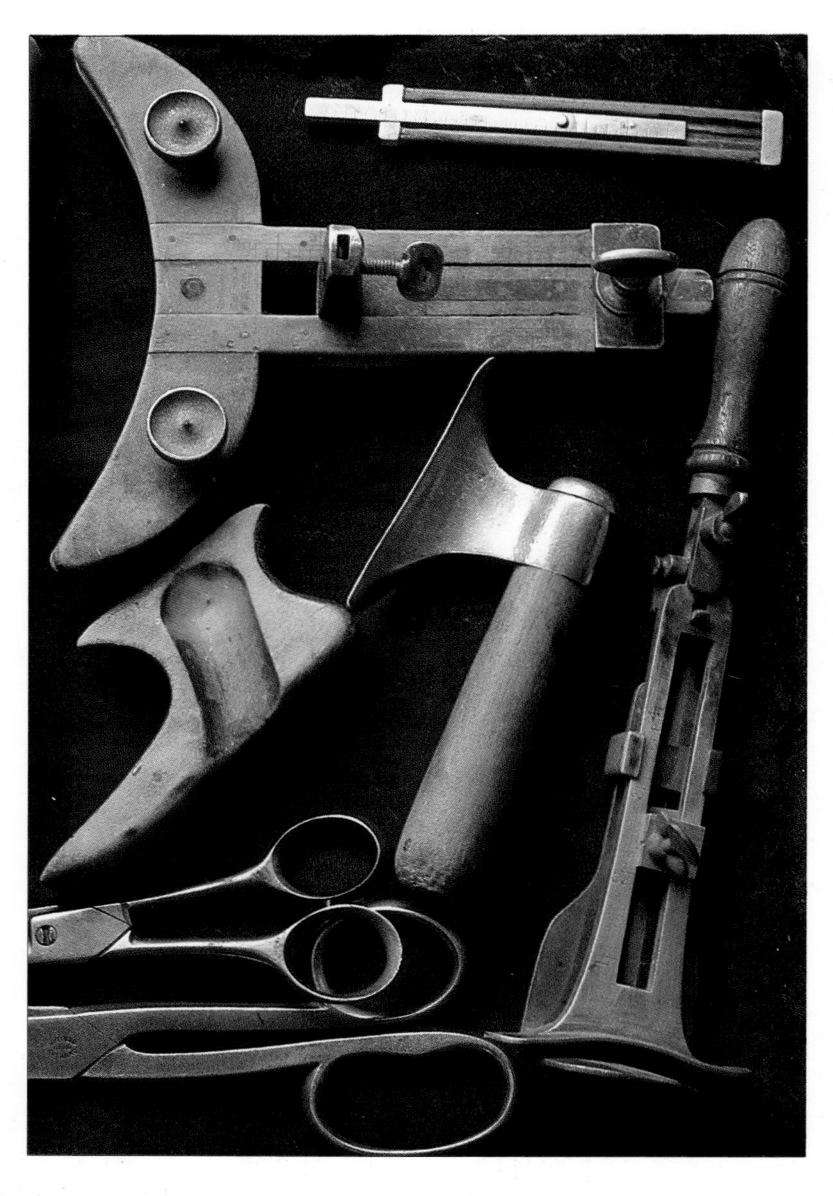

The magnificent kitchen at Castle Drogo in Devonshire (RIGHT), one of the last great English country houses in the grand style. It was designed and built by Sir Edwin Lutyens in the 1920s out of the tough local granite; Lutyens chose beech for his superb round kitchen table, designed in hexagonal 'modular' fashion. The roots of the English Windsor chairmaking industry (BELOW RIGHT) in the beech woods round High Wycombe, 1905. In the bodgers' encampment, newly felled trees were split, reduced and processed completely by hand to make components, a pile of which can be seen in the left background; the sprung pole for the lathe goes into the hut above the door.

The Gauls first used the ashes of beech to make soap, anticipating by 2,000 years its modern cultivation for the production of cellulose and artificial fibres.

Even though it is one of the most majestic trees, which occupies a prime position in European woodlands, beech is noticeably lacking in association with legend and folklore. Nor has this glorious tree ever been highly esteemed by cabinet makers; liable to warp if carelessly dried, and with an uninteresting figure, it has been used mostly where the required sections are small, and movement is less important – chairspindles, and especially upholstery frames, since it is relatively cheap and exceptionally strong.

THE 'BODGERS'

But in its own way, beech has been a constant in furniture of a less grand kind. From the 17th century onwards, because of its capacity to take stain, paint and polish, it was often used to imitate more

precious woods such as mahogany and walnut in cheaper furniture for less-discerning customers.

In the 18th century, picturesque but profitable cottage industries developed in the heart of the Chiltern hills, in Buckinghamshire, around the chair factories (which still exist) of High Wycombe. Specialist craftsmen, 'bodgers', would make the components for later assembly in the factories. These men would operate in temporary sheds in the woods, splitting the logs, roughing out the components, and then turning the spindles on a simple lathe powered by the primeval principle of bow-power – a springy pole is attached to a string which is wrapped around the spindle to be turned, and the whole device is driven by a wooden treadle.

If the ancient Egyptians developed the art of the pole lathe; it was surely perfected by the bodgers. The whole process was ecologically sound. Craftsmen could work on their own and develop great personal pride in their work; transport costs were minimal; no artificial power was needed; huge production was possible at minimal cost; the bodgers could dismantle their sheds and move on leaving neither trace nor waste. Modern factory techniques began to unbalance this natural economy in the 1920s, and today there are only a few exponents of the art.

STEAM BENDING AND THONET

Some woods,because of the constitution of the cell wall, can be rendered supple and pliable by steaming, then bent to shapes into which no tree could ever grow. When cooled and dry, they retain those shapes without damage to the fibres.

Beech and ash have just such properties, but the relative cheapness of beech gave it predominance in the 19th century in the development of 'Bentwood' furniture, associated with chairs in European cafés from the 1890s to the present day. A sophisticated steaming process was developed by the Viennese cabinetmaker Michael Thonet (1796-1871) and exploited by his descendants, which enabled structurally sound and aesthetically desirable furniture to be produced in huge quantities at economic prices without recourse to traditional expensive jointing.

Curves, scrolls, and fluid shapes were incorporated into hundreds of classic designs from hat stands to rocking chairs. In his shapes and his techniques, Thonet pioneered in beech the route for tubular steel furniture of this century.

In the 20th century, beech finds use where hard-working surfaces are required, as in laboratories, school desks and kitchens, but it is in the production of nitro-cellulose lacquers that most beech is consumed in the furniture industry. It is a touching irony that those who lament the use of modern spray finishes on furniture as being 'unnatural' are ignorant of the prime source of the raw material for these sprays

FRUITWOOD

Valued generally more for their fruit than for their timber, such trees as cherry, apple and pear have tended to be much husbanded trees, planted close to villages. Consequently, they have been attractive to local craftsmen looking for timber to use.

As a result, furniture from these woods has traditionally been recognizable as either rurally made, or merely decorated with inlays and marquetry designs using fruit timbers, because of the small timber sizes available from fruit trees. Currently, one-off custom work from designer-craftsmen has become fashionable and feasible.

APPLE

Few fruits have enjoyed such varied symbolism as the apple: as a symbol of love since Paris judged Aphrodite to be the fairest of three goddesses, and gave her a golden apple; as a symbol of deceit, it menaces readers of the tale of Snow White and her stepmother.

On the head of William Tell's son, as the symbol of hope and freedom, an apple is the target for the Swiss marksman's crossbow. It is the antidote to blissful ignorance in the biblical account of Adam and Eve's bite from the forbidden fruit in the Garden of Eden. And in modern times, New York is the Big Apple from which everyone can take a bite.

The Romans recorded as many as 29 varieties of apple, but the commonest are the orchard apple (*malus domestica*) and the crab apple (*malus sylvestris*). The woods are similar: hard, heavy and often dappled with tiny knots.

Apple was found to mesh well in timber engineering – millwheels, water wheels, and the working parts of windmills were geared with apple teeth set in oak-framed wheels. Wooden screws and golf-club heads are more esoteric uses than the turnery and carving for which the wood is especially suited.

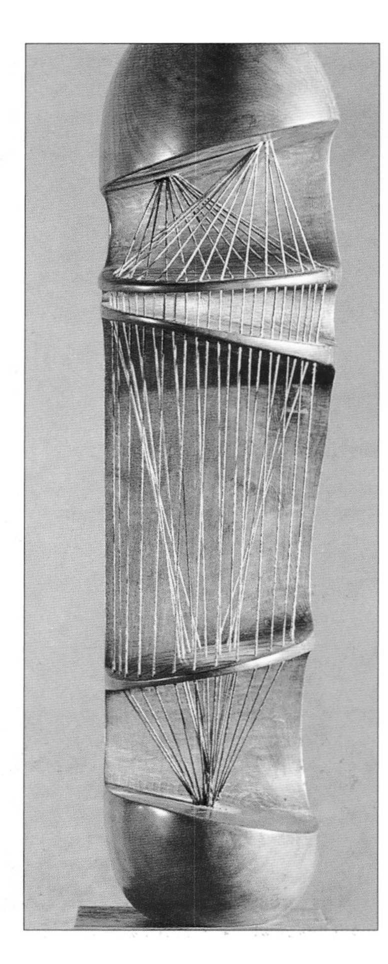

Most fruit trees are small and grow irregularly like the branched and contorted cherry (ABOVE), making them attractive to carvers and sculptors. The 1937 Henry Moore 'Stringed Figure' (LEFT), only 20in high, is in cherry. For furniture, such woods tend to appear as veneers and inlays, which demand no great size of the original tree; an Elizabethan Nonesuch chest (RIGHT) is a myriad display of different inlaid fruitwoods.

CHERRY

The association of cherry with virginity, the sweet, juicy-red fruit, plucked before the start of decay, has lured poets since the ancients. It is a recurring theme from the sonnets of Shakespeare and Dowland, to the many popular songs like 'Cherry ripe', laden with sexual innuendo as in this extract from Thomas Campion's Fourth Book of Airs, written in the 17th century:

> 'A heavenly paradise is that place
> Wherein all pleasant fruits do flow
> These cherries grow, which none may buy
> Til "Cherry ripe" themselves do cry.'

Even the blossom has a special association, this time with hope and rejuvenation; cherry trees surround memorials and cemeteries in both East and West, regardless of religious affiliation.

Cherry is probably the most popular of the fruitwoods, because it is possible to derive larger sizes of timber from it than from the orchard woods of pear, apple or plum. It has been appreciated as a fine cabinet wood in England and America because of its superficial resemblance to mahogany.

American cherry is darker and straighter than European varieties, and was especially popular with the furniture makers among the Shakers. A Puritan religious sect that flourished in 19th-century New England and of whom a tiny number still survive, their work is celebrated for subtlety of line and proportion, and refined use of timber figure in decoration.

PEAR

The archipelago of the Peloponnese, in the Mediterranean, was known to the Ancient Greeks as Apia – the land of the pear tree. With the finest texture of the fruitwoods, pear is a rival to boxwood for use in printing blocks, and for high-quality miniature carving. It takes stain well, and was popular with counterfeiters as the best ground on which to simulate ebony.

Before the advent of nylon the hardness of pear had a special use in the jacks of harpsichords, while its colour and texture have made it popular for architects' models. Varieties include alligator pear, dogwood pear, Chinese pear, white pear, Nigerian pear, and from Western Australia the 'native' pear.

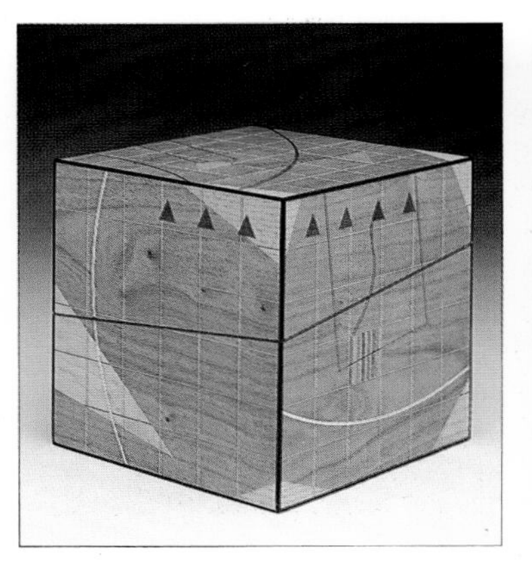

Nic Pryke's mainly cherry cube box (ABOVE) shows a modernist's approach to crafting and inlaying fruitwoods, while the blossom (RIGHT) at the Jefferson Memorial, Washington DC, acts as an eternal symbol of regeneration and hope in America as it does throughout the world. Golf clubs (ABOVE RIGHT), when they are made of wood, most often use apple for its surprising for its surprising hardness and resilience.

TEAK

In the book *Burmese Days*, in which George Orwell castigates the evils of imperialism, he chooses as his main character the manager of a teak timber yard. Teak was a peculiarly appropriate symbol – the trees being grand, noble, of supreme quality, and defenceless. The destruction of the teak forests of Indo-China, not only by the colonial powers but subsequently by national governments themselves, is one of the most painful aspects of indiscriminate forest management.

TEAK'S DURABILITY AND STRENGTH

Teak is a wood that conjures up images of elephants, sampans and straw-hats, of paddy fields and South-east Asian steam trains, of Thai tourist stalls laden with smiling teak Buddhas.

But this romance belittles one of the hardest, strongest and most durable of all timbers. In fact it is so hard that it was customary in the 19th century for European cabinetmakers to demand extra payment when converting the timber into furniture, to compensate for the blunting of their tools and the extra work involved.

However, although teak is resistant to most parasitic attacks, it does not entirely escape the attention of marine borers; though ideal for shipping (the Dutch and Portuguese assessed it superior to oak), it can be hardened further by a technique developed by the Chinese of burying it in damp earth for a few years. The logs emerge harder and totally impervious to the ravages of all insects and weather.

The Admiralty pilot books (manuals of local information published by the Admiralty in London for use by international shipping) still carry a warning for the Masters of cargo liners to avoid collision with Chinese junks, because the 'ironwood' (as such teak is known locally) is capable of overcoming steel in conflict.

It is not just in ships but in buildings too that the durability of teak is so prized. In Vijayanapur, in Southern India, the ruins of the old city include a temple of which the superstructure is supported on

Nothing in commercial quantities rivals teak's durability, which comes from its natural oiliness; it needs no maintenance, and will weather untouched to a distinguished silver grey. It is thus an obvious choice for the best garden furniture; the elegant seat (LEFT)*, made after a 17th-century Dutch design by English manufacturer Charles Verey, shows what the timber looks like when freshly cut.*

Cynthia and Margot (BELOW) may not know it, but they are lying on teak, conventionally the highest quality timber for ships' decking. Most of the world's supplies come from South-East Asia, where elephants (LEFT) are still trained to handle timber in the picturesque traditional way; teak is rare amongst tropical hardwoods in that much of it is plantation grown.

teak planks merely one and a half inches (32mm) thick! Examined in 1881, they were found to be in an excellent state of repair after 500 years exposed to the elements.

The walls of a palace of the Persian kings near Baghdad, pillaged in the seventh century, were found to contain pieces of Indian teak in perfect condition – indicating both the wood's durability and the existence of a sophisticated timber trade route at an early date. When kept under cover the timber is imperishable. Cave temples of Salsette, also in India, have 2,000 year old teak in perfect condition.

PROPERTIES AND PRODUCTS

Teak has a characteristic leathery smell and a slightly greasy feel (the oils are distilled as teak oil, which is used as a preserving finish on other timbers as well), but in the last 30 years it has only really been popular for furniture in Denmark.

It is hardly suprising that teak is favoured for making garden furniture, and for use in exterior joinery. The golden brown-colour turns to a silvery grey, but the timber needs no maintenance or further preservative treatment.

Although Burma produces as much as 70% of the world's supply, India, Thailand, Java, Sri Lanka, and Vietnam also produce teak; the priority use of tame elephants has been the extraction of the heavy trees from the forests, for which purpose elephant training schools are still maintained.

Because of its superior qualities, other woods are miscalled teak in the puffs of salesmen. There is Cape teak, Johore teak, Philippine teak, Rhodesian teak, Surinam teak, Transvaal teak, to name but a few. A relatively recent popular substitute is the African wood iroko 'Native teak' from New South Wales has similar properties but is no relation to the real thing.

EUCALYPTUS

Eucalyptus is the name given to a huge genus of trees indigenous to Australia and Tasmania. There are more than 300 species, from small shrubs to trees the size of sequoias (*Eucalyptus regnans*, the giant gum of Victoria, attains a height of over 90m or 300ft), with poetic names derived from an evocative mix of aboriginal and western vocabulary: Blackbutt, Yuba, Wandoo, Bloodwood, Cadaga, Coolibah, Gimlet, Merrit, Yapunyah, Tallow wood, Tingle-tingle, Peppermint, Tuart, Wormwood, and lest we forget the role of the Royal Navy in early British colonial history, Messmate.

FOOD, MEDICINE AND FUEL

After 1854 the eucalyptus was successfully introduced to southern Europe, Algeria, Egypt, Tahiti, Natal and India, but it has been most successful in South Africa and California. Young twigs of *E. cordata* and *E.pulverulenta* are sold by American florists as 'blue spiral'.

But to Australians the timber is as important economically as is the maple to North America. The leaf glands of many species, especially *E. salicifolia* and *E. globulus*, contain a volatile aromatic straw-coloured oil which can be extracted by steam distillation, and should contain at least 70% eucalyptus, which carries a crystalline resin. It smells like camphor and tastes spicy, and is the source of that gum found deposited underneath restaurant tables or in hidden corners of lifts.

But its chief use is as an expectorant, to encourage spitting or clearing the mucus from the throats and lungs of bronchitis and tuberculosis sufferers. It is also used to relieve colds and asthma. Botany Bay kino, an amorphous resin, is obtained from incision in the same trees; gum is readily derived from *E. gigantea*, whilst a hard opaque substance, known as manna, is procured from *E. vinnialis*.

Distilling the leaves yields a huge quantity of gas – as much as 10,000 cubic feet per ton – and it is as fuel that the tree has been extensively used, first by the Australian Aboriginals. They ascribe mystical powers to the trees, comparable with the myths of the ancient Greeks. It is not unusual for humans, or any animal, to appreciate the source of so much nourishment and protection. Even koala bears seem fond of the eucalyptus.

FRIEND AND ENEMY

Tasmanian oak is the export name for three eucalyptus trees known locally as ash, but having no connection with European oak or ash. Jarrah is the most durable of the species, and being impervious to parasites and not particularly decorative, it has been found ideal for docks, harbours and bridges. It was even specified by engineers of the

19th century for the sleepers of the London Underground. Karri and spotted gum are similar woods, though not as durable as jarrah.

Though loved in the Antipodes, eucalyptus has come to be abhorred in Madagascar; the climate appears to be especially benign, and it grows voraciously. In contrast to parts of the world where timber conservation is the aim, Madagascar is gradually being taken over by the weed-like tree, which cuts the light from indigenous flora, and rampantly colonizes wherever it can.

*The almost supernaturally rich colouring of candlebark (*LEFT*), (Eucalyptus rubida) stands out in the Snowy Mountains of New South Wales. A koala (*TOP*), enjoys a eucalyptus leaf snack; the oils are reputed to intoxicate! Railway sleepers and heavy-duty industrial timbers (*ABOVE*), are often made from jarrah, another of the many species of the genus. The trees can be enormous, like this group of karri – 'The Four Aces' – near Manjimup, W. Australia (*RIGHT*); the giant gum, which grows in Victoria, can reach 300ft.*

ELM

The rough bark of the elm (FAR LEFT), was not sufficient protection against Dutch Elm disease which has destroyed many elms throughout Europe. One of elm's longest-standing and most familiar applications is for the seats of Windsor chairs (LEFT). These examples are unusually ornate for Windsors, aping the style of the Georgian Gothic revival (c 1760). The strong, earthy figure and wide boards in which elm was obtainable made it a favourite with sculptors as well as chairmakers; this 'Head' (BELOW), by Henry Moore, 21cm (8½ in) high, dates from 1938.

The Romans dedicated the elm to Mercury, the messenger god who watched over thieves and merchants, but the Greeks, in contrast, held a more poetic belief that nymphs would plant elms in remembrance of fallen heroes. They could hardly have known that the elm itself would become a fallen hero in the 20th century, decimated by Dutch Elm disease. Although the disease has been known since the 1920s, it is a particularly virulent strain introduced from Canada to Europe in the late 1960s that has taken the most devastating toll, killing off those heroic hedgerow trees and completely altering the landscape. There are a few pockets of woodland, usually surrounded by steep hills, which have avoided the devastation, but now every European country has suffered.

The fungus *ceratocystic ulmus* was originally carried with Canadian rock elm which came to Europe for boatbuilding. A beetle eats out a path under the bark which allows the fungus to get into the phloem tissues, through the cambium and into the sapwood. The fungus then causes the cell tissues to blister, blocking the water routes and killing the tree. No lasting effective cure has been found.

CHARACTERISTICS AND USES

Strong, swirling figure gives a sensuous appearance to the polished timber, much exploited by modern sculptors such as Henry Moore, but those same swirls indicate the uneven structure of the tight

The length, strength and durability of elm made it first choice for massive structures before imported hardwoods arrived in Europe. London Bridge, (LEFT) in 1600, was originally built on elm piles. A sturdy blanket chest by Luke Hughes (ABOVE) displays typically curvilinear grain patterns.

interlocking grain. The unevenness makes elm unreliable; it has a propensity to warp and crack, and the unpredictability of the grain can make it awkward to work with hand tools, or even to machine. But decorative veneers from elm have always been sought, because it is possible to lay them on a more stable base, such as plywood.

Only relatively recently, with the advent of modern drying techniques, has it become realistic to use the solid timber for fine woodwork. However, in chairmaking there is a long-established tradition of using solid elm seats for 'Windsor' or 'Philadelphia'- style chairs, the elaborate figure lending a rustic charm to the otherwise stick-like designs, and to this day companies such as Ercol, one of the UK's best known, use what supplies of elm they have stockpiled or can still obtain.

The interlocking quality of the grain helps elm's resistance to pressure, though not to tension, which is why it may well be found as floorboards in old buildings – but rarely as supportive joists. Woodmen still use elm wedges (as a substitute for steel) to split stubborn logs of other timbers, and it was valued as the only reliable wood for the central stock of cart-wheels. Unpolished, or well weathered, elm turns a silver-grey colour, and the grain becomes deeply contoured. Aged floorboards, and the weatherboards on farm buildings are particularly attractive in this way.

But it is for its longevity when completely waterlogged that elm has been of most structural use. The Venetians built their city on elm piles, aping a technique used even by the Romans, who, indeed, built the original London Bridge on piles of this trusty timber. They were still intact when the bridge was rebuilt once more in the 1960s.

Similarly, the early engineering of London's underground rivers in 1613 incorporated elm water pipes, which were found serviceable as late as the 1930s. This durability and its symbolic associations with death caused it to become the traditional timber for coffins – the corpse might last longer, so the story goes. Ironically, those coffins put away in grand family vaults fare less well than those laid to rest under the earth. Vaults tend to be damp, but not waterlogged, and elm deteriorates astonishingly quickly if the environment is neither entirely wet nor dry.

Box

From the Bible and classical literature we learn that box hedges have been groomed in an evolving tradition of topiary that reached a peak in the studied formal gardens of Versailles and other grand European palaces – yet the Egyptians were pioneering the use of this hard, smooth, distinctive timber for decorative inlays 3,000 years ago. It is one of the densest woods of all (it barely floats in water), which, combined with the small sizes of available sections, explains its exploitation for 'treen' (literally, small wooden items from trees) – chessmen, corkscrews, snuff boxes and the like.

Box carries fine detail exceptionally well. Woodcarvers of medieval Europe achieved high standards of representation on a small scale, an art that was prized by gold and silversmiths who would employ boxwood formers against which to beat out their delicate shapes. Patternmakers had similar views, and the moulds used to cast elaborate details for gesso picture frames are still made from this wood.

ENGRAVING AND PRINTING

But it was the art of the engraver that particularly flourished with box, from the early 15th century, when printing began in Europe, through the 16th century, when the work of masters like Dürer and Holbein was transferred as engraving, to the 1920s. This was the era

Classical topiary cut from boxwood hedges at Levens Hall, Cumbria in northern England (LEFT). *The leaves and branches take sculptural character almost as well as the timber takes carving. The garden was planted in 1692 by Guy Beaumont, King William of Orange's gardener at Hampton Court. It is claimed that it is the oldest unchanged topiary garden in England.*

Boxwood moulds for casting decorative plaster features (ABOVE). *No other timber holds detail so faithfully, as any printer or engraver will testify. Boxwood cores for top hats* (ABOVE RIGHT); *the wood's density and closeness of grain are the qualitites sought by craftsmen in need of reliable mould shapes.*

when the quality of the material and the fashionable aesthetic together gave the woodblock engraving pre-eminence in book production especially in the work of private presses.

Although the development of copper and steel engraving plates temporarily eclipsed the wood engraving, it was found towards the end of the 18th century that carving fine detail into the endgrain of a boxwood block would withstand up to a million impressions, and was far longer lasting and better suited to high-speed presses than metal blocks.

INSTRUMENTS OF MUSIC AND MEASUREMENT

The Greeks discovered that when it was used for wind instruments there were few woods that possessed such a soft resonant tone as box. Although modern substitutes, such as African blackwood or ebony, have better defined qualities, the recent revival of early music and the attendant requirement for authentic instruments, such as bombards, hurdy-gurdys, and baroque flutes has revived demand.

The traditional carpenter's ruler has always been known as a boxwood rule, and early architects' and mathematicians' instruments tended to be made from the same material. Another wood, maracaibo boxwood, found in Central America, is botanically unrelated but shows similar traits to European box, and is generally used as an inferior substitute. Other practical uses include handles for chisels and mallets, and the wearing parts of carpenters' moulding planes. No better substance has been found for dressing lead (beating it into awkward shapes) in roofwork.

SYCAMORE AND MAPLE

The ancient Egyptians used to make mummy cases from a variety of fig, (Ficus sycomorus) from which the name 'sycamore' derives – contributing to the confusion in the naming of the different species of *acer*.

For what the British call sycamore (*Acer pseudoplatanus*), the Americans call maple, and what the Americans call sycamore (*Platanus occidentalis*) the British call lacewood. What the Australians, on the third hand, call maple (actually Queensland maple, *Flindersia brayleyana*) is part of the satinwood family (rutaceae). Other *acers*, it is a relief to say, do enjoy a USA/UK concord, both nations agreeing to call such species as *Acer rubru m* and *Acer saccharum* by the name of maple.

Sycamore in its UK identity lives up to five hundred years and has earned a few mythical attributes, not least that sycamore lintels and doorsteps will guard against the evil attentions of witches. Sycamore branches around doors and windows, goes the belief, are sure protection against lightning.

CHARACTERISTICS AND USES

All the *acer* timbers are principally crisp and creamy-white, sometimes with distinctive figure arising from irregular growth. This is best-known in the form of bird's eye (the little black knots come from miniature growths – a kind of canker) or fiddleback, from the traditional use in the backs of violins and 'cellos.

*The creamy delicacy of sycamore and its delightful range of decorative grain and figure are exploited by furniture makers and woodturners alike. The elliptical beauty of form in the bowl (*ABOVE*), by Cornish Anthony Bryant, is skilfully echoed in the way the growth rings appear on the curved surface; the screen (*LEFT*), by classically-inspired designer maker Luke Hughes, depends for its effect on the combination of stained bird's eye and fiddleback sycamore panels. The ancient sycamores in the Kentish countryside (*RIGHT*) would be known in the US as maples, but either way they have been cut back so hard by natural or unnatural means that their growth has been (literally) truncated; the diehard trees have sprouted new branches, which form an attractive lookout niche for young tree lovers.*

The Shaker rocking chair (LEFT) is unusual for that famous religious sect's harsh aesthetic tradition in that it uses unpainted bird's-eye maple, essentially a decorative wood. The simple beauty of Shaker work is exemplified by the hatbox (BELOW), formed from straight-grained maple cut thin and then bent into a circle. To most Americans, maple means syrup, which is boiled from the sap; here it is being tapped in Vermont (RIGHT) before processing and consumption - with the obligatory waffle, no doubt.

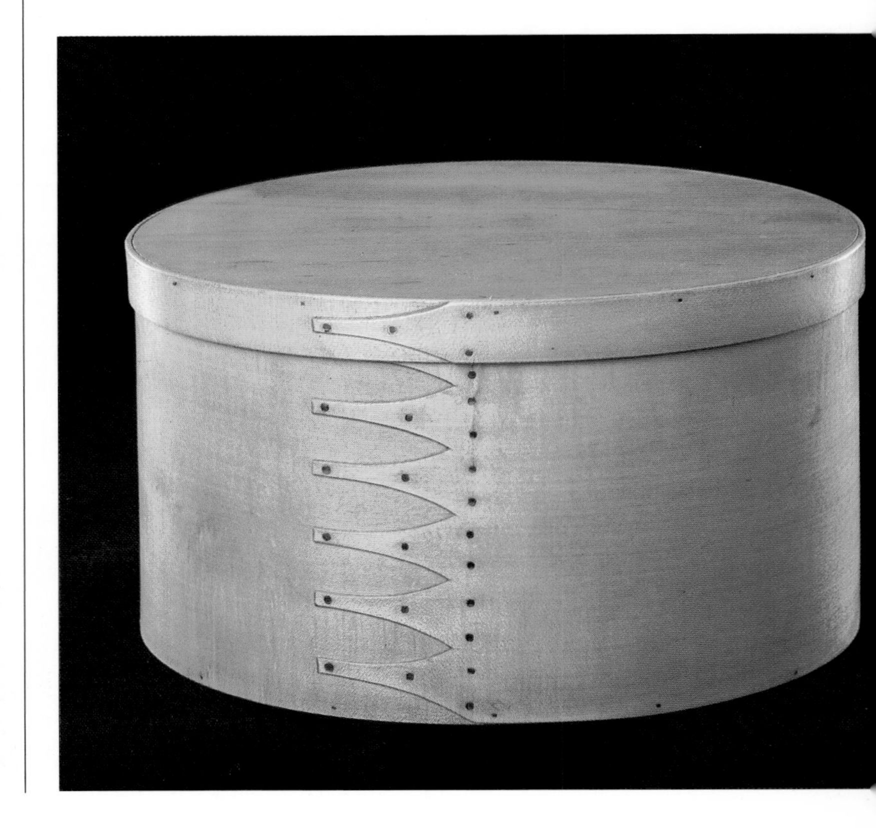

More plain-grained timbers from the *Aceraceae* are favoured for their acceptance of stain, not just in the grey-stained veneer known as harewood, but also in more dramatically bright and unusual colours; these have found popularity amongst some contemporary furniture designers, and of course they always work well in children's toys and furniture.

The propensity to stain means special care is required in extracting the timber without discoloration – it is necessary to remove the bark shortly after the tree is felled, saw immediately, sweep the planks clear of sawdust, and, unusually, stack upright as opposed to horizontal.

Acer pseudoplatanus (sycamore for the British) is particularly good for turning, and since it leaves no taint when in contact with food, was commonly used in the production of kitchen bowls and utensils. It can be obtained in wide planks, and found use in scrubbed farmhouse kitchen tables, as mentioned by Chaucer more than 600 years ago. An inferior relation is the field maple (*acer campestris*) popular for light-coloured inlays and veneers but less favoured as solid furniture.

In America, early colonists were quick to assess the many values of rock maple (acer saccharum), not just for furniture, especially turned chairs, but also for the syrup derived from boiling the sap. Each tree can yield as much as eight pounds of maple sugar each season, and it quickly became the staple sweetener of the New England diet. Although the wood is tough enough to withstand use as squash-court flooring and escalator treads, it is barely resistant to harsh weather.

Other varieties found in America include silver maple, black maple, red maple, swamp maple, Oregon maple and blister maple. The overwhelming importance of the tree to the economies of North America is perhaps best celebrated in the choice of symbol for the Canadian national flag, and it turns up, too, in the music of Scott Joplin's 'Maple Leaf Rag'.

RELATIONS

A close cousin is the London plane (*Platanus acerifolia*), a popular tree for the central parks of industrial cities. It has an appropriate ability to shed its bark frequently, and can survive the smog better than most; for this reason it features in most European capitals. The mottled, dappled figure it sometimes yields, which has led to its being known also as lacewood, has given it more use for decorative veneers than as a solid timber. Art Deco designers in particular favoured the mannered appearance.

Queensland maple, or silkwood, is pinkish, lustrous and slightly fragrant. More highly figured than any of the *acer* species, it owes its name to a superficial resemblance rather than botanical connections. It is prized for cabinet work.

The 'Butler's Tray' (ABOVE) by avant-garde English furniture designer Fred Baier makes outstandingly successful use of the ability of sycamore both to bend (in thin sections) and to absorb bright colours. Ten-pin bowling champion Cheryl Robinson (LEFT) is sure to have appreciated the springy hardness of maple, used for sports and dancing floors.

LIGNUM VITAE

Of lignum vitae, Sheraton wrote 'a very hard and ponderous wood'. When lignum vitae was 'discovered' by Europeans in the 16th century, they found that the West Indians ascribed to it the power of curing venereal disease – a property more poetic than pragmatic, but which led to a brisk trade in sawdust and shavings to credulous sufferers. Perhaps this led to some wit giving the timber the Latin name of 'wood of life'.

The best timber comes from the West Indies and the coastal regions of Colombia and Venezuela, and also from Central America. Because the trees are generally narrow and short, the timber has never found favour as a cabinet wood, although in the 20th century 'oysters' (thin veneer slices cut across the endgrain of a billet or trunk) have been applied with luxurious effect in repetitive motifs on cigar humidors and knife boxes.

Lignum vitae has been valued not only for its exceptional weight, density and toughness, but also for the high resin content in the pores, which acts as a continuous lubricant. In an age when foul-smelling animal fat greased the working parts of most wooden machinery, this quality was especially prized and found impressive uses. It was not just the sheaves of ships' pulleys, but even the bearings of ships' propellers and millwheels that sat in bushes of lignum vitae. Other mechanical applications include the gear wheels of wooden clocks, like those made by the 18th-century Lincolnshire brothers, John and James Harrison. They claimed their timepieces would need neither cleaning nor repair for fifty years. History has

Lignum vitae's legendary hardness combines with exceptional oiliness to render it suitable for functions not normally associated with wood – for example, bearings for heavy moving structures like this huge millwheel (ABOVE) in Sturbridge, Massachusetts. Its density is the reason why it is first choice for mallets, for both croquet (LEFT) and carving; a little tap with lignum vitae goes a long way!

judged, however, that they do need rather more frequent adjustment than the brothers claimed.

Lignum vitae is the heaviest timber in commercial use, and its high resistance to abrasion also means that it is difficult to work with cutting tools. Conversely the weight and density made it ideal for the pestles and mortars of pharmacists, in addition to the belief that some medicinal value might be added to the concoction.

Smart 19th-century aristocrats would carry lignum vitae coshes as 'life-preservers' against potential assailants, and there is a famous murder tale in the novels of Anthony Trollope in which the instrument of death was just such a weapon. More modern applications include the heads of croquet mallets, and the turning of 'boules' – the balls used in the French village game.

The colour varies from a deep brown through to an olive green. Despite its hardness, lignum vitae has a fine texture and it has always been popular for turning, especially goblets, loving cups (perhaps, in light of earlier medicinal attributions, with a view to prevention as well as cure) and 'wassail' bowls – the ceremonial drinking cups used on formal occasions like Christmas and Twelfth Night, principally in Scotland. For its hardness, lignum vitae was also used by the Victorians for the movable cupped bases that increase the height of billiard and snooker tables.

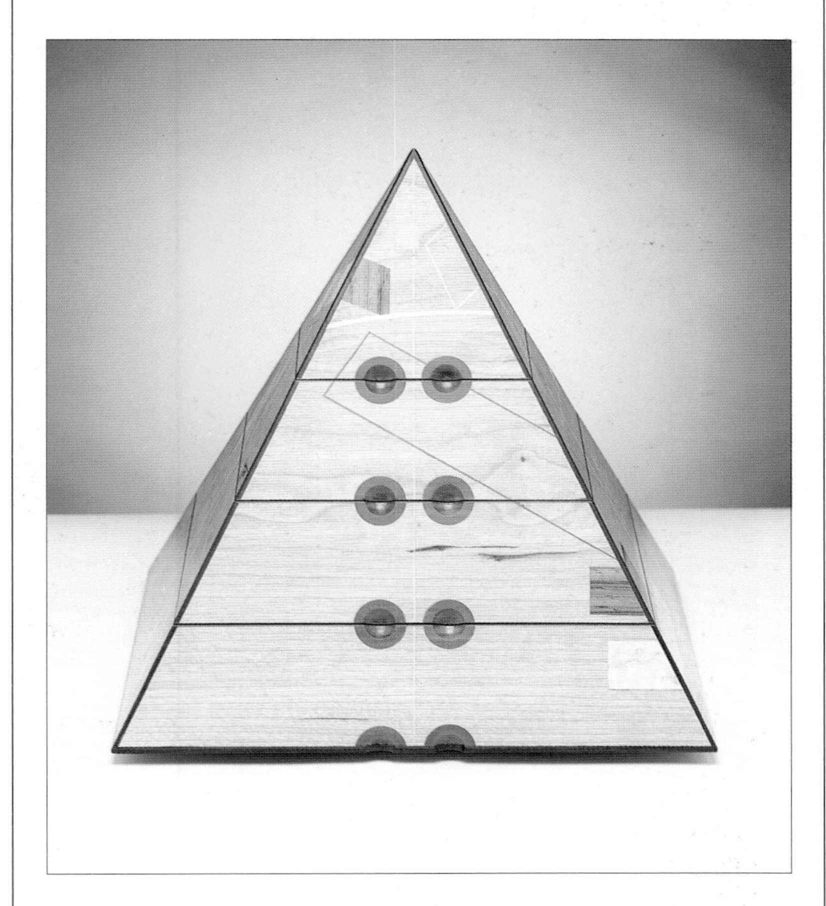

Lignum vitae can be an effective decorative wood, as the finger-pulls on Nic Pryke's pyramid box (ABOVE) demonstrate. Its durability and lubricant qualities were outstanding in pulley blocks for the rigging of tall ships like HMS Victory (ABOVE RIGHT), and its weight is ideal for blunt instruments (ABOVE FAR RIGHT, a 19th-century policeman awaits his prey); perhaps its evocation of permanence made it the traditional timber for the 'wassail bowl' with which the Scots greet Twelfth Night (RIGHT).

LIME

It has been said of the lime tree that it grows for 300 years, stands still for 300 years, and takes 300 years to die – a romantic view. However, the lime does not reach its maximum height until around the age of 150 years, and even then it goes on expanding outwards.

The sense of permanence associated with this longevity, together with the fact that it is a visually striking tree, have made the lime popular for decorative planting. It graces parklands in random fashion, and forms disciplined avenues in cities, the most impressive example of which is in Berlin. The main axis of the historic centre of the capital is marked neither by name nor geographical reference, but simply as 'Unter den Linden' – under the limes.

A customary tribute to Venus, goddess of love, is a garland of lime leaves, a feminine attribution also celebrated by the Greeks; their legends relate that the peasant Philemon and his wife Baukis earned immortality by extending their hospitality to Zeus, travelling in disguise. Philemon was transformed into an oak and Baukis a lime.

CHARACTERISTICS AND USES

Physically, lime is a soft hardwood – close-grained, creamy, and easily worked, both with and across the grain. This is why it has always been popular with carvers. Even before the Bronze Age, because it was one of the few woods soft enough to be fashioned with stone tools, it was used to build primitive houses, and subsequently weapons. The earliest bows were made from lime, and it was found that laminations of the woven phloem (the arteries beneath the bark of the tree that carry nutrients) could withstand heavy blows, which made it ideal for shields. A fine Nordic example, sheathed in leather and decorated with metal, is in the Sutton Hoo collection of the British Museum.

Although it is tough, stable and pliable, lime is susceptible to both worm and weather, so the comparatively recent discovery of a war canoe dated 300 BC in a bog at Hjortspring in Denmark, is even more exceptional than the size of the boards involved: five huge planks over 15.25m (50ft) long and more than 0.6m (2ft) wide came from a larger tree than can be found today.

Lime makes excellent 'woodwool', the thin matted shavings used conventionally for packing crockery. Less well-known uses for this by-product were in early surgical dressings, cheap upholstery, and as a substitute for hair in plaster; the shavings help to bind the lime mix and prevent it cracking or collapsing. Other utilitarian applications include the mechanisms of keyboard musical instruments, and in America, where it is also known as basswood, it is considered exemplary for beehive frames.

Carvers of the Middle Ages referred to lime as 'sacrum lignum' (the holy wood) because they associated it with the figures of Christ, the Virgin and the Apostles which they were employed to carve. But from the late 17th century, a more secular age, the lime was known simply as the carvers' tree. Grinling Gibbons (1648-1720) was one of the greatest exponents of the carver's art; although most of his creations for buildings were carved in oak, he would create exquisite miniatures in lime with which to tempt potential patrons.

Lime flowers are fecund in the extreme, as anyone who has parked a car under a lime tree for more than hour in spring will have discovered – the sticky mess is profuse. The flowers of the tree (rich in sugar, wax and oils) are reputed to be effective in reducing fever, but they also contribute to another form of suffering – hay fever. Tree pollen is a general irritant, and lime pollen particularly so. The flowers can, however, be brewed into a pleasant tea, as mentioned in 16th-century herb books.

The avenue 'Unter den Linden' (BELOW LEFT) in Berlin is a superb example of the decorative planting of lime trees . The ancient Sutton Hoo shield (BELOW) from an Anglo-Saxon burial mound illustrates the ability of lime, woven in layers, to withstand heavy blows – perfect for defensive equipment. Always the 'carver's wood', lime's greatest exponent was Grinling Gibbons, who executed this superb commission (RIGHT) for the Duke of Somerset in 1692 at Petworth House in Sussex in southern England.

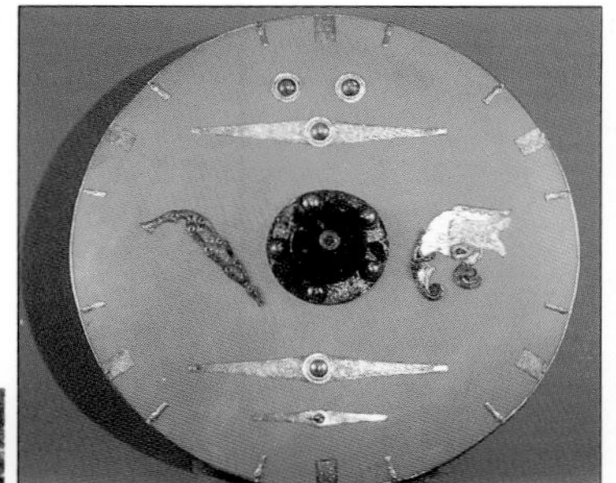

WALNUT

A native of central Asia, walnut was taken to Italy by the Greeks, then northwards by the Romans, who dedicated the tree to Jupiter, calling it 'jovis glans', or the nut of Jupiter. Hence the contemporary botanical name *juglans*. There was no wide dissemination of tree or timber in Europe before the time of Charlemagne, though the Romans had certainly introduced it to England by the first century AD.

The English name probably derives from the Old English 'wahl', meaning foreign. It is now more common in France, Italy and Spain. The native American species, known as Virginia walnut, was first planted in England as early as 1656. It is darker, straighter, seldom as finely figured, but has many similar properties. Queensland walnut (*Endiandra Palmerstonii* – it was named after the Victorian British

Contemporary Californian maker Sam Maloof uses almost nothing but black walnut for his furniture, like the sinuous rocking chair (FAR LEFT), while young Englishman Nic Pryke's cheval mirror (LEFT) contrasts the timber's dark warmth with ripple sycamore feet. Amongst all this furniture, it pays to be reminded of the nut itself; the distinctive tang of walnut oil (BELOW LEFT and RIGHT) will get any salad on its legs.

Walnut's mellow character was beautifully evoked by the 18th-century craftsmen of England's 'Age of Walnut'; the Queen Anne travelling 'bachelor's chest' (RIGHT) shows a fine aesthetic sense of proportion and restraint.

Prime Minister Lord Palmerston) has no botanical connection to the *juglans* family, but is well suited for the same purposes, namely furniture, shopfitting, panelling and domestic objects.

WALNUT IN EUROPEAN HISTORY

From the Renaissance onwards, walnut became the most popular timber for furniture, especially in the Low Countries. The paintings of Dutch interiors by Vermeer are heavily populated with elegant pieces. The English used it only rarely, until the advent of the Huguenots, who fled the religious persecution on the Continent during the second half of the 16th century. They brought with them their artisan skills and their love of walnut.

However, the stark Puritan era that followed put paid to any lusciousness in decoration, and oak remained the staple furniture timber. Indeed, Flemish and Continental styles were banned after the execution of Charles I in 1649, but the restoration of Charles II brought with it the tastes of his French and Dutch hosts-from-exile.

The years 1660-1720 became the 'Age of Walnut', a time when English furniture influenced the world, a time of exuberant design and fervent creative activity. It was also a time when the art of veneering reached new heights, principally because the timber that produced the most attractive figure tended to be structurally unsound. It was thus common to glue thin slices of attractively figured walnut on to a groundwork of plain but sound timber like oak or pine. 'Burrs' and 'curls' were and are most prized, cut from cancerous growths on the side of the tree, and where branches fork. The roots produce particularly good material, which is why it is common to fell a walnut from below ground level by digging a large pit all round the tree.

GIFTS OF THE TREE

Walnut is one of those trees as much loved for the products of its branches as for its exceptional timber. The leaves contain a tannin reputed to be a good antidote to poison, but also effective in dealing with intestinal worms and skin disorders. They produce a good mouthwash, and if mixed with linseed oil are effective as a laxative for cattle.

The nuts, which ripen in September, are protected by a green, smooth, tough, fleshy outer case wherein lies the familiar hard furrowed shell. This protects a fleshy seed. Before they are dried, the seeds contain a good deal of oil used in salad dressing, but also in soap and oil-based paints.

An old and controversial proverb recurs in a number of European languages, which recommends:

'The more you beat them, the better they be:
A woman, a dog, and a walnut tree'.

Whatever the wisdom so enshrined, it is true that in Hungary the nuts are still beaten off the tree with long poles which have metal blades mounted at the ends; the lacerations thus caused on the young shoots appear to stimulate growth in subsequent years.

CEDAR

The majestic cedars of Lebanon, with the sensual aroma of their silky timber, were symbols of wealth long before they became popular as parkland trees. They were even a source of wealth to the ancient Assyrians, in whose land the indigenous trees stood.

Egyptians, living in a land where few trees grow, are known to have been importing cedar from 2500 BC. One of their principal reasons for expanding their empire northwards was to control the cedar supply. A ship buried near the base of the Great Pyramid and unearthed in 1954 was found to have massive planks 23m (75ft) long and 125mm (5in) wide, an indication of highly developed and sophisticated trade and transport systems.

PROPERTIES, PRACTICAL AND POWERFUL

The exoticism associated with cedars of Lebanon obscures the fact that the name is applied to over 70 different kinds of wood, most of which are evergreen.

The best-known cedars are the Lebanon cedar, Atlas cedar and the Deodar, or God Tree – a native of Afghanistan and the Himalayas that still grows at heights of 12,000ft.

To the Deodar were ascribed strange religious powers. It was only at the turn of the century in the Punjab that a large old cedar was felled – a tree to which a young girl was annually sacrificed. In the Hindu Kush, the priestesses' power of prophecy was said to be acquired by inhaling the thick pungent smoke of kindled Deodar twigs. The priestesses would be possessed with convulsions and fall senseless, then rise to begin a shrill prophetic chant.

The use of burning branches is still essential to some ceremonies of Tibetan Buddhists, though they commonly use juniper (*Juniperus chinensis*), which is technically known as a 'false' cedar. The best-known 'false' cedar (those which display similar qualities in appearance and behaviour) is the cypress, but others include Bay cedar, Ceylon cedar, Honduras cedar, Nigerian cedar, Oregon cedar, Japanese cedar, and Moulmein cedar.

Virginia cedar and African cedar (both technically junipers) are those commonly used for pencils, being supportive to the delicate graphite leads but still easy to sharpen. There is, in addition, Incense cedar, a Californian tree popular for making venetian blinds.

DISTINCTIVE CHARACTERISTICS AND USES

The Renaissance Italians began the tradition of making chunky, coffin-like dowry chests from cedar; not just because the wood, being so light, could be worked in thick sturdy sections and still be lifted, but also for its resistance to termite attack. For, although structurally

similar to pine and deal, cedar is normally unaffected by woodworm; this has meant that some magnificent early pieces survive. This, in conjunction with the pleasant smell, ensured cedar's popularity.

Moths appear to avoid fabrics stored in cedar, but it is unclear whether this is because of the smell, or just the fact that cedar chests tend to be airtight (hence its use for cigar boxes). It has a tendency to swell and deform slightly, closing cracks. Cedar plugs in the mouthpieces of recorders exploit this property, being resistant to saliva, but not so strong as to split the head of the instrument when they swell from the damp.

Because of its durability, cedar has from the earliest times been valued for wooden shingle roofs, the shingles being cleft (rather than sawn) from the logs to form wedge-shaped pieces that could be overlapped like the feathers of a bird.

The Kwakiutl Indians from the north-west coast of British Columbia in Canada, who are excellent craftsmen, devised an intriguing way of utilizing both cedar's durability and its propensity to swell in making boxes, buckets and other containers. From a single plank (cut perpendicular to the tangent of the rings), they would derive all four sides with an ingenious and skilful method of scoring and steam-bending; the corners were pegged and lashed with leather thongs. The sides were then rebated into a base and similarly pegged – but the strength – and waterproofing relied on the wood being kept damp.

Pencils, now made of incense cedar, have been manufacutured in Keswick, in the Lake District, England, since the 1500s. In the 1850s, the logs would have been cut using a water-powered circular saw before the graphite strip was inserted (ABOVE LEFT). At Atterbury Estate, Shrewsbury, England (LEFT) in the beautiful parkland designed by Humphry Repton the stately Cedars of Lebanon are seen at their finest. In the USA cedar wood shingles have been used to cover roofs for hundreds of years and in the Blacker House (ABOVE) in California designed by the Greene brothers in 1906 cedar is used to recreate the look of a Swiss mountain chalet.

GLOSSARY

A

Angiosperm A class of plant whose seeds are borne inside an *ovary*.

Apical meristem The 'growing point' of tissues of living wood cells capable of repeated division, found at the tip of the stem.

B

Bast The soft, fibrous layer of *phloem* tissue which forms between the bark and the inner cell structure.

Bole The lower section of the stem or trunk of a tree, from above the root-butt to the first limb or branch.

Burrs (burls) Abnormal wart-like growths or excrescences often produced by *stooling*. Irritation or injury causes stunted growth which develops into a contorted, gnarled mass of dense woody tissue that produces a highly decorative appearance of tightly clustered dormant buds, each with a darker pith.

C

Calyx A whorl of usually green leaves, known as *sepals*, forming the outer part of a bud or flower.

Cambium An active layer of cells which divides to produce new tissues. Vascular cambium produces *xylem* on the inside of the stem and *phloem* on the outside, below the bark.

Catkin A drooping spike, or 'cats tail', of scalybracted, stalkless flowers.

Collapse Irregular and excessive shrinkage of timber during drying.

Compression wood Timber from part of a conifer that has been under compression, such as below a branch or on the concave side of a bent tree. The cells are shorter than normal and thicker walled. The wood is usually of a darker colour.

Cork A layer of dead cells on the outside of a stem or root that protects the inner tissues against damage; particularly well developed in the cork oak tree.

Corolla The ring of petals forming the conspicuously coloured part of the flower.

Cortex The part of a tree's bark between the endodermis and the *epidermis*.

Cotyledon A seed leaf that becomes the first leaf of the embryo plant.

Curls (crotches) Figure of the grain produced by suitable conversion from the junction of a branch with the stem of the tree, or between two branches. The wood fibres suffer from either compression or tension, being distored by the weight of the limb.

Cuticle The impervious outer coating of the *epidermis* that prevents damage and water loss.

D

Dicotyledon In the embryo stage dicotyledons have two cotyledons in its seed. Dicotyledons include all broad-leaved trees and most shrubs.

Diffuse-porous Where the pores are diffused or scattered across the growth ring, so that there is little difference between the pores in the *earlywood* and *latewood* zones. Also, where is a gradual change in size and distribution across the *growth ring*.

Doat or dote Localized patches of incipient decay or rot in timber.

E

Earlywood The early part of a growth ring, consisting of pale inner wood with thin walls, formed in the spring and early summer. Also known as 'springwood'.

E. M. C. Equilibrium moisture content: the moisture content at which wood is stable and in equilibrium with the humidity of its surroundings.

Endocarp The inner layer of the *pericarp*.

Endodermis The inner layer of the *cortex*.

Endosperm The nutritive tissue around the embryo of a seed.

Epidermis The layer of cells just beneath the protective cuticle.

F

False heartwood Dark innerwood caused by disease or fungal attack.

Fibres The longitudinal thick-walled wood elements in hardwoods whose function is to provide strength. The weight and hardness of wood is proportional to the amount of wood fibre in the tree.

Figure The pattern on the longitudinal surface caused by variations in the colour; the arrangement of tissues such as the grain, rays, branches, and contortions around knots; irregularity and interlacing of fibres. Abnormal figure is due to external defects, such as insect attack, decay, reaction wood, wounds, and pollarding.

Flitch A section of a log trimmed and prepared for conversion into veneers, or part of a converted log suitable for further conversion.

F.S.P. Fibre saturation point: the point at which the cell walls are fully saturated but the cells contain no 'free' moisture. Shrinkage will occur with further seasoning below this point. The physical properties of wood do not change when the moisture content is above F.S.P.

G

Girdling Cutting through the cambium layer around the circumference of a growing tree to terminate its growth before felling.

Grain The direction of the wood fibres relative to the long axis of the tree or piece of wood.

Growth ring The area of growth, comprising earlywood and latewood, by which a tree increases in diameter every year.

Gymnosperm The cone bearing softwoods, with seeds that are exposed, or 'naked' not enclosed in an ovary.

H

Honeycombing Checks in the interior of wood, invisible on the surface, caused by case hardening in seasoning.

Hypocotyl Part of the embryo seedling below the *cotyledons*.

L

Latewood The darker portion of a growth ring, consisting of denser wood with thicker walls formed in mid and late summer. Also known as 'summerwood'.

Lenticel A pore on the trunk or branch that breathes and shows a white mark on the bark.

Lignin An important chemical constituent of cellular tissue, which has a hardening, binding function in the cell walls of *xylem* and gives wood its basic character.

M

Mesocarp The middle layer of the *pericarp*.

Mesophyll The inner tissue of a leaf.

Moisture content Moisture exists in the cell walls and cell spaces of heartwood, and in the cell contents of *sapwood*. The strength and stiffness of wood varies in almost inverse ratio to the changes in moisture content.

Monocotyledon Monocotyledons have only one first seed leaf, or cotyledon, and include palms and all the grasses.

Montane forests Forests that occur in mountainous regions.
Movement The dimensional changes in timber after it has been air- or kiln-dried, caused by variations in atmospheric conditions. (See also *Shrinkage*)

N

Net veins See *Reticulate veins*.
Node The slightly swollen part of a stem from which a shoot, leaf or whorl of leaves arises.

O

Osmosis The tendency of dissimilar liquids and gases to diffuse through a membrane or porous structure.
Ovary The part of the flower containing the *ovules*.
Ovate Egg-shaped, broadest below the middle.
Ovule Young seed within the ovary.

P

Parenchyma Wood soft tissue concerned with the distribution and storage of carbohydrates. It is composed of thin-walled, brick shaped cells with pits that may be axial or radial.
Pedicel The stalk of a flower or fruit.
Peduncle The stalk of a flower or inflorescence.
Pericarp The seed vessel or wall of the developed ovary.
Periderm The tissues of the cork cambium and the outer bark.
Petiole A leaf stalk.
Phloem Vascular tissue that transports food materials made by the plant. Also known as *bast* or inner bark.
Photosynthesis The process by which carbohydrates are synthesized from carbon dioxide and water in the presence of sunlight and chlorophyll.
Pistil Female flower organs comprising ovary, style and stigma.
Pith The middle core of *parenchyma* in a stem, around which all future growth takes place, It ceases to function after the sapling stage. Also known as the 'heart of medulla'.
Pith flecks These are not connected with the pith, but are irregular, discoloured streaks caused by insect attack. Wound *parenchyma* forms on the occluding insect tunnels and cross-sections appear as flecks on converted timber.
Pits Parts of the cell wall in wood tissue which remain thin during the thickening of the cell walls and serve as a link between the cells for the transmission of sap.
Plumule The rudimentary stem in an embryo plant.
Pollarding The continuous lopping of the top, or poll, of the tree to encourage fresh growth. The succession of new shoots, or knurls, forms highly decorative *burrs* (*burls*).
Pollination The transfer of pollen from the male anther of a flower to the female stigma.

R

Rays A vertical sheet of tissue formed radially across the growth rings of trees, consisting chiefly of *parenchyma* cells. Rays allow the transmission of sap. There are two types of rays: medullary rays (also known as primary rays), which extend from the bark to the medulla, or pith, of the tree, and vascular rays which do not reach the pith.
Radicle The first tiny root sprouting from a seed.
Reaction wood The name given to either *tension wood* or *compression wood* caused by the distortion of natural growth in trees.
Respiration This is the reverse of photosynthesis. In respiration, organic matter is broken down into carbon dioxide and water with the release of energy.
Reticulate veins Veins that resemble the threads of a net (the smaller veins in most *dicotyledons*).
Ring-porous Where there is a distinct contrast in size and number of the pores in different parts of the growth ring, with larger pores in the earlywood producing a well marked boundary between the earlywood and latewood zones.

S

Sapwood The outer region of the tree trunk composed of living *xylem* cells that transport water throughout the tree.
Savannah Extensive grassy treeless plains in tropical and subtropical regions, usually covered with low vegetation.
Sepal One of the individual segments of the *calyx*.
S. G. Specific Gravity: the relative weight of a substance compared with that of an equal volume of water. The S. G. of air dry hardwoods varies between 0.45 and 1.4 approximately and the S. G. of softwoods varies from about .25 to .8 at 15% moisture content.
Shake A split or separation between adjacent layers of fibres caused by wind, thunder, frost, lack of nutrition, felling or faulty seasoning. There are compound, cross, cup, ring, heart, radial, shell, star or thunder shakes.
Shrinkage The dimensional change in wood between its 'green' state and after seasoning by air-drying or kilning to its *equilibrium moisture content* which varies with local conditions and conditions of end-use.
Stipule A growth, mostly leafy, at the base of the *petiole*, which is often shed early.
Stomata Breathing pores chiefly found on the undersurface of leaves. Each pore, or stoma, is usually controlled by two guard cells.
Stooling Throwing out shoots from a tree stump to produce a second growth from the original roots, in a similar way to *Pollarding*.

T

Tannin An acidic substance distilled from the bark, wood and excrescences of many species of trees and used for converting hide into leather.
Tension wood A term given to wood in which the cells are thin-walled and abnormally long. Thiscan occur in parts of deciduous trees that are under tension, for example on the convex side of leaning trees.
Texture The texture in hardwoods depends on the size and distribution of the wood elements and to a lesser extent on the rays. Texture varies in different species and according to the rate of growth and may be coarse, fine, medium, uniform, smooth, even or uneven.
Tracheids Often called wood fibres, these are narrow, vertically elongated tubular wood cells with rounded ends and bordered pits in their side walls. They form the bulk of the wood of conifers and correspond to the fibres in hardwoods.
Transpiration Water loss from the leaves by evaporation especially through the *stomata*.

V

Vascular tissue The living tissue that conducts water and food substances through the tree, and provides mechanical support. It is composed chiefly of *xylem*, which forms the wood, and *phloem* forming the bast.
Vessels Found only in hardwood and also known as pores, vessels are elongated cells arranged one on top of another with their ends missing, and are well perforated, forming continuous pipe-like tubes for conducting water through the tree.

W

Whorl Three or more leaves in a circle from a node; also used to describe a circle of branches or branchlets.

X

Xylem Basic wood tissue, comprising long cells with thickened walls by which water and mineral salts are transported through the tree.

INDEX

Notes: 1. References to very common uses e.g. furniture, joinery and building are omitted. Only less universal uses are specifically entered. Comprehensive lists of uses can be found on pages 64-140. 2. Common names are generally used in this index. Latin names are listed alphabetically on pages 64-140.

CREDITS

Quarto would like to thank the following for their help with this publication and for permission to reproduce copyright material.

Key = *A* - Above *B* - Below *R* - Right *L* - Left *C* - Centre

5 Farquharson Murless; **8** Parnham; **9** Parnham; **10** (*A*) Parnham, (*B*) Parnham; **11** (*A*) Parnham, (*B*) Parnham; **12** Andrew Lawson; **13** Andrew Lawson; **14** (*A*) Derek G Widdicombe Worldwide Photographic Library, (*BL*) Linda Proud, (*BR*) Andrew Lawson; **15** (*L*) Andrew Lawson, (*R*) Alan Peters, (*AR*) Alan Peters; **16** (*AL*) Andrew Lawson, (*R*) Robert Harding Picture Library, (*BL*) David Savage Furniture Makers; **17** (*L*) Robert Harding Picture Library; **18** Lucinda Leech; **25** (*A*) Peter Ryan/Science Photo Library, (*R*) Dr Jeremy Burgess/Science Photo Library, (*L*) John Howard/Science Photo Library, background Andrew Lawson; **26** (*AL*) Bruce Coleman/Eric Crichton, (*R*) Robert Harding Picture Library, (*BL*) Biophoto Associates, (*BR*) James Bell/Science Photo Library; **27** (*A*) Harry Smith Horticultural Photographic Library, (*B*) Eric Grave/Science Photo Library; **28** (*AL*) Science Photo Library, (*AC*) James Bell/Science Photo Library, (*AR*) Science Photo Library, (*BL*) Sinclair Stammers/Science Photo Library, (*BR*) James Bell/Science Photo Library, (*BR*) Science Photo Library; **29** (*AL*) Robert Harding Picture Library, (*AR*) James Bell/Science Photo Library, (*R*) John Durrell/Science Photo Library; **30** Bill Lincoln; **31** (*AL*) Lucinda Leech, (*BL*) Andrew Lawson, (*BR*) Farquharson Murless; **32** Linda Proud; **33** (*AL*) Bruce Coleman Ltd/Geoff Dare, (*AR*) Derek G Widdicombe Worldwide Photographic Library, (*BR*) Bruce Coleman Ltd/Norman Owen Tomalin; **34** Lucinda Leech; **35** (*BL*) Bruce Coleman Ltd, (*BR*) Lucinda Leech, (*AR*) Lucinda Leech; **37** (*L*) Lucinda Leech, (*R*) Lucinda Leech; **38** (*L*) Lucinda Leech, (*R*) Lucinda Leech; **40** (*A*) Lucinda Leech, (*R*) J Allan Cash Ltd; **41** (*L*) Panos Pictures, (*A*) Panos Pictures; **42** (*L*) Lucinda Leech, (*R*) Lucinda Leech, (*B*) Panos Pictures/Michael Pickstock; **43** (*L*) Lucinda Leech, (*R*) Lucinda Leech; **44** (*L*) Robert Harding Picture Library, (*R*) Michael Taylor; **45** Lucinda Leech; **46** (*L*) Lucinda Leech, (*R*) Lucinda Leech; **47** (*L*) Lucinda Leech, (*R*) Lucinda Leech; **48** (*A*) Habitat, (*C*) Robert Harding Picture Library, (*B*) Lucinda Leech; **49** Network Photographers; **50** Robert Harding Picture Library; **52** (*AL*) Robert Harding Picture Library, (*R*) Robert Harding Picture Library, (*BL*) Derek G Widdicombe Worldwide Photographic Library; **53** (*A*) Robert Harding Picture Library, (*C*) Lucinda Leech, (*B*) Robert Harding Picture Library; **54** (*R*) Michael Taylor; **57** (*A*) Trada/Danzer, (*BL*) Michael Taylor, (*BR*) Michael Taylor; **58** Sotheby's; **59** (*L*) National Canadian Aviation Museum, (*R*) British Aerospace; **60** Rycotewood College, Department of Fine Craftsmanship and Design; **61** (*L*) Rycotewood College, Department of Fine Craftsmanship and Design, (*R*) Rycotewood College, Department of Fine Craftsmanship and Design; **62** Bruce Coleman Ltd; **65-140** Bill Lincoln; **144** Parham; **145** (*L*) Anthony Blake Photo Library, (*R*) The Photo Source/E Streichan; **146** (*AR*) Andrew Lawson, (*L*) Andrew Lawson; **147** (*L*) Alan Peters, (*R*) Alan Dixon; **148**, Alan Dixon; **149** (*AR*) Nicholas Pryke/Graham Pearson, (*BL*) Morgan Motor Co Ltd, (*BR*) Wycombe Chair Museum; **150** (*R*) Andrew Lawson; **151** (*L*) Mary Evans Picture Library, (*R*) National Portrait Gallery, London; **152** (*L*) Derek G Widdicombe Worldwide Photographic Library, (*R*) John Massey Stewart; **153** (*L*) John Massey Stewart, (*R*) Bruce Coleman Ltd; **154** The Image Bank; **155** (*A*) Bruce Coleman Ltd/Norman Myers, (*R*) Andrew Lawson; (*BL*) Moira Clinch; **156** (*L*) The Bridgeman Art Library, (*R*) James Canfield; **157** (*L*), Luke Hughes and Co Ltd, (*R*) NHPA/© K Ghani; **158** (*L*) Linda Proud, (*R*) Sporting Pictures (UK) Ltd; **159** (*AL*) Derek G Widdicombe Worldwide Photographic Library; (*BL*) Hutchinson Library (*R*) David Drew; **161** (*L*) Heather Angel, (*R*) Martin Grierson, (*BR*) Bruce Coleman Ltd/© A J Stevens; **162**, Andrew Lawson; **163** (*A*) The National Trust/Charlie Waite, (*B*) © Wycombe Chair Museum; **164** (*L*) Harry Smith Horticultural Photographic Collection; **165** (*CL*) Photo Graham Pearson, (*R*) Sporting Pictures UK Ltd, (*BR*) Bruce Coleman Ltd; **166**, © Barnsley House Garden Furniture; **167** (*R*) The Image Bank/© David Brownell; **168** Robert Harding Picture Library; **169** (*AL*) © ANT, (*BL*) Jakem Ltd; **170** (*AL*) Harry Smith Horticultural Photographic Library, (*B*) Henry Moore Foundation; **171** (*L*) Mary Evans Picture Library, (*R*) Luke Hughes and Co Ltd; **172** Robert Harding Picture Library; **173** (*L*) © Jim Partridge, (*R*) Andrew Lawson; **174**, Luke Hughes and Co; **175** (*L*) © Anthony Bryant, (*R*) Andrew Lawson; **176** (*L*) Paul Rocheleau, (*AR*) Steven C Kaufman, (*B*) Paul Rocheleau; **177** (*A*) Fred Baier, (*B*) Sporting Pictures (UK) Ltd; **178** (*L*) Sporting Pictures (UK) Ltd, (*R*) Robert Harding Picture Library; **179** (*L*) Nicholas Pryke/photo, Graham Pearson, (*AC*) HMS Victory, (*AR*) Mary Evans Picture Library (*B*) Mary Evans Picture Library; **180** G & J Images; **181** (*AL*) British Museum; **182** (*L*) Sam Maloof, (*R*) Nicholas Pryke/photo, Graham Pearson, (*BC*) Anthony Blake Photo Library, (*R*) Anthony Blake Photo Library; **183** The Bridgeman Art Library; **184** (*A*) Rexel, Cumberland.

Every effort has been made to trace and acknowledge all copyright holders.
Quarto would like to apologise if any omissions have been made.